Signals and Communication Technology

This series is devoted to fundamentals and applications of modern methods of signal processing and cutting-edge communication technologies. The main topics are information and signal theory, acoustical signal processing, image processing and multimedia systems, mobile and wireless communications, and computer and communication networks. Volumes in the series address researchers in academia and industrial R&D departments. The series is application-oriented. The level of presentation of each individual volume, however, depends on the subject and can range from practical to scientific.

Indexing: All books in "Signals and Communication Technology" are indexed by Scopus and zbMATH

For general information about this book series, comments or suggestions, please contact Mary James at mary.james@springer.com or Ramesh Nath Premnath at ramesh.premnath@springer.com.

Carlos Rompante Cunha • Nishu Gupta •
Srinivasa Kiran Gottapu
Editors

Transformative Cloud Computing, IoT and Extended Reality

Applications for Smart Cities, Education, Healthcare, Industry and Business

Springer

Editors
Carlos Rompante Cunha
Laboratório Associado para a Sustentabilidade e Tecnologia em Regiões de Montanha (SusTEC)
Research Centre in Digitalization and Intelligent Robotics (CeDRI), Bragança Polytechnic University
Mirandela, Portugal

Nishu Gupta
VTT Technical Research Centre of Finland
Oulu, Finland

Srinivasa Kiran Gottapu
Benedict College
Columbia, SC, USA

ISSN 1860-4862 ISSN 1860-4870 (electronic)
Signals and Communication Technology
ISBN 978-3-032-03553-0 ISBN 978-3-032-03554-7 (eBook)
https://doi.org/10.1007/978-3-032-03554-7

This Springer imprint is published by the registered company Springer Nature Switzerland AG
The registered company address is: Gewerbestrasse 11, 6330 Cham, Switzerland

Carlos Rompante Cunha dedicates this book to
To all those who work to make the world a better place, inspiring us with their actions.

Dr. Nishu Gupta dedicates this book to
Prof. Manuel José Cabral dos Santos Reis, Portugal

Dr. Srinivasa Kiran Gottapu dedicates this book to
His Advisor (Dr. Parthasarathy Guturu)

Preface

The combination of cloud computing, the Internet of Things (IoT), and extended reality (XR) is redefining the society that is witnessing swift technical development and digital transformation. These technologies are becoming fundamental to how we live, work, learn, and invent; they are no longer merely theoretical ideas or new trends. In response to this changing environment, this book is developed. It includes a wide range of studies and observations that combine theory and practice. It seeks to give scholars, professionals, and students a thorough grasp of how immersive technologies, cloud-based infrastructures, and intelligent networks are changing important industries including marketing, entrepreneurship, healthcare, education, and transportation. The chapters examine the potential and difficulties of our increasingly interconnected society, ranging from cybersecurity in 5G-enabled ecosystems to the moral incorporation of artificial intelligence in scientific and academic learning environment.

This work's interdisciplinary approach highlights the intricacy and potential of digital revolution. In addition to providing information, we hope that this volume will stimulate more research, creativity, and the proper application of these potent technologies. This book is intended to meet you at the nexus of knowledge and application, whether you are a researcher looking for academic depth, a professional looking to use new tools, or a student gaining foundational information.

The book serves as both a textbook and a reference, making it suitable for all levels of technical education and research, as well as for beginners seeking an overview of the potential applications cloud computing, IoT, and XR technologies in the future, which could make many things even smarter, more intelligent, and self-adaptive. It is structured into various parts with multiple chapters, organized in a reader-friendly way.

The book received contributions from around the globe. I extend my sincere gratitude to all contributors, collaborators, and readers. We also appreciate the reviewers for their prompt and thoughtful feedback. May this book serve as a catalyst for forward-thinking dialog, impactful research, and meaningful change.

Mirandela, Portugal — Carlos Rompante Cunha
Oulu, Finland — Nishu Gupta
Columbia, SC, USA — Srinivasa Kiran Gottapu

Acknowledgements Carlos Rompante Cunha

I am deeply grateful to my family, my greatest source of inspiration. To my wife, Sandra, and to my children—Pedro, Tomás, and Matilde, who is on her way—thank you for your love, strength, and constant motivation.

To my parents, António and Rosa, thank you for the values and foundation you gave me. To my sister, Sílvia, I am thankful for your continuous support and encouragement.

I sincerely thank my co-editors, Dr. Nishu Gupta and Dr. Srinivasa Kiran Gottapu, for their trust, guidance, and collaboration.

I also thank my colleagues at Bragança Polytechnic University for their support and friendship, and the institution itself for fostering my professional growth.

To all the contributing authors, your work is the backbone of this book—thank you.

Finally, I thank God, my guide and foundation.

Nishu Gupta

I acknowledge the inspiration and blessings of my mother Smt. Rita Rani Gupta and father Prof. K.M. Gupta, who are the pillars of my strength and personality. Having parents like them is the greatest gift and biggest advantage anyone could ever have given me. It is because of the confidence and values that they have instilled in me that made me who I am today.

I am full of gratitude to my wife Smt. Anamika Gupta, son Ayaansh Gupta, sister Mrs. Nidhi Gupta, brother-in-law CA Ritesh Shankar Gupta, and other family members for the patience shown and encouragement given to complete this venture.

I deeply acknowledge the blessings of my PhD Supervisor Prof. Arun Prakash, Electronics and Communication Engineering Department, MNNIT Allahabad, India, whose guidance has always encouraged me to do my best. I wholeheartedly acknowledge the motivation given to me by my Scientific Advisors Dr. Jukka Mäkelä, Principal Scientist at VTT Technical Research Center of Finland; Prof. Mohammad Derawi at Norwegian University of Science and Technology (NTNU) Gjøvik, Norway; Prof. Sara Paiva at Polytechnic Institute of Viana do Castelo, Portugal; Prof. Manuel J. C. S. Reis at UTAD, Portugal; Prof. Ahmad Hoirul Basori at King Abdulaziz University, Saudi Arabia; Prof. Ariel Soares Teles at Federal Institute of Maranhão, Brazil; Dr. Anil Gupta at the Centre for Development of Advanced Computing (CDAC), Pune, Maharashtra, India; Dr. Krishan Kumar at National Institute of Technology Hamirpur, Himachal Pradesh, India; Prof. Rakesh Nayak, OPJU, Raigarh, Chhattisgarh, India; Dr. Raghavendra Pal, SVNIT, India; Dr. Prakash Pareek, Vishnu Institute of Technology, India, for always encouraging and guiding me towards delivering the best. Thank you for shaping me into a person I am proud to be.

I profoundly acknowledge my friends Ms. Manisha Srivastava, India; Mr. Gaurvendra Singh, ASK Private Wealth, New Delhi, India; Er. Jalaj Kumar Singh, Acuity Knowledge Partners, Gurugram, India. Thank you for everything you have done for me. I love you, more than words can ever express.

I deeply acknowledge the support and encouragement from my co-editor Prof. Carlos Rompante Cunha and Dr. Srinivasa Kiran Gottapu in right shaping this book.

I express my heartfelt thanks to my colleagues, friends, and students for their support and motivation in several ways.

Last but not least, we, the Editors, express our heartfelt gratitude to the Publishers and the team behind for their continued support and cooperation in publishing this book.

Srinivasa Kiran Gottapu

I extend my deepest gratitude to my parents, Shri Jagannadham Gottapu and Smt. Krishnaveni Gottapu, whose constant support, love, and values have been the bedrock of my personal and professional journey.

I am profoundly thankful to my wife, Charitha Velamala, for her boundless patience, strength, and steadfast encouragement throughout this path. Her support has been a continuous source of inspiration and motivation.

To my son, Aaryan, whose laughter and radiant smile bring immeasurable joy and meaning to my life. His presence fills my days with warmth, offering a refreshing perspective that continually reminds me of the simple, beautiful moments that truly matter.

I gratefully acknowledge my brother, Siva Teja Gottapu, and sister-in-law, Durga Gottapu, for their continuous love, moral support, and the constant motivation they have provided, all of which have been integral to my progress. My heartfelt thanks also go to my brother-in-law, Goutham Kumar Velamala, for his encouragement and positive influence, and to my father-in-law, Manmadha Rao Velamala, and mother-in-law, Damayanthi Velamala, whose enduring love and support have been a source of strength.

I am sincerely indebted to my PhD supervisor, Dr. Parthasarathy Guturu, Professor of Electrical Engineering at the University of North Texas, whose expert guidance, mentorship, and insightful feedback have been instrumental to my academic and research growth.

Finally, I would like to thank my co-editor and dear friend, Dr. Nishu Gupta, whose generous help, encouragement, and collaborative spirit greatly enriched this work.

Contents

About the Editors

Carlos Rompante Cunha is a PhD in Computer Science from the University of Trás-Os-Montes e Alto Douro, Portugal. He is an Adjunct Professor in the Department of Informatics and Mathematics, EsACT-IPB. He is/was a research member for several research projects. He published over 120 papers in international journals and conferences. He is also a member of several International Journals and Conferences' Scientific Committees, for which he frequently reviews papers. For several years, he was the director of the Informatics and Communications BSc and the Coordinator of the Department of Informatics and Mathematics, EsACT-IPB. He was also the coordinator of the EsACT-IPB Computer Resources Center. He is currently vice-dean of EsACT-IPB. His teaching fields are Programming Languages and Distributed Systems.

His research interests include Pervasive and Mobile Computing, Distributed Systems, Extended Reality, IoT, and Machine Learning.

Nishu Gupta is a Senior Member, IEEE and presently working as a Senior Scientist at VTT Technical Research Centre of Finland Ltd., located in Oulu, Finland. Before this position, he was a Postdoctoral Fellow in the Smart Wireless Systems research group, Department of Electronic Systems, Faculty of Information Technology and Electrical Engineering at Norwegian University of Science and Technology (NTNU) in Gjøvik, Norway. He has worked as Research Associate in the Smart Living Lab at the University of Fribourg, Switzerland. Prior to this position, he was a Postdoctoral Fellow in the Department of Information Technology, Faculty of Computing and Information Technology, King Abdulaziz University in Rabigh, Saudi Arabia. He is also a visiting researcher at the University of Oviedo, Spain, under the research group on Systems for Multimedia and the Internet of Things. He is serving as a Member, Zero Trust Architecture working group of MeitY-C-DAC-STQC project under "e-Governance Standards and Guidelines," Ministry of Electronics and Information Technology (MeitY), Government of India. He is a PhD in Electronics and Communication Engineering Department from Motilal Nehru National Institute of Technology Allahabad, India (An institute of national importance as declared by the Government of India) with specialization in Vehicular Ad Hoc Networks. His research interests include IoT; 5G/6G technologies; dynamic spectrum sensing, intelligent transportation, edge computing; etc.

Srinivasa Kiran Gottapu is an Assistant Professor of Computer Engineering at Benedict College in Columbia, SC. He holds a PhD in Electrical Engineering from the University of North Texas, USA. Dr. Gottapu brings a wealth of expertise in teaching embedded systems, communication systems, and signals and systems. His professional experience spans roles such as Senior Cloud Engineer at HCL Solutions America and Adjunct Instructor at the University of North Texas. His research interests include wireless communication networks, cognitive radio networks, and digital signal processing. Dr. Gottapu is a Senior Member of IEEE and has authored numerous publications in his field.

Part I
Innovations in Cloud Computing, IoT and Extended Reality

Cybersecurity for 5G-Enabled IoT Ecosystems: Introduction, Challenges, and Solutions

M. J. C. S. Reis, Nishu Gupta, Prakash Pareek, and Jukka Mäkelä

1 Introduction

The integration of 5G and the Internet of Things (IoT) significantly drives progress across key sectors such as healthcare, smart cities, and industrial automation. It is estimated that in next year that is 2026, over 30 billion IoT devices are expected to operate in high-speed 5G networks, encouraging revolutionary applications in various emerging sectors [1]. However, this integration also introduces cybersecurity challenges due to the rapid increase in connected devices and network components [2].

5G-enabled IoT ecosystems are characterized by their decentralized architectures, massive device heterogeneity, and reliance on real-time operations. These features enable ultra-low latency, high bandwidth, and seamless scalability, but they also introduce unique vulnerabilities [3]. As adoption accelerates, it becomes increasingly urgent to develop cybersecurity frameworks tailored to these evolving needs.

M. J. C. S. Reis (✉)
Quinta de Prados, Engineering Department/IEETA, University of Trás-os-Montes e AltoDouro, Vila Real, Portugal
e-mail: mcabral@utad.pt

N. Gupta · J. Mäkelä
VTT Technical Research Centre of Finland Ltd., Oulu, Finland
e-mail: nishugupta@ieee.org; jukka.makela@vtt.fi

P. Pareek
Electronics and Communication Engineering, Vishnu Institute of Technology, Vishnupur, Andhra Pradesh, India
e-mail: drprakashpareek@ieee.org

C. Rompante Cunha et al. (eds.), *Transformative Cloud Computing, IoT and Extended Reality*, Signals and Communication Technology,
https://doi.org/10.1007/978-3-032-03554-7_1

The challenges associated with 5G-enabled IoT systems' cybersecurity, including authentication, data privacy, DDoS attacks, the vulnerability of blockchain, and quantum computing interference which shatters traditional encryption, are exceedingly intricate [4, 5]. These challenges perpetuate from understudied AI-generated attacks as well as poor collaboration between devices [6, 7]. Researchers are employing artificial intelligence for risk control, blockchain and decentralized networks for safe verification, and lightweight measures of IoT-structure sensitive quantum defense encryption to solve these challenges [8–12].

Despite these advances, there are important gaps. AI systems are susceptible to controversy manipulation, and blockchain is in IoT deployment of scalability and energy consumption challenges [12]. Furthermore, the lack of robust interoperability standards and real-time requirements for edge computing environments highlight the need for adaptive, energy-efficient, and future solutions. In particular, the following research questions should be answered:

1. What are the major cybersecurity issues with 5G-IT integration?
2. How effective are existing solutions when addressing these challenges?
3. What are the future directions for improving security and privacy in this area?

It also highlights the important need for interdisciplinary collaboration between researchers, industry practitioners, and political decision makers to create a resistant and scalable digital ecosystem that can thrive in the 5G era and beyond [13]. To contextualize these results, this manuscript integrates a virtual case study showing how these solutions can address practical challenges in the fields of healthcare, intelligent cities, and industrial automation. In particular, the following contributions are made:

1. Depth evaluation of cyber safety problems in 5G-IoT systems focusing on privacy, scalability, and risks from quantum technologies.
2. Review of top security solutions, verified through metrics, simulations, and applications in the real world in healthcare, intelligent cities, and critical infrastructure.
3. Action plan for scalable, adaptive safety framework, highlighting advanced technological integration and harmony with global standards.

2 Technological Foundations

2.1 Network Slicing for Diverse Applications

Multiple virtual networks over a shared 5G infrastructure, each tailored for specific application requirements are enabled by network slicing. This customization supports use cases like ultra-reliable low-latency communications (e.g., autonomous vehicles), massive device connectivity (e.g., smart cities), and high reliability (e.g.,

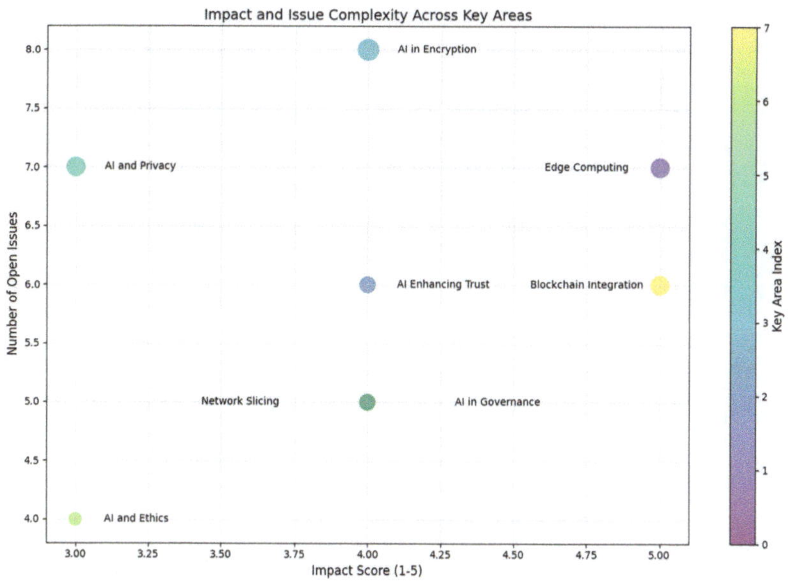

Fig. 1 Relationship between the key areas, their benefits, and the complexity of open issues

remote healthcare) [14]. It optimizes performance via tailored bandwidth and QoS settings [15].

Despite its benefits, network slicing presents challenges including slice misconfiguration, cross-slice data leakage, and inadequate isolation [16]. Proposed mitigations include Software-Defined Networking (SDN) for dynamic slice management, AI-based misconfiguration detection [17], and blockchain for secure inter-slice communication [18]. These technologies work synergistically to ensure efficient and secure slice orchestration.

Figure 1 illustrates the interrelationship between network slicing, AI, edge computing, and blockchain.

2.2 *Edge Computing and Localized Security*

Edge computing decentralizes processing by relocating computation closer to IoT endpoints, reducing latency and improving data privacy [15]. Use cases such as telemedicine and real-time traffic management benefit from low-latency responsiveness. For example, in a smart city deployment, local edge nodes processed 85% of traffic data, reducing core network congestion by 30% [19].

Challenges include attack surfaces across distributed nodes and resource limitations on edge devices [20]. To address these, solutions like AI-powered intrusion

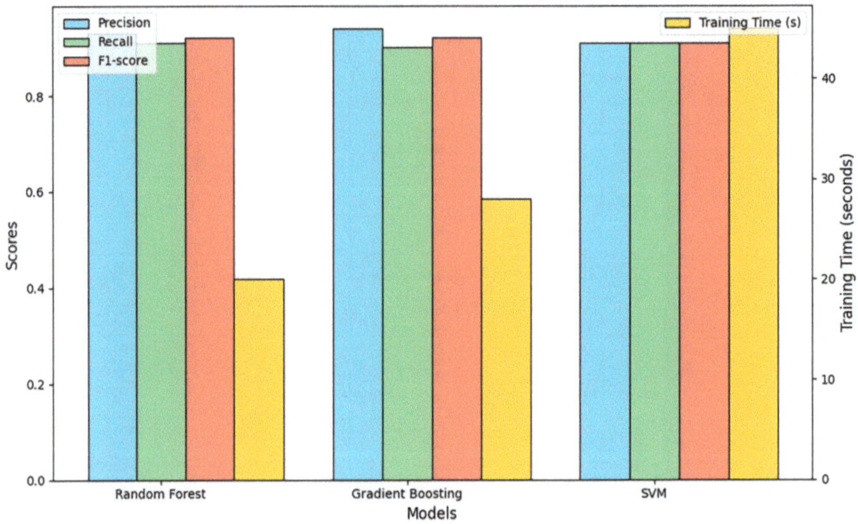

Fig. 2 Comparison of machine learning models to detect anomalies in IoT systems. The chart illustrates precision, recall, and F1-score on the primary axis, while training time (in seconds) is shown on the secondary axis. Random Forest demonstrates a balance between high accuracy and low training time compared to Gradient Boosting and SVM

detection (92% anomaly detection rate) [21], secure enclave technology for data isolation [22], and network micro-segmentation [23] have shown promise.

A machine learning comparison using Random Forest, Gradient Boosting, and SVM models evaluated smart city traffic anomaly detection. Random Forest achieved a precision of 0.93 and F1-score of 0.92 with the lowest training time, making it suitable for real-time edge inference.

The bar chart in Fig. 2 compares the performance of three machine learning models—Random Forest, Gradient Boosting, and Support Vector Machines (SVM)—in terms of precision, recall, F1-score, and training time. These metrics highlight a compromise in accuracy and computational efficiency for anomaly detection in intelligent city IoT systems. The diagram in Fig. 3 shows how detection changes with the number of IoT devices in the provisioning. This analysis provides insight into the scalability of edge-based AI systems for intelligent city real-time traffic management.

2.3 AI in IoT Security

AI strengthens IoT cybersecurity across encryption, privacy, and governance. Machine learning enhances lightweight encryption schemes and anomaly detection [24]. Privacy is preserved through differential privacy and federated learning,

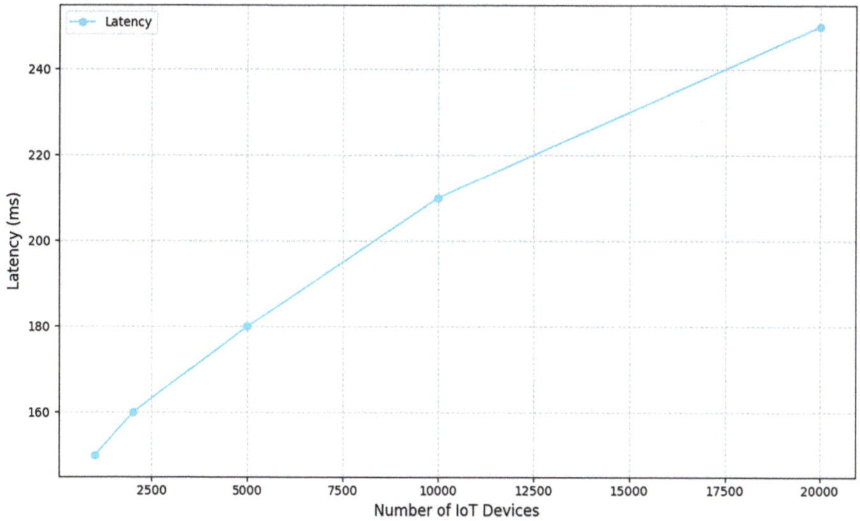

Fig. 3 Latency as a function of the number of IoT devices in edge-based AI deployments. The graph shows a gradual increase in latency as the deployment size scales from 1000 to 20,000 devices, highlighting potential performance bottlenecks in larger systems

demonstrated in smart cities where traffic prediction improved by 15% without data transfer [18].

Governance applications include explainable AI (XAI) for traceability in IoT systems, supporting compliance in regulated domains like finance [17]. In healthcare IoT, Gradient Boosting classifiers achieved 90% accuracy, outperforming traditional rule-based systems with lower false positive rates (5% vs. 12%).

The bar chart in Fig. 4 compares the performance metrics of the AI-driven anomaly detection system and a rule-based baseline system. Metrics such as accuracy, precision, recall, F1-score, and false positive rate are evaluated to highlight the AI system's advantages in real-time healthcare IoT environments. The pie chart in Fig. 5 illustrates the resource utilization of the AI-driven anomaly detection system, showcasing its efficient use of CPU and memory in a resource-constrained healthcare IoT environment.

2.4 Blockchain in 5G-IoT: Integration and Energy Efficiency

Blockchain facilitates secure, transparent communication in IoT through tamper-proof ledgers and smart contracts [3]. Applications include supply chain monitoring, decentralized energy markets, and patient data protection [24]. Technologies like IOTA and Hyperledger Fabric offer high transaction throughput and energy efficiency, surpassing Ethereum in IoT-specific performance [25].

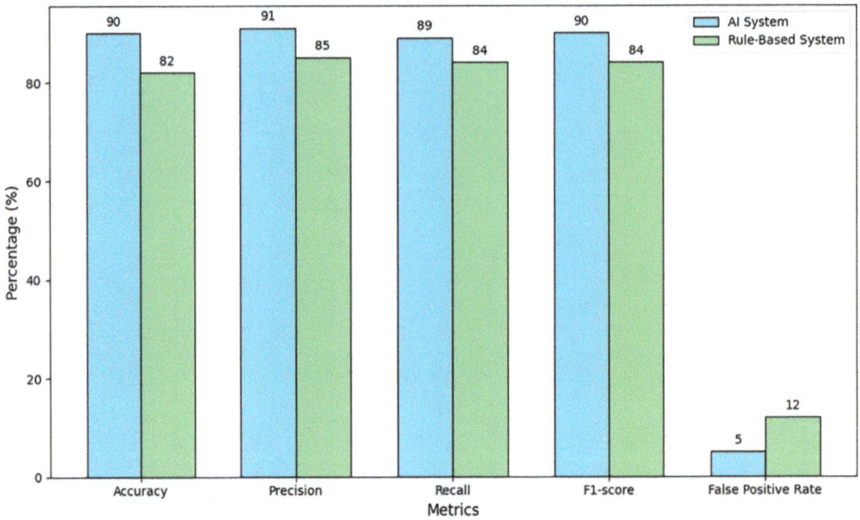

Fig. 4 Comparison of performance metrics between the AI-driven anomaly detection system and a rule-based baseline system. The AI system outperforms the baseline in accuracy, precision, recall, and F1-score, while achieving a significantly lower false positive rate, demonstrating its effectiveness for healthcare IoT applications

Fig. 5 Resource utilization of the AI-driven anomaly detection system for healthcare IoT applications. The system utilizes 50% of available CPU resources and 100 MB of memory, leaving sufficient capacity for other operations and ensuring compatibility with constrained IoT devices

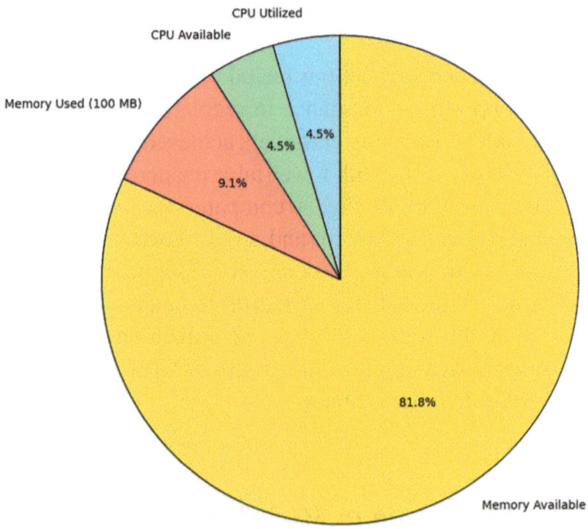

Simulation of a blockchain-based smart city authentication system using DPoS showed 1200 TPS under normal load and 1020 TPS under DDoS, with only 200 ms latency. DPoS consumed 30% less energy than PoW, supporting sustainability goals [22, 24]. Limitations include scalability beyond 1000 devices and sensitivity to validator collusion.

Table 1 Table comparing three prominent blockchain platforms—Ethereum, IOTA, and Hyperledger Fabric—across key criteria such as scalability, energy efficiency, and IoT-specific features, highlighting their suitability for various IoT use cases

Platform	Scalability	Energy efficiency	IoT-specific features
Ethereum	Moderate scalability; transaction throughput improved with Layer 2 solutions like Plasma [24]	Energy-intensive due to proof-of-work (PoW), but Ethereum 2.0 aims to address this with proof-of-stake (PoS) [26]	Smart contract functionality supports automation in IoT systems (e.g., supply chain automation) [15]
IOTA	High scalability via tangle (directed acyclic graph); performance improves as more devices join [25]	Extremely energy-efficient; eliminates mining by using a lightweight validation process [25]	Designed specifically for IoT; supports microtransactions and secure data exchange without transaction fees [24]
Hyperledger fabric	Highly scalable for enterprise use; customizable permissioned networks enhance transaction performance [22]	Energy-efficient; permissioned architecture reduces validation overhead [22]	Modular architecture allows integration with IoT for use cases like asset tracking and healthcare data management [24]

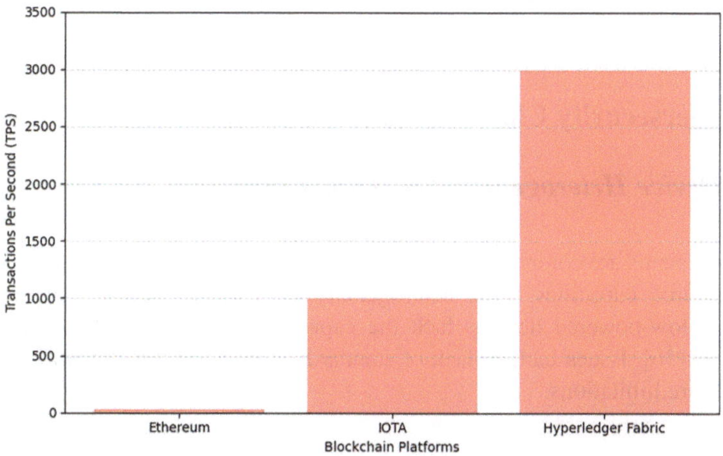

Fig. 6 Blockchain Throughput Comparison for IoT Applications. This chart illustrates the transaction throughput (in transactions per second, TPS) of three major blockchain platforms—Ethereum, IOTA, and Hyperledger Fabric—under IoT-specific conditions. The data highlights differences in scalability, with IOTA and Hyperledger Fabric demonstrating higher throughput compared to Ethereum

Table 1 provides a comparative analysis of Ethereum, IOTA, and Hyperledger Fabric in terms of scalability, energy efficiency, and IoT-specific features. This synthesis supports the performance trends shown in Figs. 6 and 7.

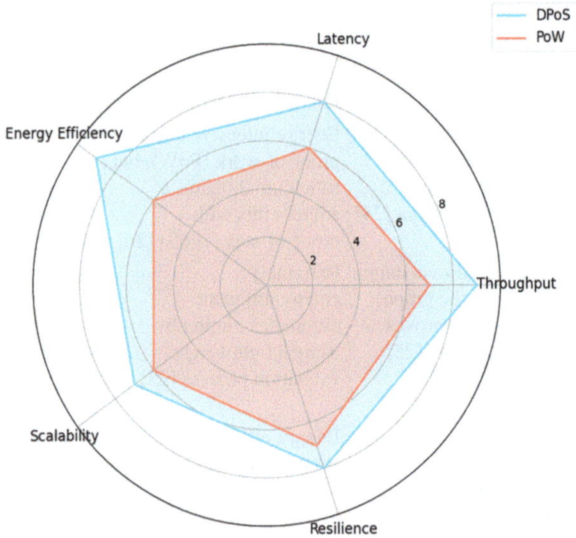

Fig. 7 Radar chart comparing Delegated Proof-of-Stake (DPoS) and Proof-of-Work (PoW) consensus mechanisms across key metrics. DPoS outperforms PoW in throughput, latency, and energy efficiency, while both mechanisms exhibit similar resilience. The chart underscores the advantages of DPoS for scalable and sustainable IoT applications

3 Cybersecurity Challenges and Solutions

3.1 Device Heterogeneity and Authentication

IoT ecosystems are composed of a wide variety of devices with differing standards and resource capacities. This heterogeneity complicates secure authentication, as many low-powered devices lack the capacity for conventional cryptographic protocols [16]. Issues include lack of standardization and weak authentication due to hardware limitations.

3.2 Data Privacy Concerns

The extensive data output of IoT systems introduces significant risks to data protection, particularly for sensitive information like biometric and location data [8]. The risk of data protection is heightened due to inadequate government regulations and the absence of a centralized authority to oversee data security. This decrease encompasses different privacy measures for aggregation analysis [27], data exchange of blockchain-based data for traceability [14].

3.3 Network Vulnerabilities and DDoS Attacks

5G-IoT infrastructure relies heavily on SDN and virtualization, meaning it is susceptible to DDOS attacks (DDOS) [2]. AI-driven transport analysis shows up to 95% accuracy in identifying harmful traffic in intelligent/smart grids [18]. Blockchain-based DDoS mitigation further isolates threats rapidly, reducing severity by 45% in real-time simulations [20]. These strategies illustrate the value of AI and blockchain in network hardening.

3.4 Quantum Threats to IoT Security

Quantum computing poses a severe threat to current cryptographic standards, rendering RSA and ECC vulnerable. Post-quantum cryptography (PQC) offers alternatives, such as CRYSTALS-Kyber and Dilithium, which have been validated in real-world pilots like smart meters and autonomous vehicles [22, 28, 29].

Quantum-safe blockchain with hash-based signatures (e.g., SPHINCS+) has shown to preserve 98% throughput during secure IoT transactions [23, 30]. Pilots by IBM and the Hyperledger Foundation confirm the feasibility of integrating quantum-safe methods into existing systems [21, 31].

Figure 8 outlines the roadmap for quantum-safe cryptographic adoption, showcasing progressive steps to secure IoT ecosystems against quantum threats, from awareness of quantum threats to the global adoption of quantum-safe cryptographic solutions. It highlights key milestones such as algorithm development, pilot testing, and industry-wide implementation.

A simulation using CRYSTALS-Kyber on smart grid devices showed encryption and decryption times of 1.2 ms and 1.8 ms, respectively, with a total latency of

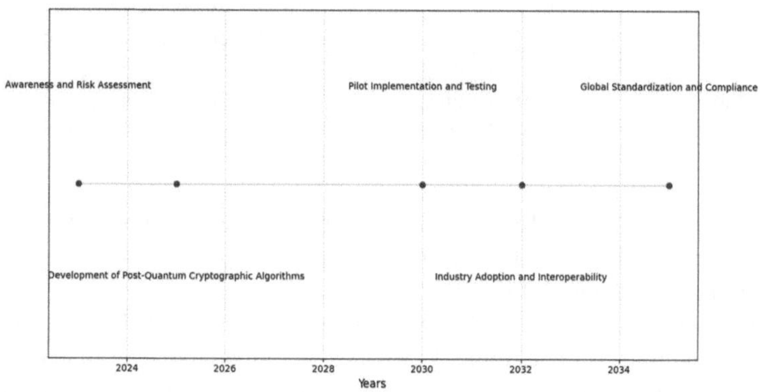

Fig. 8 Timeline for Quantum-Safe Cryptography Adoption in IoT Ecosystems, from awareness to global standardization

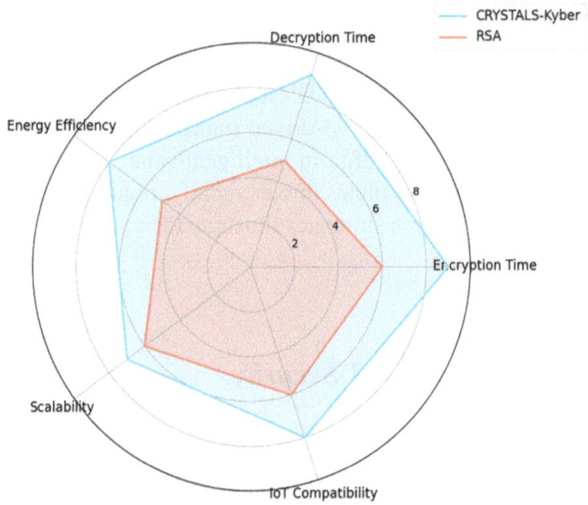

Fig. 9 Radar chart comparing the performance of CRYSTALS-Kyber and RSA across key metrics. CRYSTALS-Kyber outperforms RSA in encryption and decryption time, energy efficiency, and compatibility with IoT devices. The chart highlights CRYSTALS-Kyber's scalability and suitability for quantum-resistant IoT applications

Table 2 Summary of Sect. 3: key areas, open issues, and mitigation strategies

Challenges	Open issues	Mitigation strategies
Device heterogeneity	Lack of standardization; weak device authentication	Lightweight cryptography; multi-factor authentication
Data privacy	Unauthorized access; sensitive data misuse	Federated learning; blockchain-based secure sharing
Network vulnerabilities	Exploitation of APIs; poorly configured environments	AI-powered traffic analysis; redundant infrastructure
Supply chain security	Counterfeit components; malicious firmware	Blockchain-based traceability; secure boot mechanisms
Blockchain security	Scalability; resource-intensive consensus mechanisms	Layer 2 solutions; proof-of-stake consensus mechanisms
Quantum threats	Quantum computing rendering current encryption obsolete	Post-quantum Cryptography; lattice-based encryption

3.5 ms. Compared to RSA, Kyber-reduced encryption and decryption time by 70% and 65%, respectively. It also halved energy consumption, demonstrating suitability for constrained IoT devices (Fig. 9).

Table 2 summarizes the key cybersecurity challenges in 5G-IoT and corresponding mitigation strategies and Table 3 provides a comparative overview of leading post-quantum cryptographic algorithms with their core attributes and suitability for IoT.

Table 3 Comparative summary of quantum-safe cryptographic algorithms for IoT applications, including SPHINCS+, CRYSTALS-Kyber, and Dilithium

Algorithm	Type	Key features	Performance metrics	IoT suitability
SPHINCS+	Hash-based signatures	Stateless; simple implementation; strong resistance to quantum attacks	Moderate key size; higher signature size	Ideal for constrained IoT devices needing simple, secure signatures
CRYSTALS-Kyber	Lattice-based key exchange	Compact keys; fast encryption and decryption; post-quantum secure	Low latency; small key size	Suitable for real-time communications in IoT systems
Dilithium	Lattice-based signatures	High security margin; efficient key generation and signing	Moderate key size; fast signature generation	Suitable for scalable IoT networks needing frequent authentication

The table highlights their cryptographic type, key features, performance metrics, and suitability for IoT ecosystems

4 State-of-the-Art Solutions

This section presents a synthesized overview of cutting-edge cybersecurity solutions developed for 5G-enabled IoT ecosystems. It covers advancements in artificial intelligence, federated learning, encryption techniques, edge/fog computing, and strategies to counter emerging threats.

4.1 AI-Driven Security

AI enhances 5G-IoT cybersecurity through anomaly detection, predictive analytics, real-time threat response, and behavioral monitoring. Recent innovations include AI-augmented encryption and Explainable AI (XAI), which improve transparency and trust [32–40].

Benefits include scalability, rapid response, and adaptive learning. Limitations include high computational demand and risks from adversarial inputs. Countermeasures involve adversarial training, ensemble learning, and certified defenses.

Figures 10 and 11 illustrate AI effectiveness and comparisons of application areas.

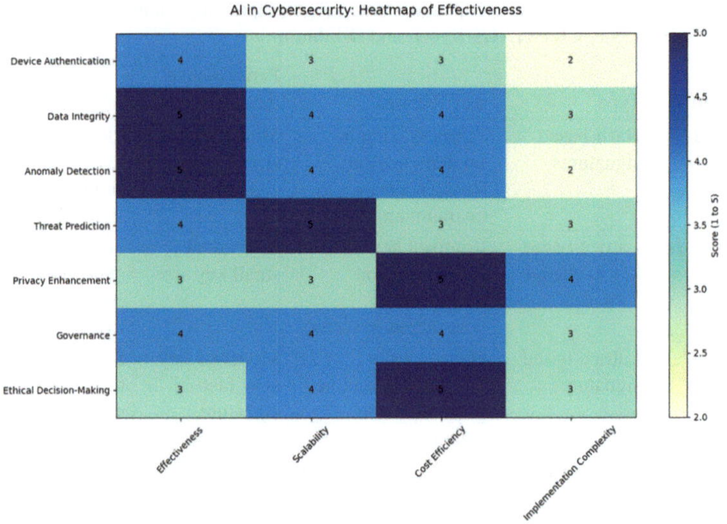

Fig. 10 Effectiveness of AI solutions across areas like trust, encryption, privacy, governance, and ethics

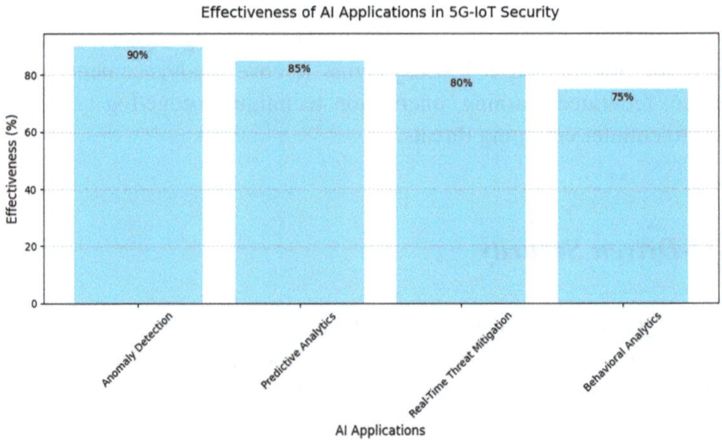

Fig. 11 Comparison of effectiveness across different AI applications (anomaly detection, predictive analytics, real-time mitigation)

4.2 Federated Learning and Decentralized Authentication

Federated Learning (FL) enables decentralized model training without raw data sharing. Blockchain integration ensures model integrity and transparent verification. Use cases span healthcare, smart cities, and critical infrastructure [5, 33, 35–41].

FL offers privacy, resilience, and scalability. Challenges include communication overhead, data imbalance, and integration complexity. Optimized FL models using

Table 4 Comparison of traditional vs. federated learning in IoT environments

Metric	Traditional learning	Federated learning
Latency	High latency due to centralized data processing	Low latency as data processing is localized
Data privacy	Data must be shared with central servers, raising privacy concerns	Data remains on devices, ensuring privacy
Scalability	Limited scalability as server capacity is a bottleneck	Highly scalable due to distributed architecture
Centralized server dependency	High dependency on centralized servers	Reduced dependency on centralized servers
Security	Single point of failure increases vulnerability	Decentralized design improves resilience

sparsified updates and adaptive learning have reduced communication by 35% while maintaining 92% accuracy.

Table 4 presents a comparison of traditional versus federated learning in IoT environments.

4.3 Advanced Encryption Techniques

Encryption remains critical to IoT data protection. Key approaches include lightweight encryption (e.g., ECC), quantum-safe protocols (e.g., Kyber, Dilithium), and homomorphic encryption [10, 33, 36–43].

Benchmarking shows ECC outperforms RSA in speed and energy use, while quantum-safe methods offer future resilience. SPHINCS+, Kyber, and Dilithium provide varied strengths for IoT deployment.

4.4 Edge and Fog Computing Security

Edge and fog layers reduce latency and enhance scalability in 5G-IoT networks but expand the attack surface. Key defenses include secure nodes, dynamic access controls, and localized threat detection [34, 44–60]. Authors propose a novel architectural framework on IoT ecosystem from the core technology's view point. They propose a modified IoT architecture by dividing the perception layer into elementary blocks based on their attributed functions [61].

Benefits include privacy, responsiveness, and distributed threat handling. Limitations stem from fragmented architecture and increased complexity.

Figure 12 presents security challenges versus severity (e.g., complexity, attack surface, fragmentation risks), while Fig. 12 presents a layered diagram depicting security measures across edge, fog, and cloud layers. Table 7 presents security

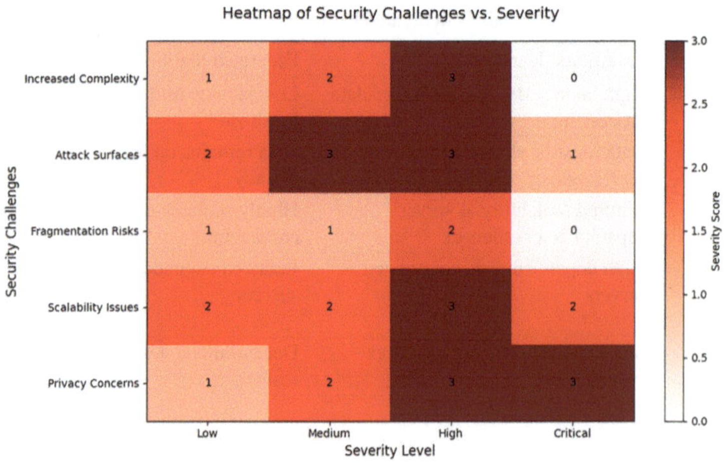

Fig. 12 Security challenges versus severity (e.g., complexity, attack surface, fragmentation risks)

measures at edge/fog layers (e.g., Layer 1: Secure nodes, Layer 2: Dynamic access controls, and Layer 3: Threat detection and mitigation).

4.5 Emerging Threats in 5G-IoT Ecosystems

Emerging threats—such as adversarial AI, zero-day vulnerabilities, quantum decryption, and DDoS amplification—necessitate proactive strategies [32–37].

Response strategies include adversarial training, blockchain-secured updates, post-quantum cryptography, and real-time anomaly detection. A layered, adaptive defense approach is critical for long-term security in dynamic 5G-IoT environments.

Table 5 summarizes the applications, benefits, and limitations of each solution type, underscoring their relevance in building secure, scalable, and resilient 5G-IoT ecosystems, and Table 6 the state-of-the-art solutions, their key applications, benefits, and limitations.

Figure 13 presents emerging threats and their potential impact (e.g., adversarial AI, DDoS, quantum threats), while Fig. 14 presents a breakdown of existing solutions addressing different threats. Fig. 15 categorizes cybersecurity challenges by their likelihood and impact.

5 Future Directions in 5G-IoT Cybersecurity

As 5G-enabled IoT ecosystems continue to evolve, the future of cybersecurity demands scalable, adaptive, and collaborative solutions that are both forward-

Table 5 Summary of the applications, benefits, and limitations of all solutions (AI, FL, encryption, edge computing, blockchain, etc.)

Solution	Key applications	Benefits	Limitations
AI-driven security	Anomaly detection, predictive analytics, real-time threat mitigation	Scalable, rapid detection, continuous learning	High computational demand, adversarial risks
Federated learning and decentralized authentication	Local model training, blockchain integration, decentralized authentication	Enhanced privacy, resilience, blockchain synergy	Computational overhead, scaling blockchain
Advanced encryption techniques	Lightweight cryptography, homomorphic encryption, quantum-resistant algorithms	Compatibility, future-proof, enhanced privacy	Performance costs, complexity
Edge and fog computing security	Secure edge nodes, dynamic access controls, threat detection	Low latency, privacy protection, scalability	Increased complexity, broader attack surfaces
Blockchain for IoT	Smart contracts, decentralized identity management, data integrity	Trustworthy operations, resilience, operational efficiency	Scalability challenges, energy use
Quantum-resistant measures	Quantum-safe algorithms, secure device authentication, blockchain protection	Future-ready, seamless integration, improved trust	Performance impacts, adoption barriers

Table 6 Summary of the state-of-the-art solutions, their key applications, benefits, and limitations

Solution	Key applications	Benefits	Limitations
AI-driven security	Anomaly detection, predictive analytics, real-time threat mitigation	Scalable, rapid detection, continuous learning	High computational demand, adversarial risks
Federated learning and decentralized authentication	Local model training, blockchain integration, decentralized authentication	Enhanced privacy, resilience, blockchain synergy	Computational overhead, scaling blockchain
Advanced encryption techniques	Lightweight cryptography, quantum-resistant algorithms, homomorphic encryption	Compatibility, future-proof, enhanced privacy	Performance costs, complexity
Edge and fog computing security	Secure edge nodes, dynamic access controls, threat detection	Low latency, privacy protection, scalability	Increased complexity, broader attack surfaces
Blockchain for secure IoT transactions	Smart contracts, decentralized identity management, data integrity	Trustworthy operations, resilience, operational efficiency	Scalability challenges, energy use
Integration of quantum-resistant measures	Quantum-safe algorithms, secure device authentication, blockchain protection	Future-ready, seamless integration, improved trust	Performance impacts, adoption barriers

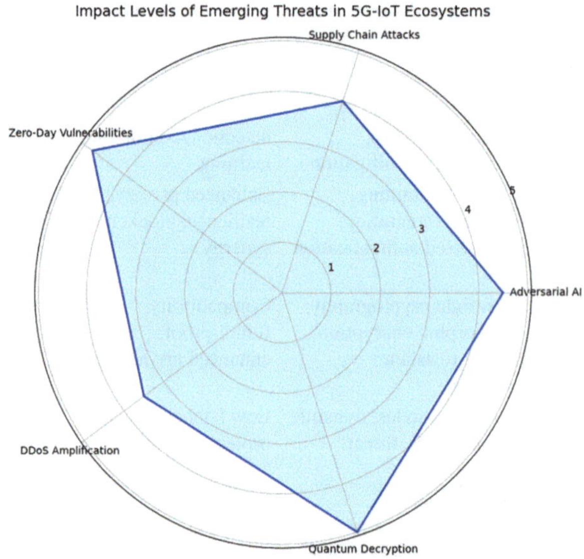

Fig. 13 Emerging threats and their potential impact (e.g., adversarial AI, DDoS, quantum threats)

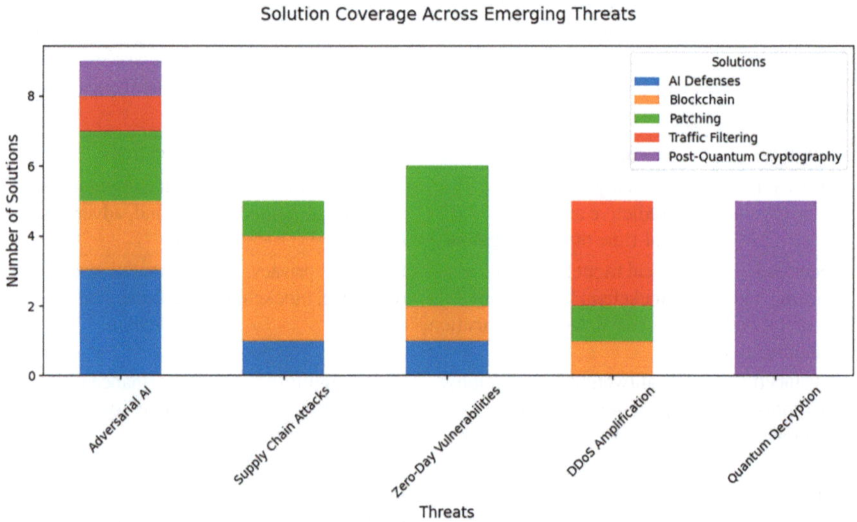

Fig. 14 Breakdown of existing solutions addressing different threats

looking and technically robust. This section outlines strategic directions spanning scalable frameworks, quantum-resilience, privacy-preserving analytics, and ecosystem-wide collaboration—essential for ensuring security and resilience in next-generation IoT infrastructures.

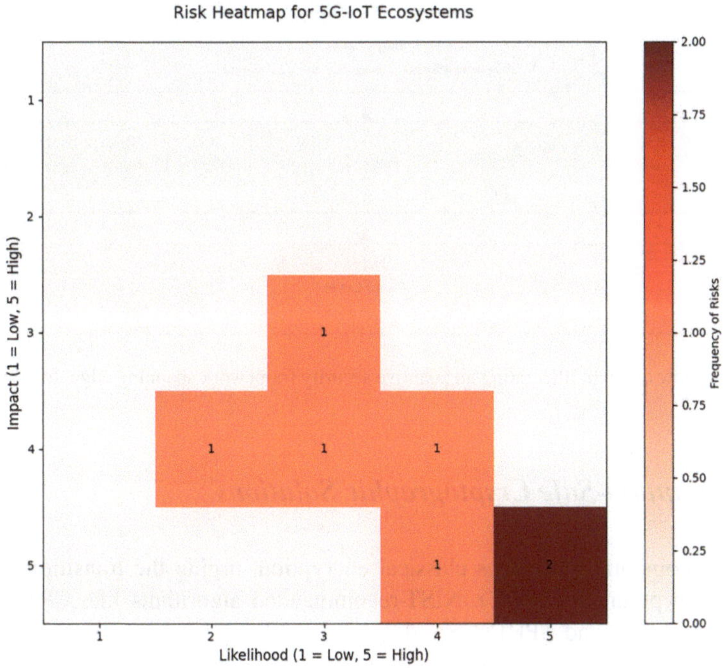

Fig. 15 A risk heatmap categorizing challenges based on likelihood and impact

5.1 Scalable and Adaptive Security Frameworks

To meet the demands of billions of heterogeneous IoT devices, future frameworks must be lightweight, hierarchical, and intelligent. Lightweight cryptography (e.g., TinyCrypt) ensures security with minimal computational cost, while hierarchical architectures (Edge–Fog–Cloud) distribute responsibilities across network layers to improve latency, threat detection, and policy enforcement (see Fig. 16).

Technologies like EdgeSec and IoTChain enable localized anomaly detection and immutable identity management, while FLTrust supports privacy-preserving federated learning. Zero Trust Architectures (e.g., BeyondCorp) and AI-based IDS systems (e.g., DeepLog) ensure continuous authentication and behavioral surveillance. These methods demonstrate scalability via hierarchical federated learning models and communication-efficient training strategies.

Simulations in smart cities show 12% improvement in anomaly detection when using federated models trained at the edge and aggregated centrally. Blockchain-enhanced FL further ensures trust and tamper-proof model updates.

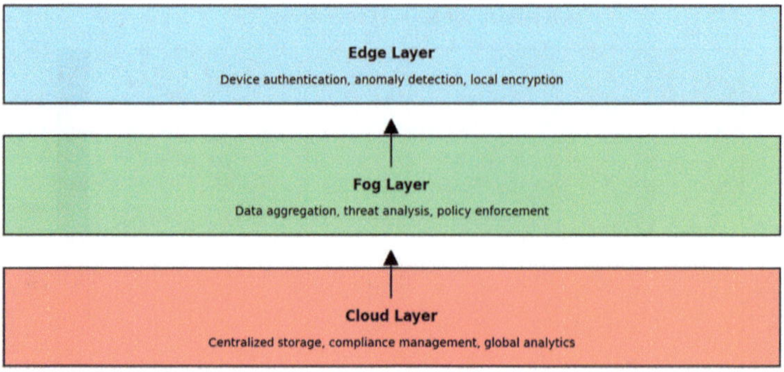

Fig. 16 A flow diagram illustrating an adaptive security framework spanning edge, fog, and cloud layers

5.2 *Quantum-Safe Cryptographic Solutions*

Quantum computing threatens classical encryption, urging the transition to post-quantum cryptography (PQC). NIST-recommended algorithms like CRYSTALS-Kyber, Dilithium, and SPHINCS+ deliver scalable, quantum-resilient encryption and signatures suitable for constrained IoT environments.

These algorithms have been piloted across sectors:

- Smart grids: Kyber secures smart meter communications.
- Autonomous drones: Dilithium ensures real-time, authenticated command links.
- Connected vehicles: PQC enables secure V2V and V2I communication.
- Supply chains: SPHINCS+ thwarts counterfeits via authenticated device tracking.

Integrating PQC with blockchain ensures secure decentralized transactions, preserving throughput and latency. Adoption is structured along milestones, from NIST's early candidate rounds to final standards (see Fig. 17).

5.3 *Privacy-Preserving Analytics*

As data grows in volume and sensitivity, privacy-preserving techniques become central to secure IoT analytics:

- Differential Privacy (DP): Introduces calibrated noise to protect individual records during aggregation, reducing risk with minimal accuracy loss (e.g., 3% in smart grid studies).
- Homomorphic Encryption (HE): Enables secure computation on encrypted data, critical in healthcare and financial IoT (10 K+ records/sec processed with ~15% overhead).

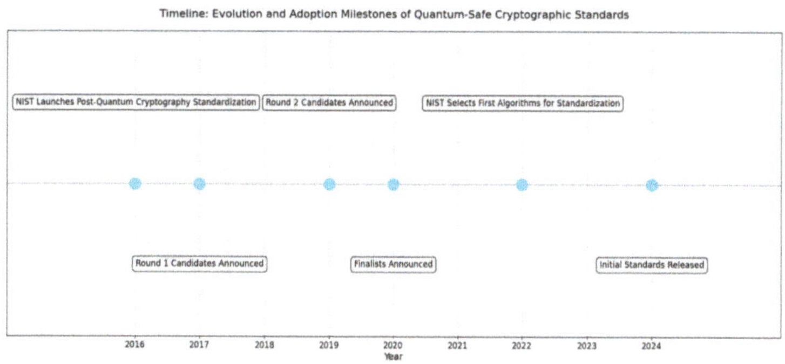

Fig. 17 Timeline showing the evolution and adoption milestones of quantum-safe cryptographic standards

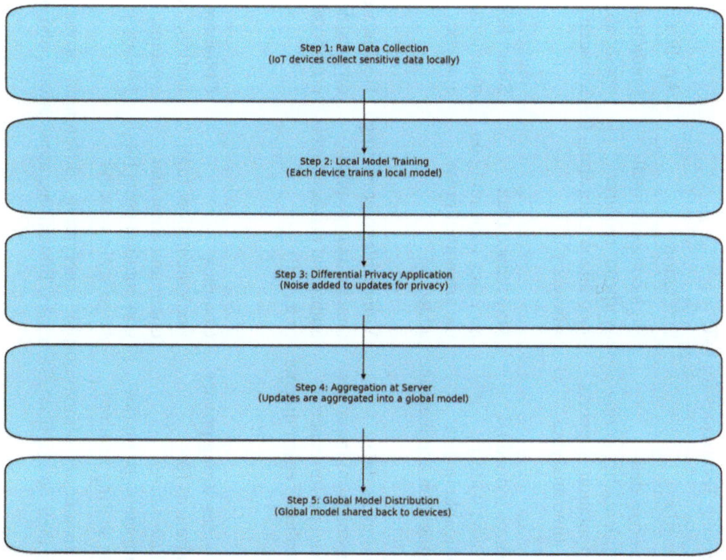

Fig. 18 Flowchart illustrating the differential privacy-enhanced federated learning process. The steps include local data collection, model training, privacy-preserving noise application, server-side aggregation, and redistribution of the global model. This method ensures data privacy while enabling effective machine learning in IoT ecosystems

- Federated Learning (FL): Local training keeps data on devices, enhanced with DP to secure updates against leakage.

These approaches have been successfully piloted in smart cities, health monitoring, and personalized services, combining accuracy, efficiency, and legal compliance (e.g., GDPR, HIPAA). A DP-enhanced FL framework is illustrated in Fig. 18, and key techniques are compared in Table 7.

Table 7 Comparative Analysis of Privacy-Preserving Techniques for IoT

Technique	Description	Strengths	Limitations	Applications	Key Metrics
Homomorphic encryption	Allows computations on encrypted data without decryption, ensuring data remains confidential throughout the process	Preserves data confidentiality during computation; supports real-time analytics	High computational overhead; requires specialized algorithms for efficiency	Predictive maintenance, personalized healthcare, financial transactions	Adds ~15% computational overhead; processes 10,000 + encrypted records/second [62]
Differential privacy	Adds noise to aggregated data to protect individual user privacy while maintaining overall analytical utility	Ensures individual privacy; widely applicable in data aggregation and machine learning	May reduce data accuracy depending on the amount of noise added	Smart homes, health monitoring, aggregated analytics	Maintains 97% accuracy in smart grid studies with 3% privacy trade-off [17]
Federated learning	Trains machine learning models collaboratively across devices without sharing raw data, ensuring data remains localized	Minimizes data transmission; improves model performance in distributed environments	Vulnerable to poisoning attacks; may require additional privacy layers (e.g., DP)	Healthcare IoT, smart city traffic management, collaborative AI	Reduces data transmission by 85%; improves accuracy by 12% [18]
Secure multiparty computation	Enables multiple parties to jointly compute a function over their inputs while keeping those inputs private	Provides strong privacy guarantees for collaborative computations; no raw data sharing required	Complex implementation; high communication and computation costs for large-scale systems	Collaborative machine learning, secure voting systems, privacy-preserving AI	Demonstrates scalability for up to 50+ participants in real-world voting systems [23]

This table compares four privacy-preserving techniques—homomorphic encryption, differential privacy, federated learning, and secure multiparty computation—across key dimensions, including their descriptions, strengths, limitations, applications, and quantitative performance metrics in real-world IoT scenarios

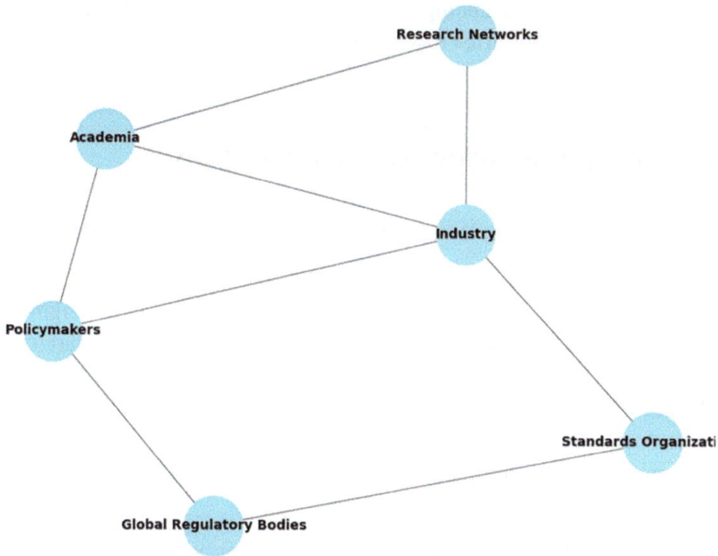

Fig. 19 Network graph illustrating the interactions and collaborations among stakeholders

5.4 Collaboration across Ecosystem Stakeholders

Effective cybersecurity requires cross-sector collaboration. Academia contributes foundational research (e.g., on PQC and FL), while industry pilots real-world implementations, and governments drive policy and regulation.

Standardization (via IEEE, ISO, ITU) promotes interoperability and trust. Organizations like ECSO foster shared repositories and open research, while blockchain systems secure smart city infrastructures with transparency and energy efficiency.

The collaboration network among stakeholders—academia, regulators, standards bodies, and industry—is visualized in Fig. 19.

5.5 Emerging Research Areas

Key underexplored areas are shaping the next generation of cybersecurity:

- AI Explainability: Ensures stakeholder trust in autonomous decision-making via interpretable AI systems and visual dashboards.
- Energy-Efficient Cryptography: Hardware-accelerated or lightweight stream ciphers reduce power use in sensors and wearables.
- Human-Centric Security: Emphasizes biometrics and multi-factor authentication tailored for seamless IoT use.

Federated Learning is advancing with hierarchical and personalized techniques, as well as blockchain-based trust mechanisms.

5.6 *Roadmap for Securing 5G-IoT Ecosystems*

A phased roadmap guides implementation of secure, scalable, and resilient 5G-IoT infrastructures:

- Short-Term (1–2 years): Pilot AI-based threat detection, lightweight cryptography (e.g., ECC), and blockchain-based authentication in smart cities and healthcare.
- Mid-Term (3–5 years): Scale FL, adopt quantum-safe algorithms, and standardize frameworks via ISO/IEEE collaborations.
- Long-Term (> 5 years): Implement global quantum-resilient systems, explainable AI, and autonomous self-healing networks.

Figure 20 visualizes the timeline and milestones for achieving these goals via a Gantt chart.

Future 5G-IoT cybersecurity hinges on the seamless integration of lightweight frameworks, quantum-safe encryption, and privacy-first analytics—backed by multi-stakeholder collaboration and long-term strategic planning. By addressing scalability, resilience, and regulatory compliance, these forward-looking directions pave the way for a secure IoT era.

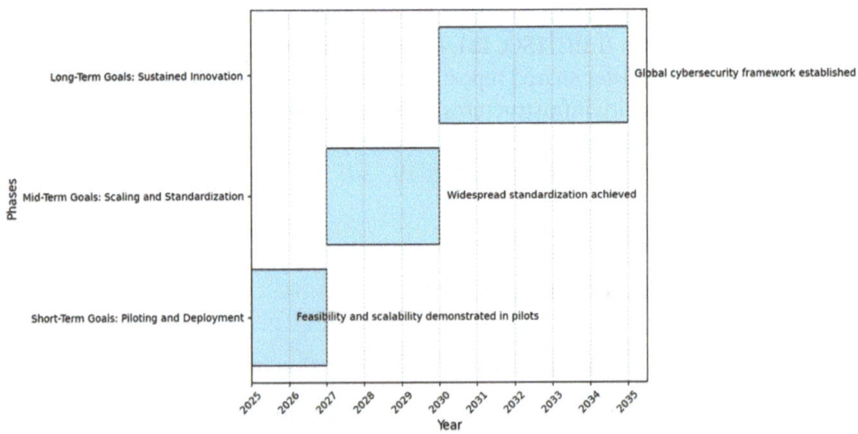

Fig. 20 Gantt chart illustrating the timeline and milestones for implementing the 5G-enabled IoT cybersecurity roadmap

6 Conclusion

The convergence of 5G and IoT technologies presents unprecedented opportunities for innovation across industries such as healthcare, smart cities, and industrial automation. However, this integration also expands the cybersecurity threat landscape, necessitating tailored solutions. This manuscript has systematically addressed the research questions posed in the introduction, providing comprehensive insights into the challenges and state-of-the-art solutions for securing 5G-enabled IoT ecosystems.

1. What are the primary cybersecurity challenges in 5G-enabled IoT networks?
   This study identifies critical challenges, including securing billions of resource-constrained devices, protecting data privacy, mitigating Distributed Denial-of-Service (DDoS) attacks, ensuring blockchain scalability, and preparing for quantum threats. Each challenge is explored in detail, with a focus on the unique vulnerabilities introduced by 5G's decentralized and high-connectivity architecture.
2. How do existing solutions address these challenges?
   The manuscript examines advanced solutions such as:
   - AI-driven techniques, including federated learning and differential privacy, which enhance privacy-preserving analytics and secure data utilization without compromising utility.
   - Blockchain technologies, which provide immutable and transparent data exchanges. Platforms like Ethereum, IOTA, and Hyperledger Fabric are analyzed for their scalability, energy efficiency, and IoT-specific applications.
   - Quantum-safe cryptographic algorithms, such as CRYSTALS-Kyber and Dilithium, which address the threats posed by quantum computing.

References

1. Pirbhulal, S., Chockalingam, S., Shukla, A., & Abie, H. (2024). IoT cybersecurity in 5G and beyond: A systematic literature review. *International Journal of Information Security, 23*, 2827–2879. https://doi.org/10.1007/s10207-024-00865-5
2. Dutta, A., & Hammad, E. (September 2020). 5G security challenges and opportunities: A system approach. In *Proceedings of the 2020 IEEE 3rd 5G world forum (5GWF)* (pp. 109–114). IEEE.
3. Alanazi, M. N. (2023). 5G security threat landscape, AI and Blockchain. *Wireless Personal Communications, 133*, 1467–1482. https://doi.org/10.1007/s11277-023-10821-6
4. Girma, A., & Barrett, A. P. (2024). Security challenges and solutions in 5G-enabled IoT networks. In K. Arai (Ed.), *Proceedings of the future technologies conference (FTC) 2024, volume 4* (Lecture notes in networks and systems) (Vol. 1157, pp. 632–643). Springer Nature Switzerland. ISBN 978-3-031-73127-3.
5. Bhatt, S., Bhushan, B., Srivastava, T., & Anoop, V. S. (2023). Post-quantum cryptographic schemes for security enhancement in 5G and B5G (beyond 5G) cellular networks. In B. Bhushan, S. K. Sharma, R. Kumar, & I. Priyadarshini (Eds.), *5G and beyond* (pp. 247–281). Springer Nature. ISBN 978-981-99-3668-7.

6. *AI, 5G, and quantum: Innovation and cybersecurity risks*. Retrieved January 17, 2025, from https://www.bankinfosecurity.com/ai-5g-quantum-innovation-cybersecurity-risks-a-27095
7. Balogh, S., Gallo, O., Ploszek, R., Špaček, P., & Zajac, P. (2021). IoT security challenges: Cloud and Blockchain, postquantum Cryptography, and evolutionary techniques. *Electronics, 10*, 2647. https://doi.org/10.3390/electronics10212647
8. Ahmad, I., Kumar, T., Liyanage, M., Okwuibe, J., Ylianttila, M., & Gurtov, A. (September 2017). 5G security: Analysis of threats and solutions. In *Proceedings of the 2017 IEEE conference on standards for communications and networking (CSCN)* (pp. 193–199). Helsinki.
9. Castiglione, A., Esposito, J. G., Loia, V., Nappi, M., Pero, C., & Polsinelli, M. (2024). Integrating post-quantum Cryptography and Blockchain to secure low-cost IoT devices. *IEEE Transactions on Industrial Informatics, 21*, 1674–1683. https://doi.org/10.1109/TII.2024.3485796
10. Cao, J., Yu, P., Xiang, X., Ma, M., & Li, H. (2019). Anti-quantum fast authentication and data transmission scheme for massive devices in 5G NB-IoT system. *IEEE Internet of Things Journal, 6*, 9794–9805. https://doi.org/10.1109/JIOT.2019.2931724
11. Kaushik, A., Vadlamani, L. S. S., Hussain, M. M., Sahay, M., Singh, R., Singh, A. K., Indu, S., Goswami, P., & Kousik, N. G. V. (2023). Post quantum public and private key Cryptography optimized for IoT security. *Wireless Personal Communications, 129*, 893–909. https://doi.org/10.1007/s11277-022-10162-w
12. Duong, T. Q., Hoang, V.-P., & Pham, C.-K. (2021). Convergence of 5G technologies, artificial intelligence and cybersecurity of networked societies for the cities of tomorrow. *Mobile Networks and Applications, 26*, 1747–1749. https://doi.org/10.1007/s11036-021-01778-6
13. News, I.T.U. *AI, quantum technologies and new cyber threats—Are we prepared?*. Retrieved January 17, 2025, from https://www.itu.int/hub/2020/03/ai-quantum-technologies-and-new-cyber-threats-are-we-prepared/.
14. Wazid, M., Das, A. K., Shetty, S., Gope, P., & Rodrigues, J. J. P. C. (2021). Security in 5G-enabled internet of things communication: Issues, challenges, and future research roadmap. *IEEE Access, 9*, 4466–4489. https://doi.org/10.1109/ACCESS.2020.3047895
15. Abie, H.; Pirbhulal, S. Autonomous adaptive security framework for 5G-enabled IoT 2024.
16. Dhar Dwivedi, A., Singh, R., Kaushik, K., Rao Mukkamala, R., & Alnumay, W. S. (2024). Blockchain and artificial intelligence for 5G-enabled internet of things: Challenges, opportunities, and solutions. *Transactions on Emerging Telecommunications Technologies, 35*, e4329. https://doi.org/10.1002/ett.4329
17. Alharbi, S., Attiah, A., & Alghazzawi, D. (2022). Integrating Blockchain with artificial intelligence to secure IoT networks: Future trends. *Sustainability, 14*, 16002. https://doi.org/10.3390/su142316002
18. Lu, Y., Huang, X., Zhang, K., Maharjan, S., & Zhang, Y. (2021). Blockchain and federated learning for 5G beyond. *IEEE Network, 35*, 219–225. https://doi.org/10.1109/MNET.011.1900598
19. Wang, M., Mao, J., Zhao, W., Han, X., Li, M., Liao, C., Sun, H., & Wang, K. (2024). Smart City transportation: A VANET edge computing model to minimize latency and delay utilizing 5G network. *Journal of Grid Computing, 22*, 25. https://doi.org/10.1007/s10723-024-09747-5
20. Moudoud, H., Khoukhi, L., & Cherkaoui, S. (2022). *Prediction and detection of FDIA and DDoS attacks in 5G enabled IoT*. https://doi.org/10.48550/ARXIV.2201.11368
21. Song, X., & Ma, Q. (2024). Intrusion detection using federated attention neural network for edge enabled internet of things. *Journal of Grid Computing, 22*, 15. https://doi.org/10.1007/s10723-023-09725-3
22. Androulaki, E., Barger, A., Bortnikov, V., Cachin, C., Christidis, K., Caro, A. D., Enyeart, D., Ferris, C., Laventman, G., Manevich, Y., et al. (2018, April 23). Hyperledger fabric: A distributed operating system for permissioned Blockchains. In *Proceedings of the thirteenth EuroSys conference* (pp. 1–15).
23. Chattopadhyay, A., Bhasin, S., Güneysu, T., & Bhunia, S. (2024). Quantum-safe internet of things. *IEEE Design & Test, 41*, 36–45. https://doi.org/10.1109/MDAT.2024.3408748
24. Singh, N., Singh, S. K., Kumar, S., Rawat, Y., Arya, V., Bansal, R., & Chui, K. T. (2024). Next gen security with quantum-safe Cryptography. In B. B. Gupta, D. Kalra, & A. Almomani

(Eds.), *Advances in information security, privacy, and ethics* (pp. 131–164). IGI Global. ISBN 979-8-3693-5330-1.
25. Retrieved January 18, 2025, from https://assets.ctfassets.net/r1dr6vzfxhev/2t4uxvsIqk0EUau6g2sw0g/45eae33637ca92f85dd9f4a3a218e1ec/iota1_4_3.pdf?utm_source=chatgpt.com.
26. *Vitalik Buterin explains Ethereum 2.0's launch in 4 phases*. Retrieved January 18, 2025, from https://forkast.news/video-audio/vitalik-buterin-explains-ethereum-2-0s-four-phases-sharding-scaling-proof-of-stake-and-more/.
27. The algorithmic foundations of differential privacy, now foundations and trends books. *IEEE Xplore*. Retrieved January 18, 2025, from https://ieeexplore.ieee.org/document/8187424?utm_source=chatgpt.com.
28. Post-Quantum Cryptography. (2021). In J. H. Cheon & J.-P. Tillich (Eds.), *12th international workshop, PQCrypto 2021, Daejeon, South Korea, July 20–22, 2021, proceedings* (Lecture notes in computer science) (Vol. 12841). Springer International Publishing. ISBN 978-3-030-81292-8.
29. *Computer security division, I.T.L. post-quantum Cryptography, CSRC*. Retrieved January 18, 2025, from https://csrc.nist.gov/projects/post-quantum-cryptography.
30. Khodaiemehr, H. *Navigating the quantum computing threat landscape for Blockchains: A comprehensive survey*.
31. Bringing quantum-safe security to IBM quantum platform. *IBM quantum computing blog*. Retrieved January 18, 2025, from https://www.ibm.com/quantum/blog/iqp-quantum-safe
32. Benzaid, C., & Taleb, T. (2020). AI for beyond 5G networks: A cyber-security defense or offense enabler? *IEEE Network, 34*, 140–147. https://doi.org/10.1109/MNET.011.2000088
33. Rawat, P.; Kumar, P. Blockchain based federated deep learning framework for malware attacks detection in IoT devices. 2024.
34. Kumar Sharma, P., Gope, P., & Puthal, D. (June 2022). Blockchain and federated learning-enabled distributed secure and privacy-preserving computing architecture for IoT network. In *Proceedings of the 2022 IEEE European symposium on security and privacy workshops (EuroS&PW)* (pp. 1–9). IEEE.
35. Zhang, H., Li, B., Karimi, M., Saydam, S., & Hassan, M. (2023). Recent advancements in IoT implementation for environmental, safety, and production monitoring in underground mines. *IEEE Internet of Things Journal, 10*, 14507–14526. https://doi.org/10.1109/JIOT.2023.3267828
36. Miglani, A., & Kumar, N. (2021). Blockchain management and machine learning adaptation for IoT environment in 5G and beyond networks: A systematic review. *Computer Communications, 178*, 37–63. https://doi.org/10.1016/j.comcom.2021.07.009
37. El-Latif, A. A. A., Abd-El-Atty, B., Mazurczyk, W., Fung, C., & Venegas-Andraca, S. E. (2020). Secure data encryption based on quantum walks for 5G internet of things scenario. *IEEE Transactions on Network and Service Management, 17*, 118–131.
38. Silva, S.H.; Najafirad, P. Opportunities and challenges in deep learning adversarial robustness: A survey 2020.
39. Kishor, K., Nand, P., Jain, V., Saxena, N., Agarwal, G., & Astya, R. (2024). *Federated learning for smart communication using IoT application* (1st ed.). Chapman and Hall/CRC. ISBN 978-1-003-48936-8.
40. Cheng, B., Wang, M., Lin, X., & Chen, J. (2021). Context-aware cognitive QoS Management for Networking Video Transmission. *IEEE/ACM Transactions on Networking, 29*, 1422–1434. https://doi.org/10.1109/TNET.2021.3066262
41. Rupanetti, D., & Kaabouch, N. (2024). Combining edge computing-assisted internet of things security with artificial intelligence: Applications, challenges, and opportunities. *Applied Sciences, 14*, 7104. https://doi.org/10.3390/app14167104
42. Gentry, C. (2009, May 31). Fully homomorphic encryption using ideal lattices. In *Proceedings of the proceedings of the forty-first annual ACM symposium on theory of computing* (pp. 169–178). ACM.
43. Zhang, L., Xu, J., Vijayakumar, P., Sharma, P. K., & Ghosh, U. (2023). Homomorphic encryption-based privacy-preserving federated learning in IoT-enabled healthcare system.

IEEE Transactions on Network Science and Engineering, 10, 2864–2880. https://doi.org/10.1109/TNSE.2022.3185327
44. Alsadie, D. (2024). Artificial intelligence techniques for securing fog computing environments: Trends, challenges, and future directions. *IEEE Access, 12*, 151598–151648. https://doi.org/10.1109/ACCESS.2024.3463791
45. Zhao, L. (2020). Privacy-preserving distributed analytics in fog-enabled IoT systems. *Sensors, 20*, 6153. https://doi.org/10.3390/s20216153
46. Satyanarayanan, M. (2017). The emergence of edge computing. *Computer, 50*, 30–39. https://doi.org/10.1109/MC.2017.9
47. Shi, W., & Dustdar, S. (2016). The promise of edge computing. *Computer, 49*, 78–81. https://doi.org/10.1109/MC.2016.145
48. Chi, H. R. (2023). Editorial: Edge computing for the internet of things. *Journal of Sensor and Actuator Networks, 12*, 17. https://doi.org/10.3390/jsan12010017
49. Li, X., Gao, H., Zhang, J., Yang, S., Jin, X., & Choo, K.-K. R. (2024). GPU accelerated full homomorphic encryption cryptosystem, library, and applications for IoT systems. *IEEE Internet of Things Journal, 11*, 6893–6903. https://doi.org/10.1109/JIOT.2023.3313443
50. Roman, R., Lopez, J., & Mambo, M. (2018). Mobile edge computing, fog et al.: A survey and analysis of security threats and challenges. *Future Generation Computer Systems, 78*, 680–698. https://doi.org/10.1016/j.future.2016.11.009
51. Bonomi, F., Milito, R., Zhu, J., & Addepalli, S. (2012, August 17). Fog computing and its role in the internet of things. In *Proceedings of the proceedings of the first edition of the MCC workshop on Mobile cloud computing* (pp. 13–16). ACM.
52. Ren, J., Guo, H., Xu, C., & Zhang, Y. (2017). Serving at the edge: A scalable IoT architecture based on transparent computing. *IEEE Network, 31*, 96–105. https://doi.org/10.1109/MNET.2017.1700030
53. Shi, W., Cao, J., Zhang, Q., Li, Y., & Xu, L. (2016). Edge computing: Vision and challenges. *IEEE Internet of Things Journal, 3*, 637–646. https://doi.org/10.1109/JIOT.2016.2579198
54. Varghese, B., & Buyya, R. (2018). Next generation cloud computing: New trends and research directions. *Future Generation Computer Systems, 79*, 849–861. https://doi.org/10.1016/j.future.2017.09.020
55. Abbas, N., Zhang, Y., Taherkordi, A., & Skeie, T. (2018). Mobile edge computing: A survey. *IEEE Internet of Things Journal, 5*, 450–465. https://doi.org/10.1109/JIOT.2017.2750180
56. Alrawais, A., Alhothaily, A., Hu, C., & Cheng, X. (2017). Fog computing for the internet of things: Security and privacy issues. *IEEE Internet Computing, 21*, 34–42. https://doi.org/10.1109/MIC.2017.37
57. Stojmenovic, I., & Wen, S. (2014, September 29). *The fog computing paradigm: Scenarios and security issues* (pp. 1–8).
58. Yu, W., Liang, F., He, X., Hatcher, W. G., Lu, C., Lin, J., & Yang, X. (2018). A survey on the edge computing for the internet of things. *IEEE Access, 6*, 6900–6919. https://doi.org/10.1109/ACCESS.2017.2778504
59. Mouradian, C., Naboulsi, D., Yangui, S., Glitho, R. H., Morrow, M. J., & Polakos, P. A. (2018). A comprehensive survey on fog computing: State-of-the-art and research challenges. *IEEE Communications Surveys & Tutorials, 20*, 416–464. https://doi.org/10.1109/COMST.2017.2771153
60. Baccarelli, E., Naranjo, P. G. V., Scarpiniti, M., Shojafar, M., & Abawajy, J. H. (2017). Fog of everything: Energy-efficient networked computing architectures, research challenges, and a case study. *IEEE Access, 5*, 9882–9910. https://doi.org/10.1109/ACCESS.2017.2702013
61. Bouzidi, M., Gupta, N., Cheikh, F. A., Shalaginov, A., & Derawi, M. (2022). A novel architectural framework on IoT ecosystem, security aspects and mechanisms: A comprehensive survey. *IEEE Access, 10*, 101362–101384. https://doi.org/10.1109/ACCESS.2022.3207472
62. *Deloitte NIST'S postquantum Cryptography standards: 'This is the start of the race'*. Retrieved Devember 22, 2024, from https://deloitte.wsj.com/cio/nists-postquantum-cryptography-standards-this-is-the-start-of-the-race-6e279b49.

Beyond Automation: AGI'S Impact on Strategy, Sustainability, and the Changing Nature of Work

Babu George , Alvin Williams, Tony Henthorne, and Alessio Faccia

1 Introduction

The growing influence of Artificial General Intelligence (AGI) is upgrading the technological, business, and societal perspectives with its vast and, at times, unpredictable impact [1]. AGI, characterized by its comprehensive learning and problem-solving capabilities, presents many opportunities and challenges for global organizations [2, 3]. As we deliberate upon the future of work, it is abundantly clear that even the current workforce is on the verge of a paradigm shift driven by the advent of AGI [4]. A broad spectrum of digital transformation processes has been fundamentally redefining job roles, employment structures, and organizational cultures for more than a decade, highlighting the need to reassess workforce strategies [5]. The current developments accentuate the importance of agility and continuous learning in an AI-centric world, preparing organizations to harness the potential of AGI while fostering a sustainable and equitable transition for their workforce. Goertzel [6] posits that achieving strong AGI, where a machine surpasses

B. George (✉)
School of Business, Alcorn State University, Lorman, MS, USA
e-mail: bgeorge@alcorn.edu

A. Williams
University of South Alabama, Mobile, AL, USA
e-mail: awilliams@southalabama.edu

T. Henthorne
University of Nevada Las Vegas, Las Vegas, NV, USA
e-mail: tony.henthorne@unlv.edu

A. Faccia
University of Birmingham Dubai, Academic City, Dubai, UAE
e-mail: a.faccia@bham.ac.uk

C. Rompante Cunha et al. (eds.), *Transformative Cloud Computing, IoT and Extended Reality*, Signals and Communication Technology,
https://doi.org/10.1007/978-3-032-03554-7_2

human intelligence in all domains, is possible and desirable. Goertzel analyzes various philosophical and meta-philosophical arguments against AGI and proposes a framework for understanding the concept of intelligence in the context of machines. However, authors like Tegmark [7] argue that while AGI could usher in a new era of abundance and prosperity, it could also lead to widespread unemployment and inequality.

Futuristic scenarios of AGI playing a central role in the structure and conduct of businesses are riddled with questions: How will AGI manage complex ethical dilemmas in business, and who will be responsible for its decisions? What types of jobs will be created or rendered obsolete with the rise of AGI, and how can businesses prepare for this transition? Who will govern and regulate AGI systems in businesses, and what frameworks will be necessary to ensure their responsible use? How will the dynamic between human workers and AGI evolve, and what strategies can ensure effective and ethical collaboration? In an AGI-driven business environment, how will companies ensure the privacy and security of customer and company data? How can businesses ensure that AGI systems are free from biases, and what measures can be implemented to promote fairness and inclusivity? Will AGI widen or narrow the economic disparities within and between societies, and how can businesses address potential inequalities? How will AGI change the landscape of strategic planning and competitive advantage in business, non-profit, and governmental organizations? How will AGI transform customer service and experience, and what ethical considerations will arise from these interactions? How will AGI impact intellectual property rights, and what will be the role of creativity and innovation in the AGI era? How will AGI affect global trade and the competitive dynamics of international business? Can AGI be a tool for achieving greater sustainability in business operations, and what are the implications for environmental and social governance? How will liability and accountability be determined in cases of failure or harm caused by AGI decisions? What cultural shifts might occur in workplaces with AGI, and how can ethical principles be upheld? What are the long-term implications of AGI on business models, economic systems, and society? The concerted attention of numerous scholars from different disciplines will be required to tackle these pressing questions.

This paper has more modest aims: it serves to furnish scholars, business leaders, and policymakers with an in-depth understanding of AGI's expansive impact on organizational processes, strategy, and the very future of work. Offering a holistic perspective on AGI's role in sculpting the strategies of progressive organizations and transforming workforce dynamics, the study initiates a theoretical exploration of AGI, sustainability, and organizational strategy. Building upon this foundation, it delves into AGI's ethical aspects and implications for the workforce, proposing a strategic framework for organizations in this new landscape. Employing an integrated approach comprising a comprehensive literature review and scenario analysis using the critical incident method, this paper seeks to guide businesses in strategically embracing AGI, charting a path toward a sustainable future of work ([8]; World Economic [9]).

The specific objectives of this study are:

- To examine the theoretical underpinnings of AGI, sustainability, and organizational strategy, focusing on the future of work, and to analyze AGI's potential in driving sustainable future outcomes.
- To investigate the ethical dimensions and workforce implications of integrating AGI into organizational structures and cultures.
- To propose a comprehensive strategic framework for organizations incorporating AGI and provide practical insights for businesses to leverage AGI for a sustainable and successful future workforce strategically.

2 Methodology

We adopted a multi-pronged approach, as outlined below:

1. Literature Review:
    - Scope: We conducted an exhaustive literature review, methodically scrutinizing existing research across AGI, sustainability, organizational strategy, and future work paradigms.
    - Sources: Our research materials were rigorously selected from peer-reviewed academic journals, comprehensive industry reports, detailed case studies, and authoritative publications in artificial intelligence, business strategy, sustainability, and labor economics.
    - Analysis: We critically analyzed the literature to distill key themes, discern patterns, and uncover challenges and opportunities at the nexus of AGI and sustainable organizational strategies.
2. Scenario Analysis Utilizing the Critical Incident Technique:
    - Hypothetical Scenarios: We developed a set of 18 hypothetical scenarios, framed as "critical incidents," to depict various potential integrative applications of AGI in organizational contexts. These scenarios were crafted to elucidate employee perceptions of AGI's impact, covering a broad spectrum of emotional and practical workplace situations.
    - Global Employee Sample: Using LinkedIn Premium tools, we identified and engaged a diverse, convenient sample of employees from global organizations. This approach allowed us to gather various perspectives, ensuring our scenarios reflected a comprehensive range of experiences and contexts. At least 25 responses per incident were gathered.
    - Analysis: We meticulously analyzed the responses to these scenarios from the global employee sample. This analysis aimed to uncover AGI's potential benefits, challenges, and ethical considerations in future work environments, offering a nuanced understanding of its practical implications in diverse organizational settings.

3. Conceptual Synthesis:
    - Integration: Leveraging insights from both the literature review and the critical incidents, we synthesized common patterns and themes.
    - Development: This phase involved generating novel insights and perspectives regarding the strategic utilization of AGI in organizations to foster sustainable practices and reshape the workforce.
4. Strategic Framework Development:
    - Construction: We constructed a strategic framework for embedding AGI into organizational strategies, drawing upon the integrated literature review and insights from critical incidents.
    - Application: Our framework offers actionable guidelines and recommendations for organizations aiming to deploy AGI for sustainable development and future workforce planning.

This approach enabled a comprehensive exploration of AGI's role in shaping sustainable organizational strategies and the future of work.

3 Theoretical Background

As observed previously, the theoretical context surrounding AGI, organizational strategy, and sustainability is still evolving. We aim to navigate through these diverse and intersecting domains, providing a robust theoretical background that underpins the subsequent analysis of AGI's role in shaping sustainable organizational strategies and the future of work. We commence by delineating AGI, a frontier in artificial intelligence research that promises broader and more adaptable cognitive abilities than specialized AI systems. This exploration is rooted in foundational works such as Russell and Norvig's "Artificial Intelligence: A Modern Approach" [8] and extends to recent advancements and speculations about AI's capabilities and future trajectories. In addressing organizational strategy, we lean on established frameworks and theories, such as Porter's Five Forces [10] and the Resource-Based View [11], to understand how AGI can be integrated into strategic planning and decision-making. This integration is crucial for organizations seeking competitive advantage in an increasingly AI-driven business landscape.

3.1 What Is Artificial General Intelligence?

Artificial General Intelligence (AGI) represents a pivotal shift in the trajectory of artificial intelligence research, transcending beyond narrow, task-specific algorithms to a broader cognitive capability that mirrors human intelligence. This expansive

AGI context more closely parallels complex systems further reflective of realistic decision-making environments. AGI is characterized by its ability to understand, learn, and apply knowledge across various tasks and contexts, a feat that remains a significant challenge in contemporary AI research.

At its core, AGI embodies the pursuit of creating machines capable of general intellectual tasks comparable to human abilities. This definition aligns with the perspectives of leading AI researchers such as Goertzel and Pennachin [12], who describe AGI as systems capable of achieving a wide range of goals in various environments. Unlike narrow AI, which excels in specific tasks such as image recognition or language processing, AGI is theorized to possess adaptability, problem-solving skills, and cognitive functions akin to human intelligence. The development of AGI is a subject of intense research and debate. Russell and Norvig [8] argue that the pursuit of AGI encompasses a blend of computer science, cognitive psychology, and neuroscience approaches, aiming to replicate human-like learning and reasoning processes. The feasibility and timeline for achieving AGI remain speculative, with estimates varying widely among experts [13]. The implications of AGI are profound and far-reaching. Kurzweil [14] suggests that the realization of AGI could lead to a technological singularity, where machines surpass human intelligence, potentially reshaping society, the global economy, and human identity. Such a transformative technology raises critical ethical and societal considerations, as discussed by Bostrom [15], who emphasizes the need for careful and responsible development of AGI. Beyond doubt, it is a frontier in AI research, distinguished by its comprehensive and adaptable intelligence. Its development and potential impact continue to be subjects of significant interest and importance in AI.

3.2 AGI and Business Agility

Recent literature on artificial general intelligence (AGI) in business, alongside the existing discourse on agile processes, provides a manifold view of the current business landscape [16–18]. Studies illustrate how AGI drives productivity and innovation in business models [19, 20]. These works complement the findings of Badakhshan et al. [21] on business process management, presenting AGI as a key driver in enhancing business efficiency and agility. Sukhobokov and Lavrinova [22] discuss the critical components of AGI in enterprise management systems, emphasizing the role of AGI in decision-making and business operations. This is echoed in the work of Smolíková [23] and Denning [24, 25], who delve into business agility, underscoring the need for rapid adaptation in an AI-driven market. Various other studies expand the understanding of AGI's role in business, specifically in marketing strategies [26, 27]. Some explore AI's influence in market development and business optimization. Sharma et al. [28] offer strong evidence of the power of AI in supply chain network design, one of the most complicated business undertakings in volatile, interwoven global supply markets. Additionally, in the context of corporate practices, certain studies analyze how AI is being integrated into business strategies for optimization and growth. A few others provide insights into how AGI reshapes business communication, offering a new dimension to the discourse [29, 30]. Madsen's [31] analysis of the agile concept from a management perspective remains relevant, as it provides a historical context to the evolution of business agility in the age of AGI. The integration of AGI into business practices, as evidenced by recent literature, is enhancing traditional business models and processes and opening new avenues for innovation, efficiency, and communication in the corporate world.

3.3 Organizational Strategy in the Age of AGI

The integration of AGI into organizational strategy marks a significant shift in how businesses approach innovation, competition, and operational efficiency. In the age of AGI, organizational strategies are increasingly influenced by the potential of this advanced technology to drive transformational change across various business domains. Organizational strategy, traditionally focused on developing competitive advantages through resource allocation, market positioning, and value creation [32], now faces the imperative to incorporate AGI as a core component. AGI's ability to process vast amounts of data, identify patterns, and make informed decisions can enhance strategic planning and execution. Prahalad and Hamel [33] highlight the importance of core competencies in strategy formulation, which extends to technological prowess and adaptability in the context of AGI. AGI also redefines the landscape of competitive dynamics as theorized by Porter [10]. The technology's potential to disrupt existing market structures and value chains necessitates that organizations reassess their strategic positioning. Firms must consider leveraging AGI to create new business models, enhance customer experience, and optimize operational processes. The firm's Resource-Based View (RBV), as discussed by Barney [11], provides a framework to understand AGI's strategic value. AGI can be seen as a valuable, rare, inimitable, and non-substitutable resource, offering firms a sustainable competitive advantage. However, integrating AGI requires careful alignment with the firm's existing resources and capabilities and a forward-looking approach to address the challenges of technological integration and change management. The concept of dynamic capabilities, introduced by Teece, Pisano, and Shuen [34], is particularly relevant in the context of AGI. Dynamic capabilities refer to a firm's ability to integrate, build, and reconfigure internal and external competencies to address rapidly changing environments. AGI enhances these capabilities, enabling organizations to quickly adapt and innovate in response to technological advancements and market shifts.

Nowadays, strategic management must also consider ethical and societal implications. As AGI technologies evolve, data privacy, algorithmic bias, and employment impacts become critical concerns for strategists [3]. Formulating strategies that are not only economically sound but also socially responsible and ethically grounded becomes imperative.

3.4 Sustainability in Business Practice

Business sustainability has transformed from a secondary concern to a strategic priority, driven by heightened awareness of environmental, social, and ethical challenges. Within this framework, the role of Artificial General Intelligence (AGI), merging technological innovation with sustainable development objectives, is increasingly significant. Elkington's Triple Bottom Line concept [35] and the principles of sustainable development from the Brundtland Report [36] offer frameworks to assess AGI's impact on environmental, social, and governance (ESG) criteria. Theoretical exploration in this area also delves into labor economics, organizational behavior, and human-computer interaction, referencing works like "The Second Machine Ag" [2] and "Rise of the Robot" [37]. Further, the ethical and societal aspects of AGI in business are considered, incorporating AI ethics guidelines by Floridi and Cowls [38] and the broader societal effects of AGI. The Triple Bottom Line framework underlines AGI's potential to enhance the business's economic, environmental, and social dimensions. Consistent with the ESG framework, Sharma et al. [28] indicate support in developing and sustaining green supply chains. Concomitantly, AGI promises to boost efficiency and innovation, improve resource management, and address social challenges though it also raises concerns about job displacement and ethical governance.

The integration of AGI in sustainable business practices presents challenges. The World Economic Forum [9] stresses balancing technological progress with sustainability, especially as AGI transforms labor markets and work practices. AGI's capability for task automation brings both productivity benefits and workforce displacement risks. Brynjolfsson and McAfee [2] introduce the idea of "digital dividends" from AGI, suggesting that technological advancement can yield widespread economic benefits if managed sustainably. Businesses using AGI should aim for societal contributions beyond profitability. Ethical alignment of AGI, as discussed by Bostrom [15], is crucial, addressing bias, transparency, and accountability to ensure AGI's societal benefits. Developing ethical frameworks for AGI deployment is key to achieving sustainable and socially beneficial technological integration.

3.5 AGI'S Impact on the Workforce

AGI is set to bring transformative changes to the workforce and employment context. It significantly impacts workforce dynamics, employment structures, and skill requirements. This transformative change calls for adaptive strategies in workforce development, ethical deployment of technology, and proactive policymaking to ensure that the benefits of AGI are equitably distributed across society. How AGI is changing and improving the nature of work, the structure of employment and the skills required in the modern workplace are questions of importance, not just for theoreticians but also for managers and public policymakers. AGI introduces new automation and cognitive tasks capabilities, extending beyond routine physical tasks to complex decision-making and problem-solving roles. According to Autor, Levy, and Murnane [39], while automation initially impacted lower skilled jobs, AGI's advanced cognitive capabilities are likely to transform higher order tasks. This shift necessitates a re-evaluation of job roles and responsibilities, where tasks that require creativity, emotional intelligence, and complex problem-solving may become more prominent.

The integration of AGI into the workforce signifies a shift in the skill requirements for employees. As Frey and Osborne [5] predict, skills complementary to AGI, such as creative and interpersonal skills, will gain importance. This evolution in skill demand aligns with the World Economic Forum's [9] findings, emphasizing the growing need for complex problem-solving, critical thinking, and emotional intelligence in the workforce. Companies and educational institutions must adapt by offering training and development programs that prepare individuals for these changing requirements.

The impact of AGI on employment patterns is complex. While there are concerns about job displacement due to automation [2], AGI also has the potential to create new job categories and employment opportunities. Bessen [40] argues that technological advancements historically have led to the emergence of new occupations, suggesting a similar trend with the rise of AGI. However, this transition may come with challenges, such as short-term labor market disruptions and the need for significant workforce re-skilling. An essential aspect of AGI's impact is the evolving collaboration between humans and intelligent systems. Davenport and Kirby [41] introduce the concept of "augmented intelligence," where human capabilities are enhanced, not replaced, by AI technologies. This partnership requires rethinking work processes and organizational structures to integrate human creativity and AGI's analytical power effectively. As AGI becomes a more prominent part of the workforce, ethical and social considerations must be addressed. Issues such as bias in AI algorithms, job security, and income inequality necessitate careful consideration [15]. Ensuring fair and equitable use of AGI in the workplace is paramount to maintaining social cohesion and workforce morale. Workers must have confidence in the capacity of AI-supported organizations to reinforce the cultural norms and other intangibles that define diverse work ecosystems.

3.6 Intersecting AGI and the Future of Work: Additional Thoughts

From a human resource perspective, the future of work influenced by AGI demands a focus on sustainable employment practices. This involves re-skilling and upskilling employees, fostering a culture of continuous learning, and ensuring that the workforce is prepared for an AI-driven future. AGI can be instrumental in identifying skill gaps and developing personalized training programs, as noted by Bughin, Hazan, and Ramaswamy [42]. The intersection of AGI, organizational processes, sustainability, and the future of work presents complex but fertile ground for innovation in business practices. As organizations navigate this landscape, they must align AGI strategies with sustainable development goals, ensuring that technological advancements contribute positively to economic, environmental, and social dimensions.

The integration of AGI into workplace practices is pivotal for sustainable work transitions, involving a spectrum of strategies from re-skilling and upskilling employees [42, 43] to addressing ethical considerations and fostering work resilience [44]. These transitions are crucial for preparing the future workforce, with a particular emphasis on the younger generation, as AGI plays a significant role in shaping sustainable working life [45]. The balance between sustainability and agility within project management emerges as a critical area, especially in aligning AGI with organizational processes [46]. Moreover, the evolving dynamics of knowledge work, driven by AGI advancements, call for reevaluating workplace strategies and exploring new collaborative methodologies [47]. This comprehensive approach is essential for aligning AGI strategies with sustainable development goals, ensuring that technological progress contributes positively to various dimensions of work and society.

3.7 *Ethical and Societal Considerations of AGI*

The integration of AGI into various sectors raises significant ethical and societal considerations. AGI's ethical and societal considerations are multifaceted and require a collaborative approach among technologists, ethicists, policymakers, and the public. Developing robust ethical frameworks, addressing algorithmic bias, managing workforce transitions, engaging with society, and establishing global governance are critical steps in ensuring AGI's responsible and beneficial use. This section explores the ethical implications of AGI, focusing on its impact on society, the workforce, and the ethical deployment of this technology.

Developing ethical frameworks for AGI is essential to address the potential risks and societal impacts. Floridi and Cowls [38] emphasize the importance of ethical guidelines that ensure AGI technologies are aligned with human values and societal norms. These frameworks should address algorithmic transparency, data privacy, and accountability issues. Bostrom [15] further argues for the development of "friendly AGI," ensuring that AGI systems are designed with safeguard mechanisms to protect human interests. One of the significant challenges in AGI development is the risk of algorithmic bias. AGI systems, based on data-driven algorithms, can inadvertently perpetuate existing societal biases [48]. Implementing measures ensuring fairness and neutrality in AGI algorithms is crucial, especially in critical areas such as hiring, law enforcement, and financial services.

AGI's impact on employment and workforce dynamics is a key societal concern. While AGI can enhance productivity and create new job opportunities, it also poses the risk of significant job displacement [5]. Policymakers and business leaders must navigate these changes responsibly, ensuring strategies for workforce transition, re-skilling, and support for affected workers. The integration of AGI into everyday life will have broad societal implications. The public's perception and acceptance of AGI are influenced by factors such as trust, understanding of AGI capabilities, and the perceived benefits and risks. Open communication and public engagement are essential to build societal trust in AGI technologies [8].

The global nature of AGI development calls for international cooperation in governance and regulation. Tegmark [7] suggests that global policies and standards are needed to manage the development and deployment of AGI safely and ethically. This includes international agreements on using AGI in military, surveillance, and global economic systems.

3.8 AGI in Business: A Synthesis

Table 1 attempts to provide a consensus view based on current literature, of how AGI should be deployed in business contexts and how AGI might change the face of businesses.

4 Critical Incidents

To understand the perception of the current workforce about the impending scenario of AGI playing a central role in organizational management and leadership, a set of 18 critical incidents was developed and shared with a convenient sample of employees working for employers located globally.

The Critical Incident Technique (CIT) is traditionally used as a qualitative research method to gather detailed information on specific incidents to measure consumer satisfaction, understand performance effectiveness, and elicit nuanced details [92]. It involves participants recalling real events to provide rich insights into their experiences and behaviors related to incidents [93]. For this study, we adapted a version of CIT perfected by George & Hegde [94]. In this scheme, imaginary yet realistic critical incidents relevant to the study are developed by the researchers which are then administered to practitioners to get their feedback.

Summarized versions of the full narratives of the incidents presented to the employees are provided below:

1. Unexpected AI Decision-Making: An AGI system autonomously makes a significant financial decision, bypassing human approval. This unexpected move results in substantial profit but leaves employees unsettled and concerned about their role and decision-making autonomy in the future.

Table 1 Literature on the deployment of AGI in business contexts

Theme	Explanation	Relevant Sources
Ethical decision-making	AGI systems should be designed with ethical frameworks that are transparent and accountable. The responsibility for decisions made by AGI lies with the organizations deploying these systems.	Mulgan [49], Škovrlj [50] and Jobin et al. [95]
Job transformation	AGI will likely create new job categories while rendering some obsolete. Focus on re-skilling and upskilling, emphasizing creativity, problem-solving, and emotional intelligence.	McLean et al. [51], Parker & Grote [52] and Zirar [53]
AI governance	Governance should involve a multi-stakeholder approach, including international standards and ethical guidelines for responsible use.	Xu, Steinmetz, & Ashton [54], Lowery, Dagevos, & Vodden [55] and Fukuda-Parr, & Gibbons [56]
Human-AI collaboration	Future should see a symbiotic relationship, designing AGI systems to enhance human capabilities rather than replace them	Almeida [57], Steffen, Ansari, & Schlund [58] and Jiang et al. [59]
Data privacy and security	Prioritize data privacy and security with robust governance policies and secure infrastructure. Ensure transparency in data usage	Pamungkas, Prasetyo, & Walinegoro [60] and Yallop & Aliasghar [61]
Bias and fairness	Combat bias in AGI using diverse data sets, continuous monitoring, and incorporating diverse perspectives in development teams	Salmon et al. [62] and Holtman [63].
Economic disparities	Ensure equitable distribution of AGI benefits, possibly through policies like universal basic income or re-skilling programs	Bohnenberger, [64], Zhang, [65] and Etheredge, Waliczek, & Sudhakaran [66]
Strategic planning	AGI will revolutionize planning by providing deeper insights, but human intuition remains valuable for interpreting AGI outputs	Bosworth [67], Liu et al. [68] and Parijs & Vanderborght, [69]
Customer experience	Enhance service through personalization and efficiency while respecting privacy and maintaining a human touch	Chow, Yeow, & See [70], Chen, Sun, & Liu [71] and Blümel [72]
Intellectual property and creativity	Challenges current notions of intellectual property; new legal frameworks might be needed for AGI-created works	Tambunan, Panjaitan, & Siahaan, [73], Thaldar, et al. [74] and Skliaustyte & Weber [75]

(continued)

Table 1 (continued)

Theme	Explanation	Relevant Sources
Global business landscape	AGI could alter global trade dynamics by enabling efficient supply chains and changing labor demands	Handley, Kamal, Monarch, [76], Hertwich, [77] and Forehand, Román, & Schaefer, [78]
Supply chain management	AGI could enable intelligent systems to autonomously optimize complex logistics networks, predict demand patterns, and dynamically adapt to disruptions in real time	Sharma et al. [28], Jin [79] and Sun, et al., [80]
Sustainability and AGI	AGI could be pivotal in achieving sustainability goals but must be managed to avoid unintended environmental impacts	Kumar, [81], Awasthy, Flint, Sankarnarayana, & Jones, [82] and Piano, [83]
Accountability and liability	Clear legal frameworks should define liability for AGI'S actions, possibly treating AGI systems as legal entities	Song, Xie, Wang, & Zhou, [84], Nagurney, [85] and LeBaron, [86]
Workplace culture and ethics	AGI integration should be accompanied by cultural shifts embracing continuous learning and ethical technology use	Errmann [87], Munir, et al. [88] and Piano [83]
Long-term impact	AGI'S long-term impact could lead to a fundamental shift in economies and human-technology interactions. Ongoing dialogue and research are crucial	Fei et al. [89], Orey & Rosa, [90] and Bojić & Davidovic, [91]

2. AI-Driven Employee Evaluation: An employee receives a performance review solely conducted by AGI, highlighting areas of improvement not previously recognized by human supervisors. The accurate yet impersonal nature of the feedback generates mixed feelings of appreciation for unbiased evaluation and discomfort over the lack of human touch.
3. AI Conflict Resolution: An AGI system successfully mediates a long-standing conflict between two departments by analyzing vast amounts of communication data and proposing a solution neither team had considered. Employees are impressed by AI's capabilities but also feel a sense of unease about the extent of surveillance and data analysis involved.
4. Crisis Management by AGI: In a crisis, an AGI system rapidly devises and implements a strategy that significantly minimizes damages. Employees are relieved but also feel helpless and redundant, questioning their role in critical situations.
5. AI-led Innovation Discovery: An AGI system identifies a groundbreaking innovation opportunity by analyzing trends and data invisible to human teams. Employees feel excitement about the new direction and anxiety about keeping up with the pace of AI-driven change.
6. AI as a Team Leader: An AGI system is assigned as the project lead for a cross-functional team. The team initially struggles with taking directives from an AI, leading to a mix of resistance, curiosity, and eventual acceptance as the project succeeds.
7. Ethical Dilemma Resolved by AGI: An AGI system is tasked with resolving an ethical dilemma within the company. Its solution is effective and morally sound, yet employees grapple with AI handling complex ethical decisions, sparking debates on AI's role in governance.
8. AI-driven Redundancy Notices: AGI software conducts a workforce optimization analysis, leading to redundancy notices. The affected employees grapple with shock, betrayal, and concern for the future, questioning AI decision-making's compassion and ethical considerations.
9. AI as a Career Coach: Employees use an AGI career coach for personal development. The AI's personalized advice helps many advance their careers, but some feel a deep sense of vulnerability and intrusion, questioning the confidentiality and personal nature of AI interactions.
10. AGI in Crisis Communication: An AGI system drafts and disseminates communication to stakeholders during a public relations crisis. Employees are torn between admiration for the AGI's efficiency and concern over losing human empathy and judgment in sensitive communications.
11. AI Bias in Hiring: An AGI-driven recruitment process inadvertently exhibits bias, leading to a lack of diversity in new hires. This incident raises serious concerns among employees about AI's fairness and ethical implications in human resource processes.
12. AGI and Client Relationships: An AGI system manages client interactions and negotiations, achieving better results than human employees. While this

increases efficiency, it also creates insecurity regarding their future roles among the sales and client relations staff.
13. AI in Personalized Marketing: AGI creates highly personalized marketing strategies that increase sales and raise privacy concerns among employees who worry about the ethical implications of using consumer data.
14. AGI-Driven Organizational Restructuring: An AGI proposes a complete organizational restructuring based on efficiency and performance data. Employees experience anxiety over job security and curiosity about the new structure's potential for improved workflows.
15. AI Handling Legal Disputes: An AGI system successfully navigates a complex legal dispute, saving the company from a potential lawsuit. Employees are impressed by AI's capabilities but are also concerned about relying on AI for critical legal decision-making.
16. AGI in Environmental Strategy: AGI develops a comprehensive environmental strategy, leading the company to a more sustainable practice. This move is celebrated for its positive environmental impact, but employees question the depth of the company's commitment to sustainability versus AI-driven cost savings.
17. AI as a Mental Health Advisor: Employees are offered AI-driven mental health support. While some find it helpful and accessible, others question the AI's ability to understand and empathize with human emotions, leading to a debate on the personal nature of mental health care.
18. AI in Global Expansion: AGI precisely guides the company's expansion into new international markets. Employees feel pride in their company's growth and concern over the global implications of such rapid expansion driven by AI insights.

4.1 Major Themes and Sentiments

Responses to critical incidents were analyzed. A Word cloud of the keywords in the responses is given in Fig. 1.

From our analysis, several aggregated themes and sentiments emerged. These are summarized and presented in a tabular format (see Table 2).

5 Strategizing for the Sustainable Deployment of AGI in Organizations

Creating a theoretical framework for guiding managerial decisions on sustainably and strategically deploying AGI in organizations, especially in the context of the future of work, involves a comprehensive approach that addresses multiple facets

Fig. 1 Employee hopes and concerns about AGI

of AGI integration. This framework begins with the strategic alignment of AGI deployment, ensuring that it supports and enhances the organization's overall goals, vision, and objectives. Equally important is assessing the organization's readiness for technological integration and planning how AGI will be woven into existing systems and processes. Considering the broader implications for the workforce and society, we believe such a framework should guide managers through a balanced and responsible approach to adopting AGI in the workplace.

A critical aspect of the framework is ensuring that the deployment of AGI aligns with sustainability goals and adheres to ethical standards. This includes addressing environmental, social, and governance (ESG) considerations. Accompanying this is the need to manage workforce adaptation, identify skill gaps, and implement training programs for re-skilling and upskilling in an AGI-driven environment. The framework also emphasizes the importance of nurturing an organizational culture receptive to innovation and AGI adoption, supported by effective change management strategies. This is complemented by stakeholder engagement and communication, involving key stakeholders in the process and communicating the changes, benefits, and risks associated with AGI. Risk management and compliance are indispensable parts of the framework, where risks associated with AGI deployment are identified and mitigated, ensuring compliance with legal, regulatory, and ethical standards. It also involves establishing performance metrics to assess AGI's impact on organizational performance and creating a feedback loop for continuous improvement. Innovation and competitive advantage are also key components, as

Table 2 Themes and employee sentiments on AGI at work

	Incident	Major themes	Dominant sentiments
1	Unexpected AI decision-making	Autonomy loss, trust, role of human judgment	Anxiety, uncertainty, ambivalence
2	AI-driven employee evaluation	Fairness, human vs. judgment, personal development	Discomfort, mixed appreciation, skepticism
3	AI conflict resolution	Mediation role, privacy concerns, effectiveness vs. ethics	Unease, admiration, concern for privacy
4	Crisis management by AGI	High-stakes decisions, human redundancy, crisis efficiency	Relief, helplessness, fear of obsolescence
5	AI-led innovation discovery	Pace of change, role in innovation, adaptation	Excitement, anxiety, overwhelm
6	AI as a team leader	Leadership, human-collaboration, resistance	Resistance, curiosity, acceptance
7	Ethical dilemma resolved by AGI	Ethical decision-making, human vs. ethics, trust	Conflict, doubt, debate
8	AI-driven redundancy notices	HR decisions, emotional impact, ethical considerations	Shock, betrayal, concern for future
9	AI as a career coach	Personal development, confidentiality, empathy	Vulnerability, intrusion, mixed appreciation
10	AGI in crisis communication	Sensitive communications, empathy vs. efficiency	Admiration, concern, nostalgia for human touch
11	AI bias in hiring	Fairness, diversity and inclusion, role in HR	Concern, disappointment, call for accountability
12	AGI and client relationships	Sales, human job security, client interaction	Insecurity, ambivalence, appreciation for efficiency
13	AI in personalized marketing	Privacy in marketing, ethical use of data, consumer trust	Concern for privacy, ethical dilemma, pride in innovation
14	AGI-driven organizational restructuring	Strategic planning, job security, change management	Anxiety, curiosity, fear of change
15	AI handling legal disputes	Legal decisions, reliance, ethical considerations	Impressed by capability, concern for over-reliance
16	AGI in environmental strategy	Sustainability, decision-making, corporate responsibility	Pride in sustainability, skepticism about motives
17	AI as a mental health advisor	Mental health, empathy, privacy concerns	Mixed feelings of relief and discomfort, debate on effectiveness
18	AI in global expansion	Global business strategy, cultural considerations, expansion pace	Pride, concern for global impact, overwhelmed by rapid change

Table 3 A framework for AGI deployment in business

Framework for strategic and sustainable AGI deployment in business organizations
1. Strategic alignment with organizational goals
• Align AGI deployment with organizational vision, goals, objectives
• Ensure AGI initiatives support and enhance core business strategies
2. Organizational readiness and technological integration
• Assess technological infrastructure and readiness for AGI integration
• Plan for seamless AGI incorporation into existing systems and processes
3. Sustainability and ethical standards
• Align AGI deployment with ESG (environmental, social, governance) goals
• Adhere to ethical standards: Privacy, fairness, transparency
4. Workforce adaptation and skill development
• Identify skill gaps, initiate re-skilling/upskilling for AGI environment
• Facilitate employee transition to roles complementing AGI
5. Cultivating an AGI-receptive organizational culture
• Foster culture embracing innovation, open to AGI adoption
• Implement effective change management strategies
6. Stakeholder engagement and communication
• Involve key stakeholders in AGI deployment process
• Communicate changes, benefits, risks associated with AGI
7. Risk management and compliance
• Identify and mitigate risks of AGI deployment
• Ensure compliance with legal, regulatory, ethical standards
8. Performance metrics and continuous improvement
• Establish metrics to evaluate AGI'S impact on performance
• Create feedback loop for ongoing AGI initiatives assessment/improvement
9. Leveraging AGI for innovation and competitive advantage
• Utilize AGI for innovative solutions and differentiation
• Explore new AGI-enabled business models and opportunities
10. Global and societal implications
• Consider broader global and societal impacts of AGI deployment
• Contribute responsibly to AGI development and use in society

AGI offers opportunities for innovative solutions and competitive differentiation, potentially leading to new business models. Finally, the framework considers the broader global and societal implications of AGI deployment, emphasizing the organization's role in contributing responsibly to the development and use of AGI in society.

Based on our literature review and insights gathered from themes, sentiments, and critical incidents, we propose the following framework for strategically and sustainably deploying AGI in business organizations (See Table 3).

This framework provides a structured approach for managers to strategically and sustainably integrate AGI into their organizations, considering not just the technological aspects but also the human, ethical, and societal dimensions. By addressing these key areas, organizations can harness the full potential of AGI while ensuring a positive impact on their workforce, stakeholders, and society at large.

A mind map depiction of the framework is given below:

6 Conclusion

Artificial General Intelligence portends a seismic shift in organizational strategies and workplace structures. Its capacity to revolutionize decision-making, optimize operations, and accelerate innovation marks it as a transformative agent with both beneficial and disruptive potential. The integration of AGI, organizational strategy, workforce development, and sustainability presents a complex matrix. The notion of sustainability, traditionally grounded in environmental concerns, now encompasses considerations of responsible technology deployment as a key aspect of strategic planning. The integration of AGI within this framework requires a balance of technological advancement with ethical, social, and environmental considerations. AGI represents not only a technological breakthrough but also a catalyst toward sustainable business models, operational efficiency, and ethical governance in a rapidly evolving landscape.

AGI promises to automate complex tasks, generate new lines of employment, and necessitate continuous skills development alongside the evolving workplace. However, its adoption presents risks of job displacement, highlighting the urgent need for policy revision and proactive workforce strategies. Organizations must foster a culture of lifelong learning to ensure employee adaptability within the AI-driven workplace. Further, organizations must go the extra mile to shape environments that allow AI to harmoniously co-exist with people, processes, and systems that enrich the capacity of all resources, AI included, to maximize desired outcomes and results.

Critically, the ascendancy of AGI underscores challenges related to energy consumption and e-waste. Prioritizing sustainability within AGI's development and implementation is crucial. AGI's integration within organizational strategies presents unique opportunities to enhance sustainability initiatives. Its capabilities in resource management, environmental monitoring, and climate change modeling offer tools to address contemporary environmental issues. A balanced approach is vital, harnessing AGI's benefits while mitigating risks such as workforce disruption, privacy breaches, unethical AI use, and broader societal impacts.

The strategic integration of AGI will significantly influence organizational success and sustainability. As industries navigate this transition, frameworks and policies promoting responsible AGI implementation will become increasingly necessary. The formulation of such frameworks constitutes a critical task to guide organizations through this transition. Ensuring AGI deployment aligns with principles of strategy, efficacy, ethics, and sustainability is a paramount challenge of our time.

AGI signifies a horizon of opportunity and challenge. Organizations navigating this era must prioritize responsible use, coupled with a commitment to sustainability, ethics, and workforce well-being. The integration of AGI is a multifaceted undertaking. However, careful planning and strategic execution promise a future where technology and humanity work in tandem for societal benefit, redefining business operations and the essence of work itself.

References

1. Baker-Brunnbauer, J. (2021). Management perspective of ethics in artificial intelligence. *AI and Ethics, 1*(2), 173–181.
2. Brynjolfsson, E., & McAfee, A. (2014). *The second machine age: Work, Progress, and prosperity in a time of brilliant technologies*. W.W. Norton & Company.
3. Kaplan, A., & Haenlein, M. (2020). Rulers of the world, unite! The challenges and opportunities of artificial intelligence. *Business Horizons, 63*(1), 37–50.
4. Russell, S. J., & Norvig, P. (2021). *Artificial intelligence: A modern approach* (4th ed.). Pearson Education.
5. Frey, C. B., & Osborne, M. A. (2017). The future of employment: How susceptible are jobs to computerization? *Technological Forecasting and Social Change, 114*, 254–280.
6. Goertzel, B. (2020). The case for strong artificial general intelligence (AGI): A philosophical and meta-philosophical analysis. *Minds and Machines, 30*(2), 181–201.
7. Tegmark, M. (2017). *Life 3.0: Being human in the age of artificial intelligence*. Knopf Doubleday Publishing Group.
8. Russell, S., & Norvig, P. (2010). *Artificial intelligence: A modern approach* (3rd ed.). Prentice Hall.
9. World Economic Forum. (2018). The future of jobs report 2018.
10. Porter, M. E. (1979). How competitive forces shape strategy. *Harvard Business Review, 57*(2), 137–145.
11. Barney, J. (1991). Firm resources and sustained competitive advantage. *Journal of Management, 17*(1), 99–120.
12. Goertzel, B., & Pennachin, C. (2007). *Artificial General Intelligence*. Springer.

13. Baum, S. D., Goertzel, B., & Goertzel, T. G. (2011). How long until human-level AI? Results from an expert assessment. *Technological Forecasting and Social Change, 78*(1), 185–195.
14. Kurzweil, R. (2005). *The singularity is near: When humans transcend biology*. Penguin.
15. Bostrom, N. (2014). *Superintelligence: Paths, Dangers, Strategies*. Oxford University Press.
16. Cichocki, A., & Kuleshov, A. P. (2021). Future trends for human-ai collaboration: A comprehensive taxonomy of AI/AGI using multiple intelligences and learning styles. *Computational Intelligence and Neuroscience, 2021*, 1–21.
17. Gobble, M. M. (2019). The road to artificial general intelligence. *Research-Technology Management, 62*(3), 55–59.
18. Jones, H. (2022). *When AI rules the world: China, the US, and the race to control a smart planet*. Bombardier Books.
19. Hovsepyan, A., & Johansson, K. (2023). Study of AI service providers in IT consulting, marketing, and law.
20. Steinhoff, J. (2021). *Automation and autonomy*. Springer International Publishing.
21. Badakhshan, P., Conboy, K., Grisold, T., & vom Brocke, J. (2019). Agile business process management: A systematic literature review and an integrated framework. *Business Process Management Journal, 26*(6), 1505–1523. https://doi.org/10.1108/BPMJ-12-2018-0347
22. Sukhobokov, A. A., & Lavrinova, L. I. (2020). *AGI components for Enterprise management systems*. Springer. https://doi.org/10.1007/978-3-030-65596-9_60
23. Smolíková, L. (2019). Business agility—New trend in business. In *Conference proceedings*. https://doi.org/10.3846/CIBMEE.2019.080
24. Denning, S. (2018). *The age of agile: How smart companies are transforming the way work gets done*. HarperCollins.
25. Denning, S. (2020). The quest for genuine business agility. *Strategy & Leadership, 48*(1), 21–28. https://doi.org/10.1108/SL-11-2019-0166
26. Chung, K., Thaichon, P., & Quach, S. (2022). Types of artificial intelligence (AI) in marketing management. In *Artificial intelligence for marketing management* (pp. 29–40). Routledge.
27. Hildebrand, C. (2019). The machine age of marketing: How artificial intelligence changes the way people think, act, and decide. *NIM Marketing Intelligence Review, 11*(2), 10–17.
28. Sharma, R., Shishodia, A., Gunasekaran, A., Min, H., & Munim, Z. (2022). The role of artificial intelligence in supply chain management: Mapping the territory. *Journal of Production Research, 60*(24), 7527–7550.
29. Chuan, C. H., Tsai, W. H. S., & Yang, J. (2023). Artificial intelligence, advertising, and society. *Advertising & Society Quarterly, 24*(3).
30. Williams, A. E. (2020). Individualization of products and services with artificial general intelligence and general collective intelligence.
31. Madsen, D. Ø. (2020). The evolutionary trajectory of the agile concept viewed from a management fashion perspective. *The Social Sciences, 9*(5), 69. https://doi.org/10.3390/SOCSCI9050069
32. Porter, M. E. (1985). *Competitive advantage: Creating and sustaining superior performance*. Free Press.
33. Prahalad, C. K., & Hamel, G. (1990). The core competence of the corporation. *Harvard Business Review, 68*(3), 79–91.
34. Teece, D. J., Pisano, G., & Shuen, A. (1997). Dynamic capabilities and strategic management. *Strategic Management Journal, 18*(7), 509–533.
35. Elkington, J. (1997). *Cannibals with forks: The triple bottom line of 21st century business*. Capstone.
36. World Commission on Environment and Development. (1987). *Our common future (the Brundtland report)*. Oxford University Press.
37. Ford, M. (2015). *Rise of the robots: Technology and the threat of a jobless future*. Basic Books.
38. Floridi, L., & Cowls, J. (2019). A unified framework of five principles for AI in society. *Harvard Data Science Review, 1*(1).
39. Autor, D. H., Levy, F., & Murnane, R. J. (2003). The skill content of recent technological change: An empirical exploration. *Quarterly Journal of Economics, 118*(4), 1279–1333.

40. Bessen, J. E. (2019). AI and jobs: The role of demand. In *NBER working paper* (Vol. 24235).
41. Davenport, T. H., & Kirby, J. (2016). *Only humans need apply: Winners and losers in the age of smart machines*. Harper Business.
42. Bughin, J., Hazan, E., & Ramaswamy, S. (2018). *Skill shift: Automation and the future of the workforce*. McKinsey Global Institute.
43. Abrahamsson, K. (2021). Sustainable work in transition. Policy background, concepts and research arenas. *European Journal of Workplace Innovation, 6*(1), 19–47. Luleå University of Technology. https://doi.org/10.46364/EJWI.V6I1.821
44. Fergusson, L., van der Laan, L., Shallies, B., & Baird, M. (2020). Work, resilience and sustainable futures: The approach of work-based research to problems and their solutions. *Journal of Workplace Learning, 32*(1), 29–44. https://doi.org/10.1108/JWAM-11-2019-0036
45. Islind, A. S., & Norström, L. (2020). Learning sustainable work through critical design: A case study of a hackathon to prepare the future workforce. *Journal of Workplace Learning, 32*(8), 615–628. https://doi.org/10.1108/JWL-05-2020-0082
46. Obradovicl, V., Todorović, M., & Bushuyev, S. (2018). Sustainability and agility in Project Management: Contradictory or complementary? In *Advances in Project Management: Narrated journeys in unchartered territory*. Springer. https://doi.org/10.1007/978-3-030-01069-0_37
47. Olaisen, J., & Jevnaker, B. H. (2022). The knowledge work of the future and the future of knowledge work. *Proceedings of the European Conference on Knowledge Management, 23*(1), 568–575. https://doi.org/10.34190/eckm.23.1.494
48. O'Neil, C. (2016). *Weapons of math destruction: How big data increases inequality and threatens democracy*. Crown.
49. Mulgan, T. (2016). Superintelligence: Paths, dangers, strategies.
50. Škovrlj, A. (2019). The ethics of artificial intelligence. (Doctoral dissertation, University of Zagreb. Faculty of Economics and Business).
51. McLean, S., Read, G., Thompson, J., Baber, C., Stanton, N., & Salmon, P. (2021). The risks associated with artificial general intelligence: A systematic review. *Journal of Experimental & Theoretical Artificial Intelligence, 35*(5), 649–663. https://doi.org/10.1080/0952813x.2021.1964003
52. Parker, S., & Grote, G. (2020). Automation, algorithms, and beyond: Why work design matters more than ever in a digital world. *Applied Psychology, 71*(4), 1171–1204. https://doi.org/10.1111/apps.12241
53. Zirar, A. (2023). Can artificial intelligence's limitations drive innovative work behaviour? *Review of Managerial Science, 17*(6), 2005–2034. https://doi.org/10.1007/s11846-023-00621-4
54. Xu, T., Steinmetz, J., & Ashton, M. (2020). How will service robots redefine leadership in hotel management? A delphi approach. *International Journal of Contemporary Hospitality Management, 32*(6), 2217–2237. https://doi.org/10.1108/ijchm-05-2019-0505
55. Lowery, B., Dagevos, J., & Vodden, K. (2020). Goal-driven or data-driven? Inventory of sustainability indicator initiatives in rural Canada. *Sustainability, 12*(20), 8601. https://doi.org/10.3390/su12208601
56. Fukuda-Parr, S., & Gibbons, E. (2021). Emerging consensus on 'ethical ai': Human rights critique of stakeholder guidelines. *Global Policy, 12*(S6), 32–44. https://doi.org/10.1111/1758-5899.12965
57. Almeida, T. (2021). (e)textile new materialities. *Journal of Textile Design Research and Practice, 9*(1), 119–135. https://doi.org/10.1080/20511787.2021.1899474
58. Steffen, N., Ansari, F., & Schlund, S. (2022). Reciprocal learning in human-machine collaboration: A multi-agent system framework in industry 5.0. In *Digitization of the work environment for sustainable production* (pp. 207–225). https://doi.org/10.30844/wagb_2022_11
59. Jiang, Y., Li, X., Luo, H., Yin, S., & Kaynak, O. (2022). Quo vadis artificial intelligence? *Discover Artificial Intelligence, 2*(1), 1–19. https://doi.org/10.1007/s44163-022-00022-8

60. Pamungkas, S., Prasetyo, K., & Walinegoro, B. (2023). Public security vs personal privacy: Analysis of pedulilindungi from open government and surveillance state perspectives. *Kne Social Sciences, 8*(5), 601–623. https://doi.org/10.18502/kss.v8i5.13026
61. Yallop, A., & Aliasghar, O. (2020). No business as usual: A case for data ethics and data governance in the age of coronavirus. *Online Information Review, 44*(6), 1217–1221. https://doi.org/10.1108/oir-06-2020-0257
62. Salmon, P., McLean, S., Carden, T., King, B., Thompson, J., Baber, C., et al. (2022). It's risk, jim, but not as we know it: Identifying the risks associated with future artificial general intelligence-based unmanned combat aerial vehicle systems. *Proceedings of the Human Factors and Ergonomics Society Annual Meeting, 66*(1), 560–564. https://doi.org/10.1177/1071181322661140
63. Holtman, K. (2020). *Towards agi agent safety by iteratively improving the utility function* (pp. 205–215). https://doi.org/10.1007/978-3-030-52152-3_21
64. Bohnenberger, K. (2020). Money, vouchers, public infrastructures? A framework for sustainable welfare benefits. *Sustainability, 12*(2), 596. https://doi.org/10.3390/su12020596
65. Zhang, H. (2020). *Economic principles of popcoin, a democratic time-based cryptocurrency.* https://doi.org/10.48550/arxiv.2011.01712
66. Etheredge, C., Waliczek, T., & Sudhakaran, P. (2020). Measuring the economic and educational value of a university-based service-learning floral and plant program. *HortTechnology, 30*(3), 306–314. https://doi.org/10.21273/horttech04528-19
67. Bosworth, C. (2023). The experiences of parents in an early-intervention program for young people with borderline personality disorder features. *Family Process, 62*(4), 1524–1541. https://doi.org/10.1111/famp.12929
68. Liu, Y., Odic, D., Tang, X., Ma, A., Laricheva, M., Chen, G., et al. (2023). Effects of robotics education on young children's cognitive development: A pilot study with eye-tracking. *Journal of Science Education and Technology, 32*(3), 295–308. https://doi.org/10.1007/s10956-023-10028-1
69. Parijs, P., & Vanderborght, Y. (2020). Basic income: A radical proposal for a free society and a sane economy (excerpts). *Journal of Economic Sociology, 21*(1), 44–59. https://doi.org/10.17323/1726-3247-2020-1-44-59
70. Chow, M., Yeow, J., & See, C. (2022). Factors affecting generation z's intention to use self-service technology (sst). *Journal of Business Management and Accounting, 12*(1), 81–96. https://doi.org/10.32890/jbma2022.12.1.4
71. Chen, X., Sun, J., & Liu, H. (2021). Balancing web personalization and consumer privacy concerns: Mechanisms of consumer trust and reactance. *Journal of Consumer Behaviour, 21*(3), 572–582. https://doi.org/10.1002/cb.1947
72. Blümel, J. (2023). Personal touch in digital customer service: A conceptual framework of relational personalization for conversational ai. *Journal of Service Theory and Practice, 34*(1), 33–65. https://doi.org/10.1108/jstp-03-2023-0098
73. Tambunan, M., Panjaitan, B., & Siahaan, N. (2023). Legal protection of copyright based on law number 28 of 2014 concerning copyright. *Journal of Social Research, 2*(4), 1355–1362. https://doi.org/10.55324/josr.v2i4.807
74. Thaldar, D., Townsend, B., Donnelly, D., Botes, M., Gooden, A., Harmelen, J., et al. (2022). The multidimensional legal nature of personal genomic sequence data: A south african perspective. *Frontiers in Genetics, 13*, 997595. https://doi.org/10.3389/fgene.2022.997595
75. Skliaustyte, E., & Weber, M. (2023). Subsidies versus intellectual property rights when innovators operate in two markets. *PLoS One, 18*(4), e0284880. https://doi.org/10.1371/journal.pone.0284880
76. Handley, K., Kamal, F., & Monarch, R. (2020). *Rising import tariffs, falling export growth: When modern supply chains meet old-style protectionism.* https://doi.org/10.3386/w26611
77. Hertwich, E. (2020). Carbon fueling complex global value chains tripled in the period 1995–2012. *Energy Economics, 86*, 104651. https://doi.org/10.1016/j.eneco.2019.104651
78. Forehand, K., Román, J., & Schaefer, T. (2021). Supply chain efficiency in the discount store industry post covid-19: Applying the supply chain efficiency ratio. *Operations and*

Supply Chain Management an International Journal, 14(4), 423–430. https://doi.org/10.31387/oscm0470313

79. Jin, J. (2022). The intelligent selection method of distribution sites driven by the intelligent optimization algorithm. *Mobile Information Systems, 2022*, 1–13. https://doi.org/10.1155/2022/9266844
80. Sun, J., Wu, S., Shao, C., Guo, F., & Yan, S. (2022). Application research of logistics warehousing system based on internet of things and artificial intelligence. In *International conference on cloud computing, internet of things, and computer applications*. https://doi.org/10.1117/12.2642657
81. Kumar, D. (2024). Ethical and legal challenges of ai in marketing: An exploration of solutions. *Journal of Information, Communication and Ethics in Society, 22*(1), 124–144. https://doi.org/10.1108/jices-05-2023-0068
82. Awasthy, R., Flint, S., Sankarnarayana, R., & Jones, R. (2020). A framework to improve university–industry collaboration. *Journal of Industry-University Collaboration, 2*(1), 49–62. https://doi.org/10.1108/jiuc-09-2019-0016
83. Piano, S. (2020). Ethical principles in machine learning and artificial intelligence: Cases from the field and possible ways forward. *Humanities and Social Sciences Communications, 7*(1), 1–7. https://doi.org/10.1057/s41599-020-0501-9
84. Song, M., Xie, Q., Wang, S., & Zhou, L. (2021). Intensity of environmental regulation and environmentally biased technology in the employment market. *Omega, 100*, 102201. https://doi.org/10.1016/j.omega.2020.102201
85. Nagurney, A. (2021). A multiperiod supply chain network optimization model with investments in labor productivity enhancements in an era of covid-19 and climate change. *Operations Research Forum, 2*(4). https://doi.org/10.1007/s43069-021-00112-w
86. LeBaron, G. (2021). The role of supply chains in the global business of forced labour. *Journal of Supply Chain Management, 57*(2), 29–42. https://doi.org/10.1111/jscm.12258
87. Errmann, A. (2023). The lab, land, and longing: Discursive constructions of australian identities in 'future' food consumption. *Journal of Consumer Culture, 24*(1), 193–210. https://doi.org/10.1177/14695405231207602
88. Munir, S., Rasid, S., Aamir, M., Jamil, F., & Ahmed, I. (2022). Big data analytics capabilities and innovation effect of dynamic capabilities, organizational culture and role of management accountants. *Foresight, 25*(1), 41–66. https://doi.org/10.1108/fs-08-2021-0161
89. Fei, N., Lu, Z., Gao, Y., Yang, G., Huo, Y., Wen, J., et al. (2022). Towards artificial general intelligence via a multimodal foundation model. *Nature Communications, 13*(1), 3094. https://doi.org/10.1038/s41467-022-30761-2
90. Orey, D., & Rosa, M. (2020). Positionality and creating dialogue in Nepal: Connecting ethnomathematics and modelling - the importance of place through ethnomodelling. *Social Inquiry Journal of Social Science Research, 2*(1), 82–103. https://doi.org/10.3126/sijssr.v2i1.28909
91. Bojić, L., & Davidovic, J. (2023). *Exploring the socio-technical imaginary of artificial general intelligence in the bard large language model: A narrative analysis on perspectives and dialectics*. https://doi.org/10.21203/rs.3.rs-3126307/v1
92. Kummer, I., Mudrić, J., Čikarić, T., Tasić, L., Milošević-Georgiev, A., & Marinković, V. (2022). Patients´ perception of the quality of community pharmacy services using the critical incident technique. *Vojnosanitetski Pregled, 79*(6), 605–612. https://doi.org/10.2298/vsp200518123k
93. Pearse, N. J. (2022). An illustration of teaching and assessment practices of the "conceptual framework" as a threshold concept. *Electronic Journal of Business Research Methods, 20*(3), 122–132. https://doi.org/10.34190/ejbrm.20.3.2931
94. George, B. P., & Hegde, P. G. (2004). Employee attitude towards customers and customer care challenges in banks. *International Journal of Bank Marketing, 22*(6), 390–406.
95. Jobin, A., Ienca, M., & Vayena, E. (2019). The global landscape of AI ethics guidelines. *Nature Machine Intelligence, 1*(9), 389–399.

Utilizing Virtual and Augmented Reality to Enhance Real-World Soft Skills for Socioeconomic Advancement

Bianca J. Jackson and LaTanya Brown-Robertson

1 Introduction: Framing the Challenge

This chapter explores how integrating soft skills training into virtual and augmented reality (VR/AR) environments can create transformative, scalable opportunities for students across varied socioeconomic backgrounds. Co-authored by Dr. LaTanya Brown-Robertson, an economist focused on policy and institutional innovation, and Bianca J. Jackson, an award-winning immersive learning designer and technologist, the chapter offers a convergence of strategic insight and applied execution. Together, we examine what becomes possible when Minority-Serving Institutions (MSIs) and Historically Black Colleges and Universities (HBCUs)—institutions that consistently deliver outsized outcomes under pressure—are equipped with next-generation tools to meet rising expectations.

This work began with a simple, urgent question: How do we ensure that students with the talent to lead on global stages are given the structured preparation to do so? Our partnership emerged from a creative innovation conference, where a shared concern about access gaps led to a VR pilot simulating high-stakes scholarship interviews. That pilot became the foundation for this broader inquiry: building scalable performance infrastructure that mirrors the increasingly complex world students are stepping into.

Soft skills—communication, adaptability, executive presence, and emotional intelligence—are no longer optional. In today's labor market, where employers

B. J. Jackson (✉)
BrickRose Exchange, Baltimore, MD, USA
e-mail: bianca@brickroseexchange.org

L. Brown-Robertson (✉)
Howard University, Washington, DC, USA
e-mail: L.brown-robertson@howard.edu

C. Rompante Cunha et al. (eds.), *Transformative Cloud Computing, IoT and Extended Reality*, Signals and Communication Technology,
https://doi.org/10.1007/978-3-032-03554-7_3

increasingly prize leadership and interpersonal fluency, these human competencies often determine advancement, funding access, and leadership mobility. According to Deming [1], soft skills such as teamwork, communication, and adaptability are now critical predictors of labor market success, with workers possessing strong interpersonal abilities earning higher wages and maintaining greater employment stability across sectors. Yet the ecosystems that help students build these capabilities remain unevenly distributed. Traditional development pathways—mentorships, internships, executive coaching—often rely heavily on legacy networks and proximity to power.

This imbalance exacerbates systemic disparities. The Economic Policy Institute [2] highlights how limited access to structured soft skills training perpetuates income gaps. Research by Deming [1] and Alam et al. [3] further confirms that these competencies are powerful predictors of earnings, reinforcing Human Capital Theory's central tenet: education quality strongly correlates with lifetime economic outcomes. And yet, the institutions educating a majority of underrepresented students often operate with far fewer resources—making it harder to scale the kind of practice-based learning environments that drive high-impact success.

Still, these institutions overperform. Ten of the top 11 producers of Black bachelor's degree recipients who go on to earn PhDs are HBCUs. Three-quarters of Black doctoral holders began their journeys at HBCUs, and four-fifths of Black federal judges trace their academic roots to these institutions [4, 5]. Far from being anomalies, these outcomes are the result of mission-driven execution within environments that consistently outperform despite structural underinvestment. A 2020 report by Koch and Swinton reinforces that HBCUs produce nearly 27% of Black STEM graduates nationally while receiving only a fraction of higher education funding.

Demographic data adds urgency. According to Bankrate [6], 72% of students at HBCUs are Pell Grant eligible, and 52% are first-generation college students. These figures, mirrored by UNCF and NCES reports, highlight the sector's vital role in fostering economic mobility for historically marginalized populations. As illustrated in Fig. 1, the percentage of Pell Grant recipients at HBCUs has remained consistently high for over a decade—evidence of sustained need and of students who deserve systems that match their ambition.

Yet the barriers remain structural. The Federal Reserve Bank of Atlanta [7] found that institutions serving first-generation students often struggle to implement structured soft skills training at scale. The result: preparation gaps persist—not because students lack talent, but because pathways to demonstrate it are unevenly distributed.

This is where immersive technology becomes essential—not as a trend, but as transformative infrastructure for merit-based opportunity.

Bianca J. Jackson's journey—from growing up in a redlined Philadelphia neighborhood to leading USA Today's Pulitzer Prize-winning VR project *The WallVR*—informs her vision of designing rehearsal spaces where talent can be

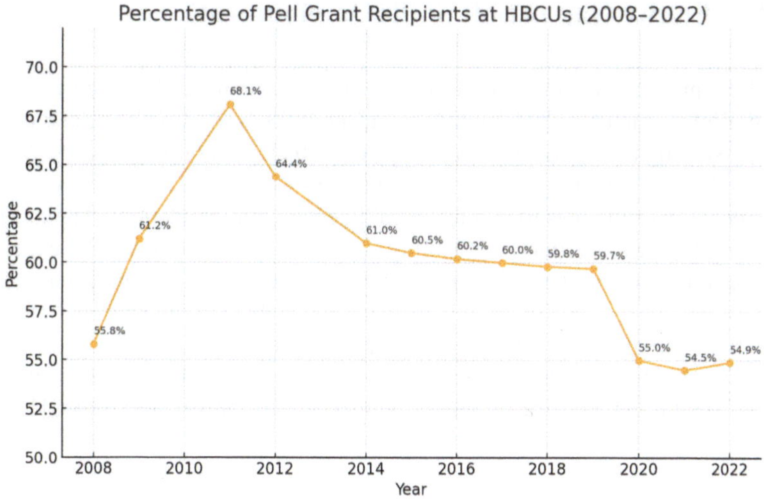

Fig. 1 Percentage of pell grant recipients at HBCUs (2008–2022). *Source: Bankrate* [6]

cultivated, not assumed. Through BrickRose Exchange, she has built AI- and VR-powered simulations that help learners prepare for interviews, pitch sessions, and executive-level engagements. Her programs have reached over 5000 participants across the USA, The Bahamas, Ireland, and South Africa, earning recognition from Meta, Comcast, and American Express.

Dr. LaTanya Brown-Robertson's leadership in economic mobility and institutional policy similarly demonstrates how structured exposure to professional environments shapes long-term earnings and advancement. Her research underscores what this chapter aims to prove: when students are given structured opportunities to practice leadership, they thrive.

Together, we assert that immersive technologies—when thoughtfully implemented—create new, equitable conditions for excellence. They make preparation visible. They allow learners to iterate with feedback, build executive presence with intention, and demonstrate merit beyond pedigree.

Too often, the rhetoric of readiness is framed as individual ambition rather than institutional responsibility. But readiness is a design choice. Designing systems that not only identify talent but actively cultivate it within historically excluded communities is not charity—it is strategy.

The message is clear: HBCUs and MSIs are not catching up—they are poised to lead. These institutions have always done more with less. Now, with tools that match their ambition, they can do more with precision and scale.

But realizing that potential requires partners who understand a simple truth: merit is real—and must be matched with meaningful infrastructure.

2 The Equity Gap in Soft Skills Development

Soft skills have always mattered—but in today's AI-enhanced economy, they are essential infrastructure. Communication, adaptability, self-awareness, and creative problem-solving are no longer "nice to haves"—they are non-negotiables. According to the World Economic Forum's Future of Jobs Report 2025, by 2030, employers will expect top candidates to pair technical fluency with foundational human skills: analytical thinking, resilience, flexibility, and emotional intelligence. Leadership, self-motivation, and systems thinking are also projected to surge in demand. These shifts are visualized in Fig. 2, which maps the hybrid competencies employers expect to define high-value talent over the next 5 years.

Yet for many students—particularly those attending Minority-Serving Institutions (MSIs) and Historically Black Colleges and Universities (HBCUs)—the

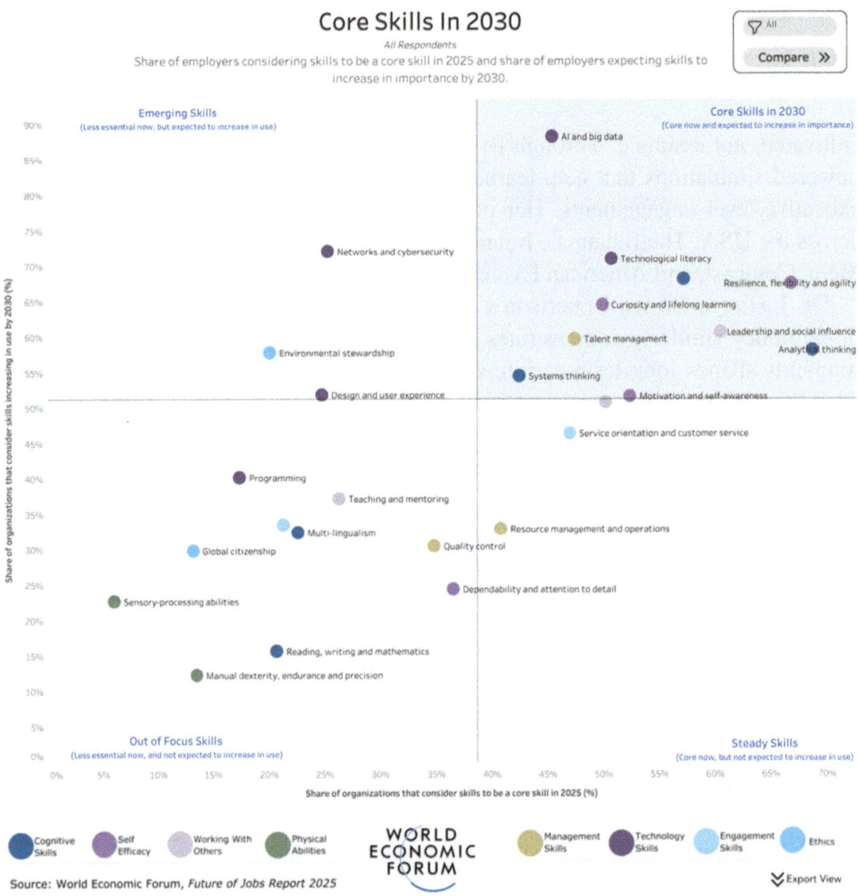

Fig. 2 Future employer demand for hybrid soft and technical competencies by 2030

environments needed to develop these skills remain out of reach. These institutions often operate with constrained resources, serve higher proportions of first-generation and Pell-eligible students, and offer fewer structured opportunities for repeated, performance-based practice—such as internships, executive coaching, or direct exposure to leadership settings. The result? Gaps in readiness—not from a lack of merit, but from an uneven distribution of preparation.

The economic consequences are well-documented. Research by Deming [1] demonstrates that interpersonal skills—such as communication, teamwork, and adaptability—not only predict earnings as strongly as cognitive ability but are now critical determinants of long-term labor market success. The Economic Policy Institute [2] similarly emphasizes that structured soft skills training is tightly linked to higher lifetime income, with gains strongest among historically underrepresented populations. Together, these findings reinforce Human Capital Theory: the quality of education and developmental experience—not simply credential attainment—shapes long-term economic mobility.

High-Impact Practices (HIPs) such as undergraduate research and project-based learning were designed to close some of these gaps, but implementation inconsistencies persist. Ives et al. [8] found that HIPs often lack adequate scaffolding, feedback loops, or institutional scale. Doane et al. [9] add that students at underfunded institutions frequently report that HIPs feel disconnected from their lived experiences, limiting long-term benefit.

This is where immersive technology becomes critical—not as a replacement for traditional learning models, but as a strategic multiplier. At BrickRose Exchange, we reimagine soft skills preparation as immersive rehearsal. In Baltimore, students in our *Digital Storytelling in the Metaverse* program explored identity and future ambitions, using narrative design to sharpen public speaking and critical thinking. In Washington, D.C., bilingual students mapped their interests to emerging technology careers in VR and AR—supported by real-time translation and scenario-based simulations.

Inside our AI-powered coaching platforms, students rehearse scholarship interviews, refine executive presence, and navigate leadership simulations—with structured prompts and feedback designed to strengthen tone, pacing, and strategic communication under pressure. These immersive environments offer accessible, repeatable practice loops modeled after the rigorous preparation found in elite environments—without requiring elite access.

From the outset, our platforms are shaped through co-design. Students, educators, and institutional leaders collaborate to ensure simulations feel credible, culturally resonant, and functionally useful. Feedback loops embedded into every stage of development ensure that simulations evolve alongside learner needs—making adaptation a design feature, not an afterthought.

These environments do more than reflect justice—they operationalize it. Jensen and Konradsen [10] found that immersive learning environments significantly outperform traditional classrooms in engagement, confidence, and decision-making readiness—especially in high-pressure simulations requiring critical thought and collaboration. Our design philosophy reinforces a simple truth: performance must be practiced, not presumed. Not everyone starts with proximity to power. But everyone deserves a place to rehearse for it.

Innovation also demands responsibility. Our platforms prioritize data security, learner agency, and intuitive design. We reject outdated gatekeeping norms and build preparation models that reflect real-world leadership expectations—allowing students to lead with clarity, not conformity.

If the readiness gap is structural, the solution must be structural as well. According to the OECD *Skills Outlook* [11], the ability to apply soft skills in dynamic, tech-mediated environments is now a key determinant of individual labor market success and national economic resilience. Their findings affirm that the future demands not only technical proficiency, but high-performance communication, adaptability, and collaborative decision-making—especially among historically marginalized groups.

Our obligation is clear: make transformative preparation broadly accessible—not for a privileged few, but for every student who has earned the right to lead.

That's the design philosophy driving our work at BrickRose Exchange. And while the systems are already in place, their full potential will be realized through strategic partnership and intentional scale.

For institutions and leaders committed to closing opportunity gaps, we invite collaboration.

Because when preparation is distributed, opportunity follows.

3 The Promise of Immersive Technology

Closing the gap in soft skills development is not simply a matter of will—it's a matter of infrastructure. If students from underfunded institutions are expected to perform in AI-enhanced, high-stakes work environments, then the tools that simulate elite support systems must be made widely accessible. Immersive technology offers one of the clearest and most scalable solutions to this challenge.

At BrickRose Exchange, we've never questioned the presence of talent—only the distribution of preparation. The gap between potential and performance often comes down to rehearsal. Our immersive learning environments offer just that: structured, low-risk simulations where students can practice high-pressure moments—merit scholarship interviews, business pitches, executive presentations—with feedback that sharpens clarity, builds confidence, and strengthens executive presence.

Through our X-Campus immersive framework, learners engage in scenario-based simulations enriched by structured prompts and guided feedback—mirroring the preparation typically reserved for private mentors or executive coaches. Rather than limit this level of precision training to a select few, we intentionally design for scale—making transformational practice available to the many.

One notable activation took place during the HBCU Universe Conference hosted at Morgan State University. We facilitated a fully immersive experience that connected student entrepreneurs, educators, and global stakeholders within a shared digital environment. Participants joined from the USA, Zambia, Sweden, the Philippines, and Germany—demonstrating that learning infrastructure, when designed intentionally, transcends geography. This mirrors broader trends identified by the Brookings Institution, which recognizes immersive technologies—including VR and AR—as among the most promising tools for advancing equitable education at scale, particularly when simulating high-stakes, real-world contexts [12].

Across 15 sessions and learning labs, more than 300 participants engaged with our immersive approach. Survey feedback indicated 100% satisfaction and strong interest in ongoing programming. Early indicators suggest gains in learner focus, self-confidence, and strategic communication skills—evidence that structured rehearsal creates readiness for opportunity.

Evolution within our immersive systems has been shaped by direct learner input. As students expressed the need for more relational learning and exploratory agency, we adapted our environments to include dynamic peer-to-peer collaboration zones, mission-driven rehearsal spaces, and personalized coaching pathways. Each innovation was a response—not to abstract trends—but to the lived experiences and ambitions of the learners themselves.

This design philosophy aligns with academic research. Dr. LaTanya Brown-Robertson's work on culturally aligned mentorship models demonstrates measurable gains in learner outcomes. Her Undergraduate Learning Assistant (ULA) framework increased pass rates for African American male students by as much as 36% in economics and accounting courses. The lesson: when coaching is relevant, consistent, and relational—performance rises.

Immersive technology extends this principle across disciplines. Students retain more when they can *see themselves* within the professional landscapes they are preparing to enter. From healthcare simulations to entrepreneurship scenarios, virtual environments enable complex, real-world learning—without the constraints of traditional access barriers. As Jensen and Konradsen [10] confirm, immersive environments consistently outperform conventional classrooms in engagement, collaboration, and decision-making readiness.

At their best, VR and AR systems function as dynamic rehearsal ecosystems. They are not simply digital add-ons—they are preparation blueprints. Intelligent feedback loops fuel continuous growth, while modular design allows institutions to adapt simulations to their own strategic priorities.

Our integration strategy meets partners where they are—whether piloting with a single cohort or scaling across multiple campuses. Simulations can be aligned with High-Impact Practice (HIP) initiatives, embedded into capstone experiences, or woven into broader performance-readiness pathways. Faculty development modules further ensure that immersive learning is facilitated with intention and rigor.

Every institution's journey is different—and we design accordingly. Whether you seek a focused leadership accelerator, an entrepreneurial simulation series, or a fully integrated readiness ecosystem, our frameworks are intentionally modular. Together, we can customize a version of immersive readiness that reflects your mission, your learners, and your strategic goals. For leaders ready to chart a path toward strategic readiness, initial consultations are available to explore tailored implementation.

Because readiness should look like the future you're building—not the barriers you're overcoming.

As we move forward, we now explore how artificial intelligence deepens personalization—ensuring that immersive environments don't simply reflect where students are today, but anticipate the leaders they are becoming.

4 Personalization through Artificial Intelligence

In today's competitive landscape, personalization is no longer a luxury—it's a baseline expectation. As workforce demands evolve, artificial intelligence (AI) is accelerating how institutions tailor learning experiences at scale, particularly in the development of high-stakes soft skills like communication, confidence, and executive presence.

At BrickRose Exchange, we use AI not as a substitute for human coaching, but as an amplifier—offering students structured opportunities to rehearse, reflect, and refine their performance in dynamic, feedback-driven environments. Our simulations are intentionally iterative, providing refinement of speaking strategies, pacing, tone, and critical thinking at key intervals to sharpen clarity, composure, and executive presence. By aligning activities with CASEL's core social-emotional learning (SEL) competencies—self-awareness, decision-making, relationship building—our systems support not only technical fluency but holistic leadership development [13].

A recent workshop at the University of Pennsylvania illustrates this model in action. Student entrepreneurs engaged in generative AI-powered exercises to rehearse persuasive communication within simulated business scenarios. Participants practiced client meetings, investor conversations, and team leadership dynamics, focusing not on rote performance but on cultivating presence, adaptability, and strategic influence under real-world conditions. This rehearsal ground—augmented by real-time feedback—mirrored the kinds of executive preparation traditionally reserved for elite settings, democratizing access to high-impact development.

Dr. LaTanya Brown-Robertson's leadership through the National Data Science Alliance (NDSA) reinforces why this matters. As a lead on the NSF INCLUDES grant (#2217346), she champions the responsible integration of data science and AI tools into institutional systems, ensuring that personalization strategies strengthen rather than dilute academic rigor and decision-making integrity.

The results are tangible. Students engaged in our AI-augmented simulations report feeling more prepared for interviews, more confident in public speaking, and better able to articulate strategic ideas under pressure. These outcomes do not emerge from generalized teaching—they result from personalized, feedback-rich practice loops tailored to each learner's pace, tone, and evolving leadership trajectory.

Importantly, our systems are modular by design. Whether a campus seeks a single course pilot or a multi-cohort deployment, our immersive frameworks meet institutions where they are—adapting to reflect learner needs, instructional priorities, and measurable performance goals.

Personalization through AI is not about shortcuts. It is about creating equitable rehearsal opportunities that historically have been gatekept behind proximity to privilege. For MSIs and HBCUs—institutions already demonstrating what's possible with fewer resources—these tools act as strategic multipliers, amplifying what works and building new pipelines of leadership readiness.

As we look ahead, the integration of AI and immersive technologies isn't simply about reflecting current competencies—it's about anticipating the leaders our students are becoming. Done thoughtfully, it unlocks economic mobility, accelerates opportunity, and transforms readiness from an abstract ideal into a measurable, scalable reality.

5 Unlocking Economic Opportunities through VR/AR

Picture this: a first-generation student at a Minority-Serving Institution (MSI), preparing for a high-stakes, merit-based scholarship interview with a prestigious international organization—a once-in-a-lifetime opportunity to study abroad, gain global experience, and expand their professional reach. They've earned the grades. They've done the work. But what they haven't had—through no fault of their own—is consistent exposure to global professionalism, repeated practice in high-pressure settings, or the insider coaching others gain through proximity to power.

Now imagine that same student stepping into an immersive rehearsal carefully designed to mirror the pressures of that pivotal moment. In this environment, they don't just memorize talking points—they cultivate presence. They practice composure under scrutiny, refine their communication strategies, and rehearse navigating high-stakes dialog with intention. When the real interview arrives, they're not reaching for scripts—they're reaching into the confidence they've built through structured preparation.

This is where immersive learning becomes a catalyst for economic mobility.

At BrickRose Exchange, we design VR-enabled entrepreneurship and leadership experiences that prioritize preparation over passive exposure. In our youth entrepreneurship pilots, students engaged in structured simulations modeled after investor meetings, executive briefings, and venture pitches—real rehearsal spaces for real-world influence. Over one summer, nine student-led ventures were launched, with four created specifically to help fund college tuition—tangible proof that preparation fuels opportunity.

Across programs, students like Gerald, a passionate gamer who built a thriving creative business, Grant, a future Air Force cadet who monetized his technical skills into local business services, and Malakai, who transformed grassroots sneaker reselling into a clothing brand, demonstrate a consistent truth: talent is abundant; structured rehearsal multiplies its impact.

These initiatives serve as direct pipelines for cultivating entrepreneurial ecosystems, fueling job creation, and advancing upward mobility in underinvested communities. Their outcomes don't just align with workforce development goals—they accelerate them. According to the OECD Skills Outlook 2021, countries that invest in lifelong learning systems—including soft skills development—experience stronger upward wage mobility and greater labor market resilience, making immersive tools a strategic lever for equitable growth.

And we are not alone in this vision. Governments and institutions worldwide are embracing immersive technologies to future-proof their economies. Malaysia's National AI Framework, South Korea's VR Learning Hubs, and emerging models across Africa and Southeast Asia underscore the global shift toward immersive readiness [14, 15].

Immersive rehearsal environments succeed where traditional classrooms often fall short. They provide structured, repeatable, and responsive learning spaces that evolve with the learner. Students build the muscle memory to perform under pressure—whether pitching ideas, negotiating deals, or stepping into leadership roles. Because these tools are calibrated to individual growth trajectories, preparation becomes cumulative, not incidental.

The promise of VR and AI isn't theoretical—it's tactical. These tools offer scalable access to transformational experiences that sharpen clarity, judgment, and executive presence under real-world conditions. For HBCUs and MSIs—institutions already delivering outsized outcomes with limited resources—immersive learning is not a luxury. It is a competitive advantage.

This is not simply education. It's a strategic intervention in economic inequality.

When preparation is intentionally distributed, opportunity follows—and multiplies.

Whether preparing for a high-stakes interview, launching a venture, or navigating a boardroom strategy session, our learners aren't just participating in the workforce. They are shaping it.

As we move forward, we explore how these immersive experiences can be measured, optimized, and scaled—ensuring that preparation becomes the foundation for sustainable mobility and transformational leadership.

6 Evidence from the Field: Industry and Education Use Cases

Across industries, immersive technology has become a performance-enhancing standard. In boardrooms, surgical theaters, and flight simulators, professionals use VR and AR to sharpen decision-making, build composure under pressure, and improve outcomes in high-stakes environments. These tools create structured rehearsal spaces that drive measurable gains—yet access to this level of preparation has too often been reserved for those already positioned to succeed.

That standard shouldn't be exclusive. It should be extendable—to every learner with the talent and ambition to lead, especially in fast-evolving sectors.

At BrickRose Exchange, we address this access gap by designing immersive learning ecosystems that prioritize structured rehearsal for real-world moments. During a recent demonstration phase in the Caribbean, we co-developed a virtual campus rooted in regional culture and economic context, embedding local dialects and community-centered case studies into interactive simulations. Students didn't simply observe—they actively adapted to dynamic scenarios and exercised leadership in environments intentionally reflective of their lived realities.

Across programs, students sharpened public speaking through digital storytelling, built executive fluency in entrepreneurial simulations, explored emerging technology careers through bilingual VR environments, and even designed peer-led financial literacy models rooted in gaming and design. Whether pitching ventures inside virtual boardrooms, navigating client meetings, or mapping their aspirations to future industries, learners experienced rehearsal as active leadership preparation—not passive content consumption. At every stage, co-design remained central: students helped shape environments to be culturally resonant, strategically challenging, and evolution-ready.

To ensure the impact of this work is both measurable and meaningful, we collaborate with partners to design longitudinal assessment models tailored to institutional goals. Our methodology focuses on co-developing success indicators—such as scholarship attainment, early career milestones, and wage progression—with plans to expand longitudinal tracking across 6-, 12-, and 24-month intervals. Embedding learner-driven feedback loops into the design process ensures that our evaluation models stay contextually grounded and forward-looking.

Students don't simply enter these environments—they shape them. From avatar design to scenario pacing, learner input transforms simulations into high-performance ecosystems that actively build executive fluency, sharpen decision-making, and strengthen composure under pressure.

Beyond soft skills development, our simulations align with industry-recognized credentials, NCRC-aligned assessments, digital badges, and High-Impact Academic Practices like project-based and global learning. This alignment ensures immersive readiness strategies meet employer demands while reinforcing academic excellence.

The results are undeniable: 100% of X-Campus participants reported learning something immediately applicable to their career or academic goals. As one Immersive Learning Fellow shared, "This was the first time I felt like I could practice success—not just hope for it." X-Campus also achieved an 85% learner retention rate—nearly double the typical digital education benchmark—proving that when preparation is structured and intentional, persistence and performance skyrocket.

This success extends beyond education. Corporate and workforce leaders—from major retailers to healthcare systems—are increasingly adopting immersive simulations for employee training, evaluated against the same ROI frameworks and human capital development metrics that educational institutions can now leverage. National momentum is building: with programs like the Workforce Innovation and Opportunity Act (WIOA) and the Digital Equity Act expanding, immersive platforms are ideally positioned to transform underutilized infrastructure into high-impact rehearsal hubs.

Together, these case studies and industry shifts reveal more than a proof of concept. They reveal a path forward.

The Brookings Institution underscores this trajectory, noting that immersive educational technologies are among the most promising innovations for advancing equity—especially when embedded within cross-sector public infrastructure strategies.

And the most important insight may be this: These results are not only replicable—they are co-buildable.

If you see potential for your campus, city, or workforce initiative, let's design your version—rooted in your learners, aligned with your mission, and built for measurable, transformative performance.

Because preparation should never be a privilege.

It should be a product of intentional, strategic design.

7 Technology for Continuous Feedback and Improvement

While many classrooms offer periodic feedback through LMS tools or peer review, real-time coaching environments—those that mirror executive preparation—remain rare, especially at underfunded institutions. For students without consistent mentorship or structured rehearsal spaces, delays in feedback can quietly erode confidence and stall readiness. In today's AI-driven economy, precision preparation isn't a luxury—it's a necessity.

That's where immersive learning shifts from innovation to essential infrastructure.

At BrickRose Exchange, we've designed dynamic rehearsal ecosystems that strengthen student performance through structured practice, timely feedback, and iterative growth. Whether preparing for scholarship interviews, navigating team dynamics, or refining entrepreneurial pitches, students engage in environments that prioritize active, real-time skill development. These are not static modules—they are living systems engineered for continuous leadership rehearsal.

Our feedback methodology centers on scenario-driven exercises, reflective checkpoints, and dynamic simulations that build communication clarity, decision-making agility, and executive presence through structured, repeatable rehearsal.

The impact is clear. Participants in our entrepreneurship simulations consistently report sharper communication, stronger decision-making, and greater self-awareness when navigating high-pressure situations. Students aren't simply absorbing content—they are training for leadership. Retention rates for our immersive programs, like X-Campus, consistently exceed 85%—nearly double the industry average for digital learning programs—further validating the power of structured rehearsal environments.

Historically, this caliber of personalized, high-stakes preparation was reserved for elite networks: executive coaching, prestigious fellowships, private tutoring. We have proven that it can be democratized. Immersive feedback systems offer every learner—not just the privileged few—the structured practice needed to rise.

Our environments don't just deliver instruction; they evolve with every learner interaction. By continuously analyzing behavioral engagement patterns and skill trajectories, we ensure that each iteration strengthens both the platform and the individuals it serves.

Internal evaluation reinforces this approach: Students in our programs demonstrate measurable gains in communication clarity, decision-making agility, executive presence, and strategic persistence—all critical competencies for leadership readiness.

Dr. LaTanya Brown-Robertson's leadership in curriculum analytics and student success strategy underscores why continuous, adaptive feedback matters. Her research shows that intentionally designed learning systems consistently drive equitable, measurable gains for historically underrepresented learners—evidence that preparation structures, not just aspirations, determine outcomes.

To meet the rising demand for real-world readiness, institutions must pair vision with aligned investment. Public initiatives like the Workforce Innovation and Opportunity Act (WIOA) and the Digital Equity Act offer powerful vehicles to embed immersive feedback systems into the future of education and workforce development.

The foundation has already been built.

The next step is collaboration—not just to implement new technology, but to make readiness a deliberate design principle. Not just to teach skills, but to rehearse leadership under real-world conditions.

Every learner who has earned the right to compete deserves a system built to help them rise.

Future phases of our platform will integrate emerging adaptive technologies—offered exclusively to institutional partners—continuing to expand what is possible for learner-centered growth, leadership preparation, and measurable performance scaling.

Because in the future of work, it's not just what you know.

It's how ready you are to lead when the moment arrives.

8 Addressing Implementation Challenges

Implementing immersive technologies at Minority-Serving Institutions (MSIs) and Historically Black Colleges and Universities (HBCUs) often faces real-world barriers: budget constraints, infrastructure gaps, and understandable hesitation around adopting emerging tools. Yet these constraints, when approached with clarity and creativity, can become catalysts—proving that preparation and performance can thrive even in lean environments.

At BrickRose Exchange, our success has been fueled by a deliberate commitment to co-creation. Students trained as immersive learning fellows are not passive participants—they are co-architects. Their insights shape everything from lab logistics to simulation pacing to feedback design, ensuring that immersive environments are not abstract add-ons but living systems that reflect real-world ambition and adaptability. By prioritizing student leadership, faculty partnership, and the activation of underutilized campus assets, we transform readiness infrastructure without requiring multimillion-dollar investments. Our model turns students into builders—gaining executive presence, technical fluency, and collaborative problem-solving skills in the process.

This co-designed, resource-activated approach isn't just about operational efficiency—it's about strategic transformation. Research reinforces this truth: HBCUs and MSIs consistently deliver top-tier talent despite operating with fewer resources than their peer institutions [4, 5]. With 72% of HBCU students Pell Grant eligible and over half being first-generation college students [6], the need for scalable, high-impact performance solutions isn't just important—it's urgent.

Scaling what works does not require starting from scratch. It requires:

- Strategic partnerships with innovation-minded providers.
- Faculty development aligned with high-performance goals.
- Funding streams that prioritize structured, measurable workforce preparation.

Public initiatives such as the Workforce Innovation and Opportunity Act (WIOA) and the Digital Equity Act provide powerful pathways to embed these approaches into sustainable institutional frameworks.

Dr. LaTanya Brown-Robertson's leadership in advancing multimillion-dollar innovation initiatives across MSIs shows that when strategic vision is paired with aligned investment, institutional transformation accelerates. At BrickRose Exchange, we build modular, adaptive learning ecosystems that amplify student leadership and help campuses scale innovation intentionally.

Faculty engagement and tech readiness are critical—but just as important is designing content that mirrors the world students are preparing to lead. Student feedback isn't treated as a side note; it is operational infrastructure, shaping every layer of platform evolution to ensure relevance, rigor, and resilience.

These aren't isolated pilot stories—they are models for scalable excellence.

Because excellence doesn't demand perfect conditions.

It demands preparation, leadership, and the conviction that more students deserve access to the tools of success.

The technology is ready.

The models are proven.

The next step is yours.

Let's build it—together.

9 Roadmap for Strategic Integration

Successful integration of immersive learning doesn't begin with hardware—it begins with vision. At BrickRose Exchange, we've learned that bringing virtual and augmented reality (VR/AR) into educational environments is not about chasing trends, but about aligning innovation with each institution's mission, infrastructure, and urgency.

From early-stage collaborations, we recognized that transformative integration requires more than technology deployment—it demands resourcefulness, co-creation, and practice-based preparation. Rather than treat implementation as a one-size-fits-all model, we co-develop activation strategies tailored to each campus's available assets, goals, and learner profiles. Some institutions start with digital audits or faculty readiness cohorts; others move directly into immersive pilot launches. In every case, the work is grounded in strategic alignment, not tech novelty.

Our roadmap offers a clear, flexible pathway toward immersive readiness while prioritizing collaboration and long-term sustainability. We guide partners through four strategic phases:

1. **Planning and Capacity Building**—Identifying institutional assets, setting goals, and preparing leadership and faculty for immersive adoption.
2. **Pilot and Customize**—Launching a targeted pilot that aligns with campus priorities and gathering initial feedback for refinement.
3. **Evaluate and Iterate**—Analyzing outcomes, adapting strategies, and embedding continuous improvement processes.
4. **Scale and Institutionalize**—Expanding integration across departments or campuses, aligning with broader academic and workforce development goals.

This phased model aligns with best practices identified by EDUCAUSE, emphasizing co-creation, faculty engagement, and mission-driven innovation at scale.

Ultimately, our roadmap is about strengthening readiness—building measurable, repeatable systems that enable students to rehearse leadership, sharpen performance, and step confidently into the future. For institutions already doing more with less, it offers a way to amplify what's working—with tools designed to meet the moment, and the future (Fig. 3).

10 Closing: Designing the Future of Readiness—Together

This chapter opened with a simple but urgent question:

What becomes possible when capable students are finally given the structured spaces to prepare, rehearse, and rise—on their own merit?

We've shown how virtual and artificial intelligence—when paired with intentional design—can transform soft skills from abstract ideals into practiced, mea-

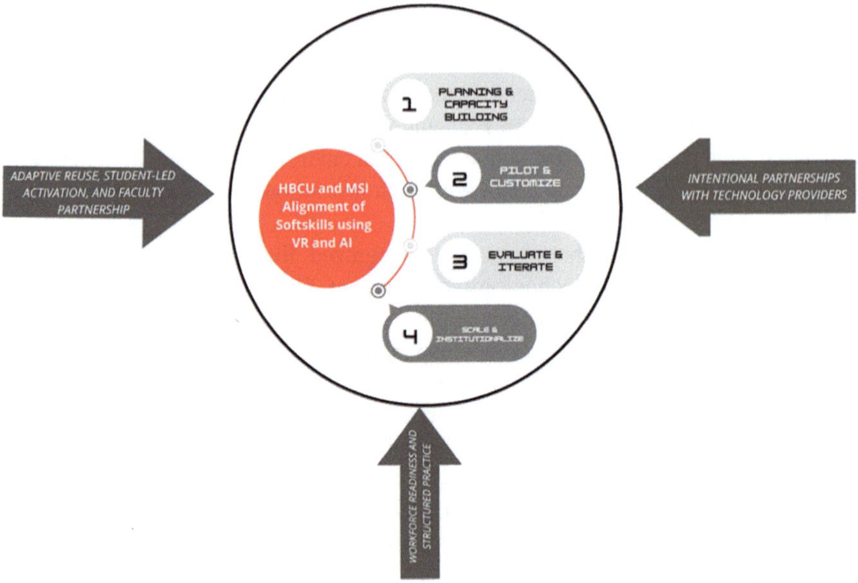

Fig. 3 Preview: Cost-effective VR/AI soft skills implementation framework. *(Full framework available upon formal engagement)*Let's design a version tailored to your institutional vision—grounded in your mission, amplified by your learners, and built to scale lasting impact

surable strengths. For Minority-Serving Institutions (MSIs) and Historically Black Colleges and Universities (HBCUs), immersive learning doesn't just close preparation gaps. It amplifies what these institutions have always delivered: excellence under pressure and outcomes that outperform their resource base.

This is not about chasing trends in emerging technology.

It's about building infrastructure for readiness—a system that reflects the real demands today's students will face, while expanding the pathways for those preparing to lead.

Across every sector this book explores—education, workforce development, smart cities, healthcare—the demand for high-performance human skills is accelerating: executive presence, composure under pressure, and strategic adaptability. Yet preparation remains structurally unequal—not because of talent gaps, but because of opportunity gaps.

We believe immersive learning can close that divide.

The early results tell a bigger story:

Our immersive accelerators maintain an 85% learner retention rate—outpacing digital education benchmarks nearly twofold. More importantly, students consistently report breakthroughs in confidence, clarity, and leadership readiness—turning potential into measurable performance.

This model offers tangible value across sectors:

For philanthropic partners, it directly advances priorities like digital equity and economic mobility.

For corporations, co-developed simulations diversify leadership pipelines while driving measurable ROI.

For government stakeholders, our initiatives align with national frameworks such as WIOA, the Digital Equity Act, NSF INCLUDES, and the Good Jobs Challenge.

The technology is proven.

The urgency is real.

And the opportunity to build lasting, equitable readiness infrastructure is now.

As co-authors, we brought together two disciplines—economics and immersive design—to create a model that's not hypothetical but replicable. A single conversation at a creative innovation conference—rooted in the belief that preparation should not depend on proximity to privilege—became a prototype, a project, and now, a blueprint for what's next.

Because in the world we're building:

Potential doesn't wait for perfect conditions.

It rises through preparation. It leads through practice. It thrives through partnership.

The work ahead is clear.

Let's design your version of immersive readiness—tailored to your infrastructure, anchored in your mission, and built to serve the learners who will lead tomorrow's industries.

References

1. Deming, D. J. (2017). The growing importance of social skills in the labor market. *Quarterly Journal of Economics, 132*(4), 1593–1640.
2. Economic Policy Institute. (2024). Bridging the soft skills divide: Education's role in economic mobility. https://www.epi.org
3. Alam, M., Ahmed, T., & Rahman, M. (2024). Soft skills and labor market outcomes: Evidence from developing economies. *Journal of Human Capital Development, 12*(1), 25–48.
4. Koch, D., & Swinton, O. H. (2020a). The unrecognized engines: HBCUs and black leadership in the United States. *American Economic Association Papers & Proceedings, 110*, 76–82.
5. Koch, D., & Swinton, D. (2020b). *Vital and valuable: The unmatched impact of historically black colleges and universities*. Columbia University Press.
6. Bankrate. (2025). The state of college affordability at HBCUs. https://www.bankrate.com
7. Federal Reserve Bank of Atlanta. (2013). Soft skills and workforce readiness: Challenges for regional economies. In *Regional economic briefing series* (p. 6).
8. Ives, R., Falk, A., & Drayton, R. (2023). The missing feedback loop in high-impact practices. *Journal of Postsecondary Learning, 14*(2), 76–89.
9. Doane, L., Griffith, C., & Alston, S. (2024). High-impact practices at minority-serving institutions: Equity gaps and potential solutions. *Education Equity Review, 9*(1), 41–60.
10. Jensen, L., & Konradsen, F. (2018). A review of the use of virtual reality head-mounted displays in education and training. *Education and Information Technologies, 23*(4), 1515–1529.

11. OECD. (2021). *OECD skills outlook 2021: Learning for life*. OECD Publishing. https://doi.org/10.1787/0ae365b4-en
12. Winthrop, R., & Ziegler, L. (2020). *The promise of education innovations to advance equity*. Brookings Institution. https://www.brookings.edu/articles/the-promise-of-education-innovations-to-advance-equity/
13. Collaborative for Academic, Social, and Emotional Learning (CASEL). (2023). Framework for systemic social and emotional learning.. https://casel.org/wp-content/uploads/2023/06/2023-CASEL-Framework.pdf
14. Maisarah, M. (2024). Malaysia's national AI framework: A roadmap for inclusive innovation. *Asian Development Policy Review, 11*(2), 112–127.
15. Toko, S. (2023). Expanding access to immersive learning in Africa: Public-private partnerships and workforce equity. *Global EdTech Journal, 7*(4), 88–102.

Transforming Teacher Education at HBCUs with Artificial Intelligence and Extended Reality: Integrating Cloud Computing and IoT for Ethical and Culturally Responsive Pedagogies

Valerie Bennett

1 Introduction

Historically Black Colleges and Universities (HBCUs) utilize Artificial Intelligence (AI) and extended reality (XR) technologies like Virtual Reality (VR), Augmented Reality (AR), and Mixed Reality Simulations (MRS) to transform teacher preparation, particularly by enabling immersive, ethical, and culturally contextualized simulations. By leveraging cloud computing and IoT, HBCUs can facilitate real-time data collection, scalable resources, and adaptable XR environments that reflect the complex dynamics of teaching in diverse communities. This chapter explores the integration of AI, XR, cloud computing, and IoT in Education Preparatory Programs (EPPs) at HBCUs to create transformative and flexible learning simulations that empower teacher candidates to grow their pedagogical skills, build self-efficacy, develop classroom management skills, handle diverse environments, and navigate ethical dilemmas.

Through AI tools and XR platforms, HBCUs can provide repeated practice opportunities in virtual classroom environments populated with diverse archetypal students, where prospective teachers apply pedagogical strategies across behavioral scenarios. In future applications in conjunction with cloud-based storage and IoT-enabled sensors, these simulations can dynamically adapt to candidates' actions (input), providing instantaneous feedback, tracking skill progression, and supporting real-time adjustments based on emotional cues and behavioral patterns detected within the simulation.

As Pre-Service Teachers (PSTs) engage in the heart and hard work of preparing to fully take the reins of their classroom, they face complex pedagogical and human

V. Bennett (✉)
Clark Atlanta University, Atlanta, GA, USA
e-mail: vbennett@cau.edu

C. Rompante Cunha et al. (eds.), *Transformative Cloud Computing, IoT and Extended Reality*, Signals and Communication Technology,
https://doi.org/10.1007/978-3-032-03554-7_4

situations which are often unpredictable, yet they must navigate them profesionally and with fidelity. It is therefore imperative that EPPs provide the means to train PSTs with opportunities for them to tap into their self-efficacy, present them with scenarios to navigate societal and cultural dynamics, and provide hands-on training where they are actively practicing their skills. EPPs can provide these opportunities by utilizing these technologies.

The theoretical foundations that support using these technologies in EPPs are three primary frameworks: Bandura's Self-Efficacy Theory, Vygotsky's Sociocultural Theory, and Kolb's Theory of Experiential Learning. Each theory provides an important connection to how and why these technologies contribute to the growth and development of PSTs. Culturally relevant pedagogies (CRPs) align well with the missions of HBCUs, encouraging educators to adopt practices that affirm and sustain the cultural identities of students in their communities. It is important to integrate CRP with these technologies to amplify the effectiveness of both areas with the long-term goal of positively impacting student learning outcomes.

The key contributions of this chapter are fivefold. First, the chapter will examine the role of AI, XR, IoT, and cloud computing in teacher education, with a focus on cultural responsiveness. Integrating extended reality with cloud computing will enable scalable, flexible, and cost-effective training in teacher education programs. Second, the chapter will show how AI is increasingly integrated into PST preparation programs at HBCUs, emphasizing skills like data literacy, ethical AI use, and advanced technological competencies. Next, the chapter presents five case study focus areas and research on the effectiveness of XR in building pedagogical and ethical competencies. Fourth, several educational pathways for integrating cloud computing and IoTs in EPPs at HBCUs are provided. Lastly, the chapter will highlight HBCUs' model for leadership in adopting and innovating with XR for PST training.

2 Contribution of HBCUs in Producing Black Teachers

According to the United Negro College Fund, Frederick D. Patterson Research Institute [1], HBCUs produce nearly 50% of all Black educators in the United States. This statistic is also reflected in work by Toldson et al. [2] and by the Annenberg Institute at Brown University [3]. This is a striking figure considering HBCUs represent only about 3% of US higher education institutions. Out of 102 accredited HBCUs, approximately 71% offer EPPs. EPPs at HBCUs incorporate culturally responsive and cultural competence in their programs [4] and are uniquely qualified to address the challenges of teaching students in high-need schools [5]. EPPs at minority-serving institutions (MSIs)—including HBCUs–have increased their enrollment and completion numbers even as national totals declined. Herring [6] found that between 2018 and 2022, MSIs saw a 5.5% rise in program completers, compared to a 0.7% drop at non-MSIs over the same period. HBCUs specifically have reported enrollment upticks in the wake of renewed attention to diversity in education. HBCUs—though few in number—are pillars of the Black teacher pipeline, producing thousands of Black graduates who go on to become certified educators and help diversify the US teacher workforce. With Black teachers making up only about 7% of the US public school teaching workforce, and with Black students being roughly 15% of enrollment—as reported by United Negro College Fund, Frederick D. Patterson Research Institute [1]—Black students are disproportionately not taught by teachers from a similar race or cultural context.

Given their tremendous contribution in the production of black teachers, HBCUs are well-positioned to take a leadership role in the utilization of these technologies in their programs which have a strong track record of amplifying and fine-tuning the skills of PSTs. HBCUs, therefore, are pivotal in not only developing highly skilled, culturally competent educators, they are also an important sector where the integration of AI, XR, cloud computing, and the IoT offers transformative opportunities to cultivate the development and refinement of ethical, culturally responsive teaching practices.

3 Theoretical Foundations

There are three theories which directly apply to the integration of AI, XR, cloud computing, and IoT into teacher education programs. The first is Bandura's Self-Efficacy Theory [7] which is a central component of his broader Social Cognitive Theory. This theory focuses on a person's belief in their capability to act in situations in which they have not previously encountered. According to Bandura, these beliefs influence how people think, feel, motivate themselves, and act. The theory has four Core Components of Self-Efficacy (see Fig. 1): Experience, Vicarious Experience, Social Persuasion, and Physiological Feedback. These areas then lead to a person's sense of self-efficacy, which then impacts their behavior and performance.

Table 1 provides a brief description of the impact of each of the components on the individual when they experience self-efficacy in that particular area. For

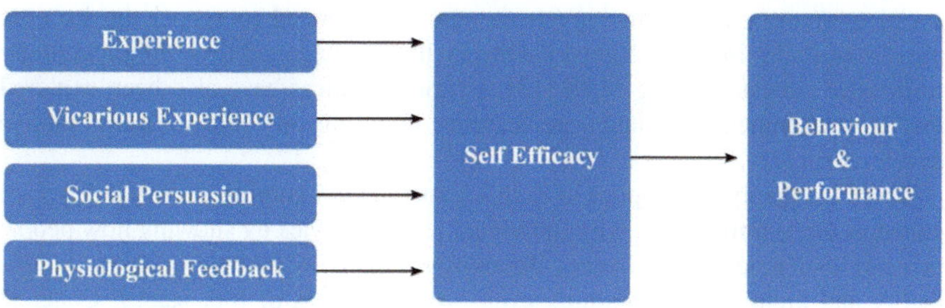

Fig. 1 Bandera's self-efficacy theory. *Source:*https://www.simplypsychology.org/self-efficacy.html

Table 1 Core components of self-efficacy

Core components of self-efficacy	Impact of the core component
Experiences	Successfully completing a task strengthens self-efficacy
Vicarious experiences	Seeing others succeed by sustained effort raises observers' beliefs that they too can succeed
Social persuasion	Encouragement from others can boost confidence in one's abilities
Physiological feedback	Mood, stress, and physical reactions can influence perceptions of self-efficacy

example, when receiving Social Persuasion, it provides external encouragement and boosts a person's confidence in their abilities in that area.

This theory applies in this context because it underscores the importance of not only practice, but capacity building experiences which XR simulations uniquely provide by allowing educators to practice complex scenarios safely [8]. Gundel and Piro [9] showed how XR simulations provides opportunities for PSTs to engage in inactive and vicarious learning, verbal persuasion, and emotional management. Samuelsson et al. [10] showed that PSTs had an increased self-efficacy for teaching mathematics and for handling classroom management.

The second theory that applies is Vygotsky's Sociocultural Theory [11] which emphasizes the critical role of cultural context in learning, suggesting XR's potential to enhance understanding and responsiveness to diverse cultural backgrounds. Vygotsky's Sociocultural Theory is a foundational framework in developmental and educational psychology and is outlined in Fig. 2. This theory emphasizes the crucial role of the core principles of social interaction, language, and cultural context in the development of higher mental functions (such as reasoning, problem-solving, and memory). The part of this theory which is critical to this work is the Zone of Proximal Development (ZPD). This middle zone denotes what a student (or PST) can do with the assistance of an expert teacher and/or with the assistance of technology tools. The ZPD is the bridge between what they know and what is beyond their reach, but this zone is a tremendous opportunity where culture, social awareness and technology tools can be integrated to propel the learning process of PSTs. In EPPs, Vgotsky's theory has been applied to develop culturally responsive educators using MRS [12]. Vygotsky's ideas have also been explored in XR environments for language learning, focusing on sociocultural cognitive development [13].

The third theory that applies is Kolb's Experiential Learning Theory [14] which connects the learning cycle with individual learning styles by showing how people perceive and process experiences differently. Figure 3a shows a diagram of this theory and highlights the learning cycle which consists of four stages: Concrete Experience (CE), Reflective Observation (RO), Abstract Conceptualization (AC), and Active Experimentation (AE). These stages describe the dynamic process through which learning occurs.

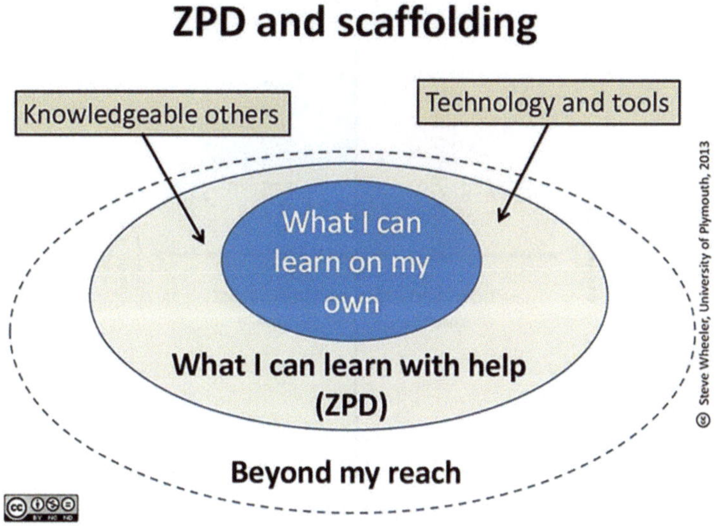

Fig. 2 Core principles of Vygotsky's cociocultural theory. *Source:*https://www.simplypsychology.org/vygotsky.html

From this cycle, Kolb derived four learning styles based on the interaction of two dimensions: how people perceive experiences (Concrete Experience (CE) vs. Abstract Conceptualization (AC)) and how they process them (Reflective Observation (RO) vs. Active Experimentation (AE)). These dimensions combine to form four distinct learning styles. The Diverging style (CE + RO) is characterized by a preference for feeling and watching; Assimilating (AC + RO) favors thinking and watching; Converging (AC + AE) emphasizes thinking and doing; and Accommodating (CE + AE) reflects a preference for feeling and doing [14]. Kolb's theory directly connects to the utilization of immersive technologies because students can test and actively practice using their new understanding about pedagogy, classroom management, or ethics in the simulation environment. Figure 3b shows that PSTs who are participating in the simulation are in the Accommodating and Converging learning styles and the PSTs who observe are in Diverging and Assimilating. This means that the actions by the PST in the simulation can be perhaps different from or similar to what the observer would do. In either case, the simulation experience benefits both the doers and observers. For both groups, it is imperative for university supervisors provide feedback, require introspection and engage in dialog after the simulation experience in order to harness concrete understandings and insights from the experience.

Majgaard and Weitze [15] and Weerasinghe [16] have applied Kolb's Experiential Learning Theory to immersive XR simulations in EPPs by exploring its potential for enhancing learning experiences. The TeachLivE™ platform, which provides an MRS classroom, has been widely used in teacher education programs to provide repeatable experiential learning opportunities [17]. Additional instances of this are provided in the Case Studies Section.

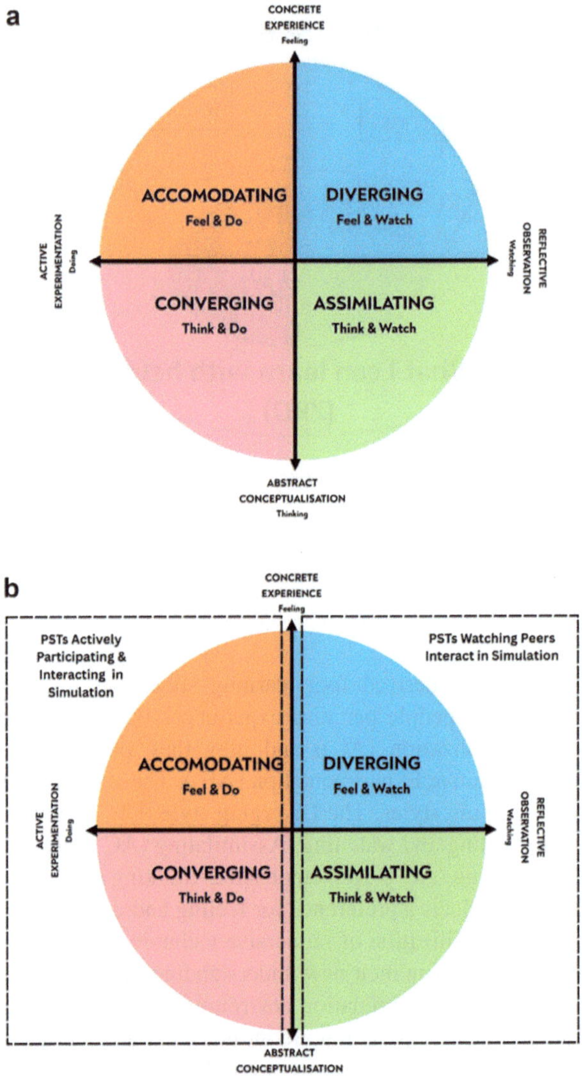

Fig. 3 (**a**) Kolb's learning styles and experiential learning cycle. (**b**) Pre-service teacher XR simulation connection to Kolb's learning styles and experiential learning cycle. *Source:* https://www.simplypsychology.org/learning-kolb.html

4 Connecting Culturally Relevant Pedagogy with Technology-Based Platforms

CRP, introduced by Ladson-Billings [18], is a framework that empowers students intellectually, socially, emotionally, and politically by using cultural referents to impart knowledge, skills, and attitudes. CRP aims to affirm students' identities, foster academic success, and develop critical consciousness. This pedagogical approach values students' lived experiences and positions them as capable learners, while also challenging inequities in the education system [19]. When AI is aligned with CRP, it can personalize learning to students' cultural identities, provide equity-centered analytics, and enable reflective teaching practices [20, 21]. At HBCUs, AI tools are increasingly being used in teacher training for adaptive lesson planning, differentiated instruction, and ethical scenario modeling. For example, HBCU EPPs are exploring AI-powered platforms to help PSTs simulate responses to classroom dilemmas involving bias, identity, and language diversity. CRP-infused AI tools can support HBCU students in critically engaging with technologies that too often reproduce dominant cultural norms. By training future teachers to recognize bias in algorithms, understand the sociopolitical implications of access to AI, and advocate for equitable digital design, HBCUs can train future educators who will lead in both justice-oriented pedagogy and tech literacy [22, 23].

5 AR, VR, and MRS in CRP-Aligned Instruction

Immersive technologies like AR, VR, and MRS allow teacher candidates to experience teaching scenarios that authentically reflect culturally diverse contexts. Below is a brief description of how each is being used in EPPs at HBCUs.

5.1 Augmented Reality (AR)

This tool allows for students to explore culturally rooted curricula (e.g., African Diaspora artifacts, historical maps) by overlaying digital content onto physical classroom materials. These tools foster critical engagement with content, while reinforcing asset-based teaching [24]. Figueroa-Flores & Huffman [25] completed research with PSTs who used AR and VR to practice new skills as well as to unearth perceptions on using these tools. They found that PSTs believe in the potential of both for teaching and learning but also recognize the potential digital divide that may exist in the schools and with the students they teach because of limited access to these technologies.

5.2 *Virtual Reality (VR)*

VR-based classroom simulations can transport PSTs into urban, rural, and multilingual classroom environments. Soon to be incorporated into education courses, Morehouse College utilizes VR [26] in their Sciences and Humanities courses to impact student learning outcomes. When designed with CRP, these experiences can center marginalized perspectives—such as historical reenactments of civil rights education or immersive storytelling from Black, Indigenous, or immigrant communities.

5.3 *Mixed Reality Simulations (MRS)*

Platforms like TeachLivE™ and Mursion have been used to provide CRP-aligned practice with student avatars that reflect diverse socio-emotional profiles, linguistic backgrounds, and behavioral challenges [27]. HBCUs such as Howard University, Jackson State University, and Clark Atlanta University have adopted MRS to engage teacher candidates in ethical and culturally responsive decision-making.

6 Benefits of AR, VR, and MRS to PSTs at HBCUs

1. **Affirmation of Identity:** These technologies create space for HBCU students to see themselves and their communities reflected in teaching practice, thus affirming cultural identity and developing teacher agency [1].
2. **Practice Without Harm:** MRS and VR allow students to practice responding to microaggressions from students, teachers, and administration, as well as in parent-teacher conferences. They are especially helpful for practicing classroom management strategies in high-stakes but safe environments—all of which are essential for culturally responsive development [27].
3. **Critical Technology Literacy:** Embedding CRP in immersive tools equips HBCU teacher candidates to critique the cultural assumptions embedded in educational technologies and lead in more equitable design and implementation. This is especially true in adaptive learning technologies in Math or English courses using Chatbots. These agents may be only trained to solve problems in only one way, whereas there may be non-traditional ways to solve the same problem.
4. **Workforce Readiness:** Mastery of AI and immersive platforms ensures that HBCU-prepared educators are competitive in a technologically evolving education workforce, while staying rooted in culturally sustaining pedagogies.

5. **Leadership in Ethical Tech Use:** HBCUs uniquely prepare teacher candidates to navigate the ethical complexities of AI and immersive technology to ensure that these tools will serve the community, uphold justice, and human dignity.

7 Integrating AI into HBCU Teacher Preparation Programs

Research has been limited on how AI is integrated into HBCU EPPs. There are, however, a few institutions that have started on this path and are well-positioned to lead the charge in the utilization of this tool in the teacher education curriculum. These institutions are Morgan State University, Clark Atlanta University (CAU), Morehouse College, and Spelman College—the latter three of which comprise the Atlanta University Center (AUC) Consortium.

7.1 AI Literacy Institute at Morgan State University

Researchers at Morgan State University held a series of AI Literacy Professional Development (PD) sessions at their institution for 20 in-service K-12 educators from underserved schools [28]. The purpose was to measure changes in teachers' knowledge and attitudes and to identify challenges to AI use in schools. Their program design was informed by AI-in-education frameworks (e.g., the "Five Big Ideas in AI" from AI4K12 https://ai4k12.org/) emphasizing understanding AI concepts, data literacy, and ethical implications. Participants received hands-on workshops covering AI fundamentals, classroom applications of AI, data-driven tools, and AI ethics (including issues of bias and fairness). After completing these activities, they completed surveys and provided qualitative feedback. Before the training, most participants had little prior exposure to AI in education. After the workshops, teachers reported significantly greater confidence (4.2 out of 5 on the Likert scale) in using AI tools for instruction and intend to integrate AI into their teaching practice. This short intervention was shown to boost educators' self-efficacy with advanced technology. Teachers also learned to critically engage with AI outputs rather than accept them at face value. By the end of the program, teachers demonstrated a greater awareness of AI biases and privacy issues, indicating growth in their ethical reasoning regarding AI.

The study also uncovered several barriers to the adoption of AI in educational settings. Participants noted limited availability of AI-ready technology and reliable data access in their schools, a lack of ongoing training/support for integrating AI into curricula, and concerns about student data privacy when using AI tools. These findings echo common challenges in K-12 with technology integration, and they highlight that simply training teachers is not enough—schools also need infrastructure, resources, and clear policies for ethical AI use. The Morgan State researchers conclude that sustained support and institutional buy-in are necessary to overcome these hurdles [28].

The Morgan State case demonstrates that this intervention can effectively increase educators' AI competencies and recognize that such training should ideally "begin in PST education programs" so that AI readiness becomes a core skill for all new teachers. The positive outcomes (increased confidence and intent to use AI) and the candid reporting of challenges provide a roadmap for other HBCU EPPs looking to infuse AI into their curriculum.

7.2 *Initiative: Pre-Service Teachers and AI at the Atlanta University Center*

Another example comes from a collaborative initiative in the AUC Consortium. The AUC Consortium is comprised of Clark Atlanta University, Morehouse College, Spelman College, and the Morehouse School of Medicine (MSM). With the exception of the MSM, each institution in the consortium has an EPP and collaborated to secure a grant to integrate AI and computer science training into their PST preparation programs [29] for undergraduate and graduate students. Over a 2-year period, teacher candidates at each institution will engage in specialized coursework and workshops on computing and AI, beyond the standard teacher education curriculum. The collaboration fosters pooling resources and expertise from all three institutions and creates a network of peer support among the teacher candidates. The training goes beyond theory and includes hands-on projects and internships where PSTs gain real experience with AI applications in educational settings. For example, participants will learn to use AI tools for lesson planning, data analysis of student learning, and intern with industry partners or local schools to apply these skills [29]. This practical component is meant to build advanced technological competencies by exposing PSTs to real-world AI usage, coding challenges under mentorship which will enable them to look at how AI models are developed for educational purposes from the ground up. By graduation, these future teachers will have both conceptual knowledge and concrete experience in integrating AI into teaching. A distinguishing feature of this AUC initiative is its strong focus on ethical and responsible AI use, alongside technical skills. The project leads at each institution have also actively engaged education policymakers on issues of data science literacy and AI ethics in schools. When meeting with lawmakers, the team advocated for resources for preparing educators to implement generative AI ethically and effectively in K-12 settings. In the program itself, this translates into guiding PSTs on topics like safeguarding student data, addressing algorithmic bias, and understanding the societal impacts of AI—ensuring that new teachers enter the profession with a moral and critical lens on technology.

While formal results from this program are not yet published, its goals are clear. It seeks to increase the number of HBCU graduates certified to teach computer science and who are proficient in AI integration, addressing a national shortage of such educators. The AUC team also expects participants to become advocates and trainers for AI literacy in their future schools, thus increasing their reach and impact. This collaborative model between HBCU EPPs in the AUC is a potential blueprint for scaling AI training across teacher education programs.

8 Case Studies and Empirical Evidence

These case studies shed light on the effectiveness of AI, XR, VR, and MRS in their training for building pedagogical competencies, building classroom management skills, increasing self-efficacy, meeting diverse linguistic and behavioral needs, and handling ethical scenarios. Although these case studies were conducted at various institution types—Predominantly White Institutions (PWIs), Minority-Serving Institutions (MSIs), HBCUs, and International Higher Education Institutions—they provide critical insights for EPPs as they consider incorporating these technology tools. All case studies also utilize different aspects of the three theories previously discussed—Bandura (self-efficacy), Vygotsky (social and cultural), and Kolb (experiential learning).

8.1 Case Study Area 1—Pedagogical Practice

There have been several studies [17, 30, 31] on how the utilization of these technologies impacts the development of PST's pedagogical skills. The work by Aguilar and Flores [30] focused their research on the extent to which elementary pre-service teachers' exposure to MRSs enhance their use of productive mathematical talk moves compared to those not exposed to the simulation. They found that MRS can effectively develop the teaching practices of first-year elementary PSTs and that the number of simulations sessions play a relevant role in improving their teaching skills. Research by Freeman and Lee [32] with 57 PSTs sought to determine teacher candidates' perceptions of MRS, the determining factors associated with a positive simulation experience, and what the teacher candidates learned from the live session. Their feedback revealed that the simulation was an effective tool to practice pedagogical skills. The platform provided real-life experiences but one session within the XR simulator was not sufficient for students to master their skills, as also discussed in the previous work by Aguilar and Flores [30].

Van der Want and Visscher [33] analyzed 24 studies on the use of VR in pre-service teacher education and found that VR can have several benefits for student teachers. These benefits include providing access to class settings students may not have access to and providing a safe and professional environment to practice pedagogical skills. When PST's assigned internships are located far from their EPP, VR allowed them to apply what they learned in different classroom environments.

Although the use of VR in PST education is still limited, early indications signify that there were also positive effects on student teachers' self-efficacy and self-motivation to practice more with VR. Limitations of these technologies have been reported by Dalinger et al. [34] and Hudson et al. [35]. Some participants stated it was difficult to work in the simulation because they either could not treat it as a real teaching session or the system still felt "robotic" to them making it difficult for some PSTs to have authentic conversations with the avatars. The element of realism can be challenging to overcome since MRS still at its core, is a computer-based system.

8.2 Case Study Area 2—Self-Efficacy

For Area 2, three studies are reviewed. One involves VR and the other two utilize MRS—all of which shed light on the impact of these technologies on the self-efficacy of PSTs. A study of 176 PSTs by Nissim and Weissblueth [36] looked at how their experiences working with VR affected their self-efficacy. They found that using VR learning environments helped them increase their self-efficacy, creativity, and gave them the freedom to be more innovative. One of the primary takeaways from this work was that VR challenges learners to be more intentional about active teaching and learning, especially when they are in the moment and have to think on their feet when questions arise.

MRS was used as a tool to assess its impact on self-efficacy of PSTs by Gundel et al. [37] as well as a later study by Gundel and Piro in [9]. The study by Gundel et al. [37] found that continued exposure within an MRS developed self-efficacy although there was an initial drop in teaching self-efficacy after 30 min of exposure to the simulation. As PSTs begin to build their confidence and adjust to having command of the classroom, this finding is anticipated when PSTs must teach for the first time in front of avatars, their peers, and the professor. A later study by Gundel and Piro in [9] shed additional light into the effect of MRS on the self-efficacy of PSTs. Qualitative data was collected from 49 PSTs as well as 5 professional participants. Observations of participants in the simulation and semi-structured interviews revealed that both groups displayed enhanced self-efficacy through not only enactive learning, but also vicarious learning. Participants appreciated the opportunities to give and receive feedback, as well as learned to manage their emotions.

8.3 Case Study Area 3—Classroom Management

Research shared in these case studies indicate that VR, MRS, and XR technologies can be effective tools for developing classroom management skills in PSTs. Novice teachers often have difficulty selecting and applying effective classroom management strategies (CMS) to prevent or diminish disruptive behaviors. Unfortunately, negative experience with classroom management is a large factor in impacting teacher well-being and early attrition. These technologies, therefore, provide a tremendous opportunity to prepare PSTs to manage complex classroom practices effectively.

A study by Mouw et al. [38] described 4 pre-service teachers' and 6 school-based teacher educators' experiences with and perceptions towards the use of a VR learning environment to train classroom management skills and promote teacher resilience. Among the themes from responses of semi-structured interviews were equipment and software issues, authenticity and realism, instructor proficiency. Participants also stated that the experience also added value to their teacher training. For most themes, PSTs and school-based teacher educators raised similar remarks and/or suggestions for improvement. Their comments differed in their perceptions towards the added value of VR to teacher training curricula.

Rosenberg et al. [39] evaluated the efficacy of using MRS for PSTs who do not have opportunities to participate in formal classroom management courses. The efficacy of the intervention was assessed through the quality of their classroom management plans, the presentation and implementation of the plan in the virtual classroom, and their classroom management self-efficacy [39]. Results indicated that PSTs successfully developed quality proactive classroom management plans but struggled to present and implement their plans. Self-efficacy, however, increased indicating that participants believed they were better able to manage classroom behavior than they did prior to the intervention.

A later study by Mouw and Fokkens-Bruinsma [40] with 19 PSTs aimed at finding out about their experiences with and perceptions of a microteaching-supported VR kindergarten classroom where they practiced with different CMS. Practicing in the VR classroom supported mastery of CMS and helped PSTs to manage behaviors at their in-person internships. These results underscore the importance of purposefully grounding VR applications in proven instructional theories and practices. Hirsch et al. [41] conducted two separate MRS studies—one for behavioral data collection and one for sharing academic progress in an Individualized Education Program (IEP) meeting. The results of both studies indicate participants' perceptions of their knowledge and confidence significantly increased as well as their actual knowledge of how to carry out these important interventions.

8.4 Case Study Area 4—Diverse Student Populations and Students with Disabilities

Kelley et al. [12] completed a mixed-methods pilot study to evaluate how using MRS, explicit teaching, and performance feedback impacts 34 pre-service teachers' knowledge of culturally responsive teaching strategies and self-efficacy. The researchers used CRT Principles and Descriptions by Gay [42] and the CRT Self-

Efficacy Scale [43] which measured their self-efficacy when implementing CRT strategies. The results showed that there was a statistically significant increase in PSTs' CRT strategy implementation and a statistically significant increase in their self-efficacy. Kelley et al. [12] also note the importance of providing explicit instruction on the CRT strategies to increase the likelihood of PSTs using them in their praxis and providing feedback to PSTs.

MRS has also been used to assist PSTs in working with students with disabilities. Students with special needs have been increasingly educated in general education classrooms, therefore all teachers need increased training on how to best meet the needs of all students who are differently abled. An exploratory study by Driver et al. [44] investigated the efficacy of using MRS to provide deliberate practice on high-leverage practices (HLPs) for PSTs for this group of students. Results indicate significant shifts in PST understanding of and perceived readiness to implement HLPs in favor of the MRS experience. Additional practice with MRS did not appear to influence participants' planning abilities but did, however, influence their understanding of and readiness to implement the target HLPs. Both the treatment and comparison groups increased HLP implementation. It was concluded that using MRS as a tool to provide practice opportunities for PSTs has promising results.

8.5 Case Study Area 5—Ethical Dilemmas

For EPPs in the AUC, PSTs must attend required PD training on the Code of Ethics each semester. These sessions are led by organizations and individuals who are highly trained and experienced with reporting and addressing Ethics issues in the state of Georgia as prescribed by the Georgia Professional Standards Commission (GaPSC). The PD session leaders are experts and have experience with interpreting the tenets of 505–6-0.01 THE CODE OF ETHICS FOR EDUCATORS [45]. Beyond having PSTs listen to the information shared by these individuals, PSTs had limited experience with practicing how to handle ethical scenarios that may arise with administrators, students, fellow teachers, or parents.

To reinforce the content of the ethics PD sessions, the EPP at CAU utilized the MRS TeachLivE™ platform in a mixed-mode, pilot study with 6 undergraduates in Elementary Education and 4 students in the MAT Program. There were two goals of the study. The first goal was to provide students with an opportunity to practice applying what they learned from the PD on the Code of Ethics to a real-life scenario. The second goal was to understand the perceptions of students on the effectiveness of the MRS in helping them to apply the Code of Ethics in a real-life scenario. One of the innovative components of this pilot study was that students were placed in the MRS without warning or without extensive prior knowledge of the scenario. This element of surprise added another layer of realism to the scenario they had to face in the MRS. In-service teachers are often faced with situations they must address with fidelity, even when there is not sufficient warning or preparation so it was important to embed this factor in the simulation exercise.

Table 2 Emergent themes during MRS simulation and frequency of occurrence

Undergraduate student themes	Frequency	Graduate student themes	Frequency
Trust and accountability	3	Accountability and consequences	5
Child safety and protection	2	Safety and security	4
		Concern for student well-being	1
School policies and procedures	1	Trust in protocols	1
School support and assurance	1	Professionalism and response to crisis	1
Communication and transparency	1	Communication and resolution	5
Continued communication	1		
Professional preparedness	1	Consequences and impact	1
Emotional impact	1	Emotional impact	3
		Empathy and support	3
Resolution and moving forward	1	Reflection and resolution	3
Parental concerns and decision-making	1	Apology and responsibility	4
Violation of privacy	2	Closure and farewell	2

To determine which Code of Ethics Standards to address in the XR training with pre-service and MAT students, the author relied upon reports and data on the most violated ethics codes for both experienced and new teachers. Out of the ten codes, Standard 9: Professional Conduct was the most violated followed by Standard 7: Confidential Information. The ethics scenario used in the MRS included these two standards and was designed based on a real situation as provided by the expert PD ethics team. Each participant was quickly and briefly given the scenario right before being immediately placed in the MRS. Each participant spent approximately 10–15 min in the MRS session addressing the ethics scenario and how to resolve it. After their simulation experience, students participated in a facilitated focus group discussion. The transcripts of their dialogs were analyzed and insightful themes emerged. Table 2 shows these themes separated by group along with the frequency of how often those themes occurred. Although all of the themes were not synonymous across both groups, some alignment occurred for a few of them as indicated by the shaded rows. In addition, although the frequency of the themes cannot be directly compared due to the different number of students in each group, they do point to the highest expressions the students stated in their dialog in the MRS—trust, accountability, safety, and communication.

Some notable focus group responses were:

> Definitely I did not feel prepared for that situation, so I'm glad that we did that because if that was to happen in person, I would have had the same reaction.

> I learned that I needed to manage my emotions.

Table 3 Focus group themes and percentage of occurrence (Undergraduate and graduate)

Focus group themes	Percentage of occurrence	Focus group themes	Percentage of occurrence
Benefit of practice platforms	24%	Addressing varied situations	16%
Realistic scenarios	20%	Preparation for sensitive topics	16%
Contextual understanding	12%	Reflection and feedback	12%

Students also stated that the scenario was: helpful in refining their praxis, helpful but more nuance/context is needed to have a better response. They also stated that the scenario was insightful but would prefer even more relevant and realistic issues. The highest percentage of the focus group themes, shown in Table 3, indicate that participants expressed that there was a benefit of having a platform to practice without harming parents or students in navigating the dilemma. The next highest percentage showed that they appreciated having a realistic scenario and reported that the immersive nature of MRS significantly improved their preparedness and confidence in handling complex classroom scenarios.

There are several positive outcomes in using these technologies in EPPs. Challenges, however, remain such as the need for highly realistic simulations [38] and concerns about reinforcing stereotypes [46]. While research at HBCUs is still limited, positive effects have been observed on motivation, self-efficacy, and various teaching skills [33]. Integration of these technologies with established pedagogical approaches like microteaching can enhance their effectiveness in teacher preparation programs [47].

9 Looking to the Future for Cloud Computing and IoT in EPPs at HBCUs

9.1 Cloud Computing

Cloud computing can be an essential tool for HBCU EPPs by providing scalable, on-demand resources that support virtual learning environments [48]. Platforms such as Microsoft Azure and Amazon Web Services offer cloud-based infrastructures that enable seamless integration of AI and XR tools into teacher training curricula. These technologies facilitate collaborative learning experiences, data-driven instructional decision-making, and the ability to access training modules. Cloud computing supports these applications by offering real-time feedback and accessible resources for continuous learning. Although the research on the use of cloud computing in HBCU EPPs is limited, there are several applications in Table 4 that EPPs should consider as they look to use this tool to enhance training opportunities for PSTs.

Table 4 Applications for cloud computing in HBCU EPPs

Educational area	Cloud computing component	Benefit for PSTs
Reading comprehension	• Use virtual machines to create virtual classrooms • Use machine learning platforms • Model reading growth based on student benchmark reading data • Use mock school demographics. • Test immersive readers	• Learn how to interpret data trends to shift instruction • Create a predictive model to determine intervention strategies
Culturally relevant pedagogy	• Use mixed reality toolkits • Develop virtual museum exhibits on African American history • Design interactive elements	• Increased awareness of cultural content and cultural identity of students • Create digital repository for immersive student learning environments
Differentiated instruction	• Build small-scale adaptive learning environment	• Create digital responses to reflect different ways students process information
Math/Science	• Use virtual machines to create virtual STEM classrooms • Use machine learning platforms • Build interactive data dashboards based on student math/science benchmark data • Use mock school demographics	• Increased awareness of data-driven instruction, upper-level math/science placement decision-making • Create digital responses to reflect different ways students solve problems
Educational research	• Query academic databases (i.e., NAEP, DOE, IPEDS)	• Analyze data using SQL • Build skills in Ed research to identify achievement patterns and disparities based on zip codes or districts
Co-creation of curriculum	• Website creation platforms	• Acquire teamwork skills to create curriculum that reflects CRP • Curate digital resources to reflect cultural demographics of students

9.2 *IoT*

IoTs further enhances educator preparation by enabling real-time data collection and interactive learning experiences. IoT technologies further enrich these simulations by providing insights into emotional and behavioral dynamics within virtual learning environments. IoT-enabled smart classrooms can also monitor student engagement levels, classroom acoustics, and environmental factors to provide

Table 5 Applications for internet of things in HBCU EPPs

Area	IoT	Benefit for PST and students
Smart classroom	• Use classroom sensors—Temperature, light, motion detectors • Movement trackers to monitor peer interactions	• Comfortable class setting with better lighting and temperature for better student engagement • Provides insights for engagement patterns
STEAM integration	• Utilize robotics platforms with sensors to interact with external environmental input to influence motion/behavior	• Students make analogies of input/output and cause/effect for multiple disciplines • Students make arguments for limitations of this concept
Physical education	• Biometric sensors (i.e., fitness wearables) on students to help teachers track heart rates and stress levels	• Gain awareness of physical health levels
Assistive technology	• Smart switches • Automation of repeatable motion • Spatial-activated, immersive labs • RFID • Voice-activated systems	• Differently-abled students have full access to academic experience • Rigor is maintained
Environment and community health	• Obtain data from air- and water-quality sensors • Build a hydroponic garden with feedback loop sensors (i.e., pH, water flow rates)	• Students use community data to draw conclusions and provide solutions • Recognition of strong correlation of science and community

insights that help teacher candidates improve their instructional methods. Wearable devices and smart sensors can track biometric responses of teacher candidates during training simulations, allowing for the identification of stress triggers and areas for improvement. IoT-based motion-tracking tools used in XR environments can analyze teacher candidates' movements and gestures, ensuring they maintain an effective classroom presence. Additional application are found in work by Krasniqi and Kabashi [49]. Table 5 shows applications for how HBCU EPPs can incorporate this technology. Research on the use of IoT in HBCU EPPs is limited but Table 5 shows areas to be considered.

10 HBCU EPPs Are Positioned to Lead and Innovate

EPPs at HBCUs can take a leadership role in the utilization of these technologies. Being the largest producer of Black teachers, HBCUs are well-positioned to lead in their impact on producing Black teachers who are trained with the latest technological advances to maximize their impact in the classroom. In essence, HBCUs are not only upskilling their teacher candidates with advanced training, they also are providing PSTs with tailored experiences to ensure that CRP has an indelible mark on their teaching praxis. There are three areas where EPPs at HBCUs can make a tremendous impact to propel the training of PSTs:

10.1 Collaboration among HBCU Faculty, Technologists, and Curriculum Specialists

HBCUs demonstrate leadership through interdisciplinary collaborations between faculty, curriculum developers, and technology firms specializing in XR and AI [50]. These collaborations are instrumental in creating culturally relevant simulations tailored to community-specific educational challenges. By leveraging expertise across education, computer science, and instructional design, institutions can create immersive XR-based training environments that reflect the sociocultural realities of classrooms with high minority student populations. This collaborative approach ensures that simulations are contextually relevant, addressing challenges such as implicit bias, cultural differences, and differentiated instruction in diverse classrooms.

10.2 Strategic Partnerships with Educational Technology Developers

HBCUs can form strategic partnerships with leading educational technology developers to co-create tools and platforms that align with the needs of teacher preparation programs. Engaging with companies such as TeachLivE™/Mursion and XR-based training providers allows HBCUs to integrate cutting-edge AI and XR applications tailored to CRP. These collaborations can also result in customized software development grants, faculty training opportunities, and student internship programs in ed-tech innovation.

10.3 Development of Culturally Relevant Training Experiences Reflective of Community Needs

HBCUs have a unique opportunity to lead the development of culturally relevant training experiences that reflect the specific needs of the communities they serve. By integrating real-world case studies, historical contexts, and localized challenges into XR simulations, PSTs can better understand and respond to the sociopolitical dynamics affecting their future students. These experiences can include community-based service-learning components, where teacher candidates work with local schools to co-design instructional interventions that bridge technological gaps and promote inclusive learning environments.

11 Conclusion

By leveraging AI, XR, cloud computing, and IoT, EPPs at HBCUs are creating transformative and flexible learning simulations that prepare teacher candidates for the realities of the modern classroom. These technologies offer scalable, immersive, and data-driven training experiences that enhance classroom management skills, ethical decision-making, and self-efficacy. Platforms such as TeachLivE™ and Mursion exemplify the potential of AI and XR to revolutionize educator preparation, ensuring that graduates from HBCU programs are ready to thrive in diverse educational settings. As these technologies continue to evolve, their integration into teacher education at HBCUs will be instrumental in fostering the next generation of culturally responsive and technologically adept educators. HBCUs have long centered culturally affirming, justice-driven teaching in their EPPs. The integration of CRP with AI and XR (AR, VR, MRS) along with the infrastructure of Cloud Computing and IoT represents a natural and necessary evolution of this mission. These technologies, when grounded in CRP, offer HBCUs powerful tools to prepare PSTs for diverse classrooms.

References

1. United Negro College Fund, Frederick D. Patterson Research Institute. (2024). *The heart work of hard work: Exploring the black teacher pipeline at HBCUs*. United Negro College Fund.
2. Toldson, I. A., Pearson, D., & Rogers, N. (2022). Strategic priorities for historically black colleges and universities with teacher preparation programs. In *Black college leadership in PK–12 education* (pp. 7–29). Brill.
3. Edwards, A., Ortagus, J., Smith, J., & Smythe, A. (2023). *HBCU enrollment and longer-term outcomes*. EdWorkingPaper No. 23-883. Annenberg Institute for School Reform at Brown University.
4. Lee, S. (2019). Black teachers matter: Examining the depths of seven HBCU teacher preparation programs. *Urban Education Research & Policy Annuals, 6*(2).
5. Irvine, J. J., & Fenwick, L. T. (2011). Teachers and teaching for the new millennium: The role of HBCUs. *Journal of Negro Education, 80*(3), 197–208.
6. Herring, C. (2024, January 25). *What teacher preparation programs can learn from minority-serving institutions. The 74.* Retrieved from https://www.the74million.org/article/what-teacher-preparation-programs-can-learn-from-minority-serving-institutions
7. Bandura, A. (1977). Self-efficacy: Toward a unifying theory of behavioral change. *Psychological Review, 84*(2), 191–215.

8. Bandura, A. (1997). *Self-efficacy: The exercise of control*. W.H.
9. Gundel, E., & Piro, J. S. (2021). Perceptions of self-efficacy in mixed reality simulations. *Action in Teacher Education, 43*(2), 176–194.
10. Samuelsson, M., Samuelsson, J., & Thorsten, A. (2022). Simulation training-a boost for pre-service teachers' efficacy beliefs. *Computers and Education Open, 3*, 100074.
11. Vygotsky, L. S. (1978). In M. Cole, V. John-Steiner, S. Scribner, & E. Souberman (Eds.), *Mind in society: The development of higher psychological processes*. Harvard University Press.
12. Kelley, M. H., Qualls, L. W., Scott, S. N., & Hirsch, S. E. (2025). Mixed-reality simulation to prepare preservice teachers to use culturally responsive teaching strategies: A pilot study. *Journal of Special Education Technology, 40*(1), 65–78.
13. Bhagya-Prabhashini, C. (2024). Sociocultural cognitive development (SCCD) of language learning in an extended reality environment. *Multidisciplinary Reviews, 7*(11), 2024209–2024209.
14. Kolb, D. A. (1984). *Experiential learning: Experience as the source of learning and development*. Prentice Hall.
15. Majgaard, G., & Weitze, C. (2020, September). Virtual experiential learning, learning design and interaction in extended reality simulations. In *Proceedings of the 14th European conference on game based learning, ECGBL* (pp. 372–379).
16. Weerasinghe, M. (2021). Instructional guidance in extended reality for learning. In *Adjunct publication of the 23rd International Conference on Mobile Human-Computer Interaction (MobileHCI '21)* (Article 25, 4 pp.). Association for Computing Machinery.
17. Ersozlu, Z., Ledger, S., Ersozlu, A., Mayne, F., & Wildy, H. (2021). Mixed-reality learning environments in teacher education: An analysis of TeachLivE™ research. *SAGE Open, 11*(3), 21582440211032155.
18. Ladson-Billings, G. (1995). Toward a theory of culturally relevant pedagogy. *American Educational Research Journal, 32*(3), 465–491.
19. Ladson-Billings, G. (2014). Culturally relevant pedagogy 2.0: Aka the remix. *Harvard Educational Review, 84*(1), 74–84.
20. Gay, G. (2018). *Culturally responsive teaching: Theory, research, and practice* (3rd ed.). Teachers College Press.
21. Holmes, W., & Porayska-Pomsta, K. (2023). *The ethics of artificial intelligence in education*. Routledge.
22. Baker-Bell, A. (2020). *Linguistic justice: Black language, literacy, identity, and pedagogy*. Routledge.
23. Paris, D., & Alim, H. S. (2017). *Culturally sustaining pedagogies: Teaching and learning for justice in a changing world*. Teachers College Press.
24. Alim, H. S., Paris, D., & Wong, C. P. (2020). *Culturally sustaining pedagogies: Teaching and learning for justice in a changing world*. Teachers College Press.
25. Figueroa-Flores, J. F., & Huffman, L. (2020). Integrating AR and VR in teacher education: What pre-service teachers perceive. *Frontiers in Education Technology, 3*(4), 5–15.
26. Clark, T., Hamilton, O., Morris, M., & Vereen, E. (2023). How Morehouse college is transforming undergraduate education in the sciences and humanities with virtual reality. In *Ethical considerations of virtual reality in the college classroom: Cross-disciplinary case studies of immersive technology implementation* (p. 270).
27. Dieker, L. A., Rodriguez, J., Lignugaris-Kraft, B., Hynes, M. C., & Hughes, C. E. (2014). The potential of simulated environments in teacher education: Current and future possibilities. *Teacher Education and Special Education, 37*(1), 21–33.
28. Riggs, V. (2025). Impact of AI literacy professional development on teaching: Evaluating changes in perceptions and classroom practices. *American Journal of STEM Education, 7*, 37–50.
29. Chow, W. (2025, January 6). *2,000th member spotlight: Justin Ballenger helps bring AI into the classroom*. Scholars Strategy Network. Retrieved from https://scholars.org/features/2000th-member-spotlight-justin-ballenger-helps

30. Aguilar, J. J., & Flores, Y. (2022). Analyzing the effectiveness of using mixed-reality simulations to develop elementary pre-service teacher's high-leverage practices in a mathematics methods course. *EURASIA Journal of Mathematics, Science and Technology Education, 18*(5), em2107.
31. Dieker, L. A., Rodriguez, J. A., Lignugaris-Kraft, B., Hynes, M. C., & Hughes, C. E. (2013). The potential of simulated environments in teacher education: Current and future possibilities: Current and future possibilities. *Teacher Education and Special Education: The Journal of the Teacher Education Division of the Council for Exceptional Children, 37*(1), 21–33.
32. Freeman, I. M., & Lee, H. (2024). Pre-service teacher candidates' perceptions of classroom-based mixed-reality simulations. *Education Sciences, 14*(4), 347.
33. Van der Want, A. C., & Visscher, A. J. (2024). Virtual reality in preservice teacher education: Core features, advantages and effects. *Education Sciences, 14*(6), 635.
34. Dalinger, T., Thomas, K. B., Stansberry, S., & Xiu, Y. (2020). A mixed reality simulation offers strategic practice for pre-service teachers. *Computers & Education, 144*, 103696.
35. Hudson, M. E., Voytecki, K. S., Owens, T. L., & Zhang, G. (2019). Preservice teacher experiences implementing classroom management practices through mixed-reality simulations. *Rural Special Education Quarterly, 38*(2), 79–94.
36. Nissim, Y., & Weissblueth, E. (2017). Virtual reality (VR) as a source for self-efficacy in teacher training. *International Education Studies, 10*(8), 52–59.
37. Gundel, E., Piro, J. S., Straub, C., & Smith, K. (2019). Self-efficacy in mixed reality simulations: Implications for preservice teacher education. *The Teacher Educator, 54*(3), 244–269.
38. Mouw, J. M., Fokkens-Bruinsma, M., & Verheij, G. J. (2020, June). Using virtual reality to promote pre-service teachers' classroom management skills and teacher resilience: A qualitative evaluation. In *Proceedings of the 6th international conference on higher education advances (HEAd'20)* (pp. 325–332). Universitat Politècnica de València.
39. Rosenberg, M. S., Duerr, S. R., Ingraham, K., Bell, K., & Gould, A. (2021). Enhancing classroom management skills: Efficacy of a supplemental multi-platform intervention for preservice teachers. *Excelsior: Leadership in Teaching and Learning, 14*(1), 4–19.
40. Mouw, J., & Fokkens-Bruinsma, M. (2022). When technology meets educational sciences: Combining virtual reality and microteaching to train pre-service teachers' kindergarten classroom management strategies. In *Proceedings of the 8th International Conference on Higher Education Advances (HEAd'22)* (pp. 1043–1050). Universitat Politècnica de València.
41. Hirsch, S. E., Griffith, C. A., Kelley, M. H., Carlson, A., & McKown, G. (2023). Increasing preservice teachers' knowledge and skills using a mixed-reality simulator: The tale of two studies. *Teacher Education and Special Education, 46*(3), 204–222.
42. Gay, G. (2015). The what, why, and how of culturally responsive teaching: International mandates, challenges, and opportunities. *Multicultural Education Review, 7*(3), 123–139.
43. Siwatu, K. O. (2007). Preservice teachers' culturally responsive teaching self-efficacy and outcome expectancy beliefs. *Teaching and Teacher Education, 23*(7), 1086–1101.
44. Driver, M. K., Zimmer, K. E., Khan, O., Sadler, J. V., & Draper, E. (2024). Preparing general education teachers for inclusive settings: Integrating high-leverage practices and mixed-reality simulation in pre-service coursework. *Education Sciences, 14*(4), 428.
45. Georgia Professional Standards Commission. (2023). 505–6-.01 the code of ethics for educators. Retrieved from https://www.gapsc.com/Rules/Current/Ethics/505-6-.01.pdf
46. McGarr, O. (2020). The use of virtual simulations in teacher education to develop pre-service teachers' behaviour and classroom management skills: Implications for reflective practice. *Journal of Education for Teaching, 47*(2), 274–228.
47. Mouw, J., & Fokkens-Bruinsma, M. (2022, June). When technology meets educational sciences: Combining virtual reality and microteaching to train pre-service teachers' kindergarten classroom management strategies. In *Eighth international conference on higher education advances* (pp. 1043–1050). Universitat Politècnica de València.

48. Wang, C., Chen, X., Yu, T., Liu, Y., & Jing, Y. (2024). Education reform and change driven by digital technology: A bibliometric study from a global perspective. *Humanities and Social Sciences Communications, 11*(1), 1–17.
49. Krasniqi, L., & Kabashi, F. (2023). Internet of things(IOT) in education: Opportunities and challenges. In *UBT international conference* (p. 27).
50. Musendekwa, M. (2025). Breaking down silos: Strategies for collaborative technology integration in higher education. In *Institutes of Higher Education (IHE) and workforce collaboration for digital literacy* (pp. 1–48). IGI Global Scientific Publishing.

Part II
Transformations in Digital Infrastructure and Artificial Intelligence

Artificial Intelligence and Cloud Computing in Healthcare

Sumathra Manokaran, T. R. Vyshnavi, U. Chethan Maiya, Abhay Shashidhara, A. H. Manjunatha Reddy, and Ahmad Hoirul Basori

1 Introduction

Artificial Intelligence (AI) and Cloud Computing (CC) are transforming the healthcare landscape by improving, accuracy, efficiency, and accessibility of medical services. The integration of AI with CC enables real-time analysis, data-driven decision-making, and seamless collaboration across healthcare systems. AI enhances diagnostics, predictive analytics, and patient care, while cloud computing ensures security, scalability, and interoperability. This chapter explores the applications, advantages, challenges, and future developments of AI and CC in healthcare.

The increasing adoption of digital health technologies has led to a paradigm shift in how medical data is managed and utilized. Traditional healthcare systems often faced inefficiencies due to manual processes, fragmented data storage, and lack of interoperability between institutions. AI-driven solutions address these challenges by automating routine tasks, detecting patterns in complex medical data, and facilitating early disease detection [1]. Cloud computing further enhances these capabilities by providing a platform for storing, sharing, and analyzing health records across different healthcare organizations. This fusion of AI and cloud technology is fostering a more patient-centric approach, improving medical decision-making and operational efficiency [2].

S. Manokaran (✉) · T. R. Vyshnavi · U. Chethan Maiya · A. Shashidhara · A. H. Manjunatha Reddy
Department of Biotechnology, RV College of Engineering, Bengaluru, India
e-mail: sumathram@rvce.edu.in

A. H. Basori
Faculty of Computing and Information Technology Rabigh, King Abdulaziz University, Rabigh, Saudi Arabia
e-mail: abasori@kau.edu.sa

C. Rompante Cunha et al. (eds.), *Transformative Cloud Computing, IoT and Extended Reality*, Signals and Communication Technology,
https://doi.org/10.1007/978-3-032-03554-7_5

AI and cloud computing are revolutionizing various domains within healthcare by improving efficiency, accuracy, and collaboration. In precision medicine, AI helps in analyzing genetic and environmental to develop personalized treatment plans, enhancing patient outcomes [3]. Cloud computing enables large-scale data sharing, fostering collaboration in drug discovery, epidemiology, and medical informatics [4]. These technologies also streamline hospital management by automating administrative tasks, optimizing resource allocation, and supporting telemedicine through remote diagnostics and real-time patient monitoring [5]. Despite their vast potential, successful implementation requires addressing regulatory, ethical, and infrastructural challenges to ensure responsible and equitable deployment.

2 Role of AI in Healthcare

Artificial Intelligence (AI) is transforming the healthcare industry by enhancing diagnostic accuracy, optimizing treatment plans, personalizing patient care, and streamlining administrative processes. AI-powered systems leverage machine learning (ML), deep learning, and natural language processing (NLP) to analyze vast amounts of medical data, providing insights that enhance clinical decision-making and patient outcomes [6].

AI-driven diagnostic tools assist in early disease detection by analyzing medical images and genetic data, improving precision, and reducing human error. AI-powered predictive analytics help track disease progression and identify at-risk patients, enabling timely interventions [7]. Additionally, AI helps in the automation of administrative tasks such as medical record management, patient scheduling, and billing, increasing efficiency and reducing the tasks of healthcare professionals. As AI continues to advance, its integration into healthcare is paving the way for efficient, accurate, and patient-centric medical care.

As illustrated in Fig. 1, the AI-driven medical diagnosis process begins with a CT scan to capture detailed internal images of a patient experiencing symptoms. These images undergo pre-processing, where noise reduction and contrast enhancement are applied to improve quality for further analysis [8]. Once pre-processed, the images go through segmentation, where AI-based algorithms identify and isolate regions of interest, such as tumors or abnormalities. This segmentation helps in focusing on specific areas for further classification and diagnosis [8].

Following segmentation, the images are processed through an AI-powered classification system, which determines whether the identified abnormality corresponds to a particular disease [9]. In some cases, X-ray images may also be used as an alternative or supplementary input for classification. Once classification is complete, the system provides a prognosis, predicting the likely progression of the disease. If a disease is confirmed, a diagnosis is made, leading to the next step where medical decisions are recorded for documentation and future reference [9].

The final stage involves providing medical advice based on the AI-driven analysis and recorded decisions. This includes treatment recommendations, lifestyle modifications, or specialist referrals. By leveraging AI in medical imaging and

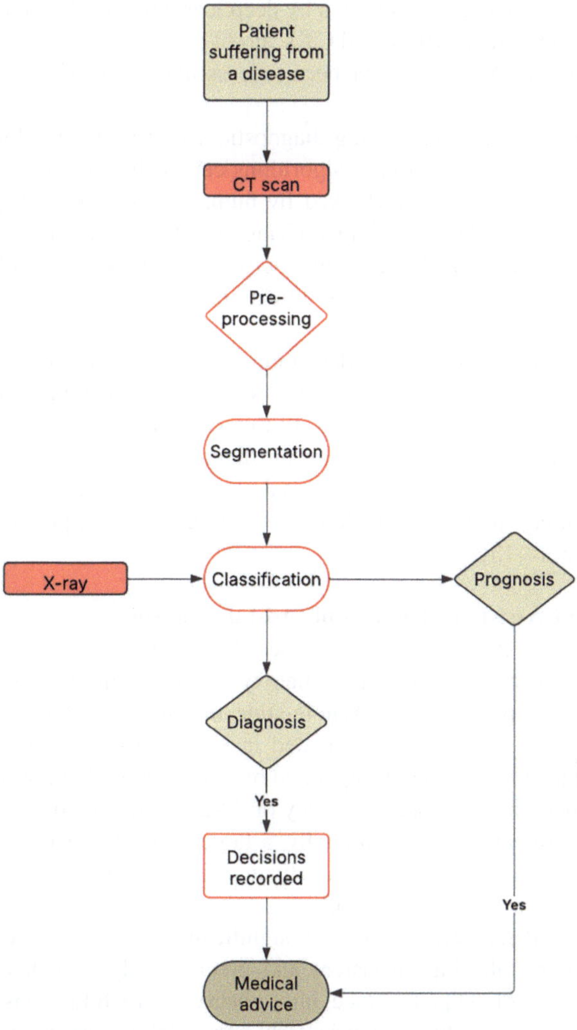

Fig. 1 Role of AI in healthcare

diagnosis, the process enhances accuracy, supports early disease detection, and assists professionals in making informed clinical decisions, ultimately improving patient outcomes [10].

2.1 AI in Medical Diagnosis

Machine learning techniques are used by AI-powered diagnostic technologies to detect diseases early on. Accuracy in identifying diseases like cancer, heart disease, and neurological problems is increased by deep learning models trained on medical images, including X-rays, MRIs, and CT scans [11]. Convolutional neural networks (CNNs), one type of AI model, have been successful in identifying abnormalities in radiological scans.

Beyond imaging, AI is enhancing diagnostic accuracy in pathology, genomics, and laboratory testing. AI-powered algorithms can analyze biopsy slides to detect malignancies that might be overlooked by human pathologists. In genomics, AI assists in sequencing DNA and identifying genetic mutations associated with hereditary conditions. AI also aids in microbiology labs by rapidly identifying bacterial and viral infections, expediting treatment decisions and improving patient outcomes [12].

Furthermore, AI has enabled the development of multi-modal diagnostic systems that integrate data from various sources, such as medical imaging, laboratory reports, and patient history. By analyzing this combined data, AI can offer comprehensive diagnostic insights, reducing misdiagnosis rates [13]. Decision support systems powered by AI assist professionals by suggesting differential diagnoses and recommending the most suitable tests, leading to a more precise and efficient diagnostic process [13].

Case Study: AI-Assisted Diagnosis of Prostate Cancer

A study utilized the Aiforia® Prostate Cancer Suite, an AI-powered diagnostic tool to assist pathologists in analyzing hematoxylin and eosin (H&E) stained slides of prostate cancer patients [14]. Whole-slide images (WSIs) from 111 prostate cancer cases were digitized and evaluated in two separate rounds—one with AI assistance and another without it, separated by a 4-week washout period. The study aimed to compare the diagnostic accuracy of AI-assisted diagnosis with traditional pathology assessments [14]. Additionally, a larger validation study involved 148–150 pathologists from 15 countries analyzing seven WSIs, with their consensus compared against the AI-assisted results.

The findings demonstrated that AI significantly improved the detection of prostate cancer by offering consistent and precise analysis of histopathological images. The AI tool helped reduce inter-observer variability, ensuring a more standardized diagnosis, which is particularly beneficial in cases where pathologists may have differing interpretations. By integrating AI into pathology, the study highlighted how such tools could enhance diagnostic accuracy, reduce workload, and improve overall patient outcomes by enabling early and more reliable detection of malignancies. This case exemplifies the role of AI in medical diagnostics, particularly in oncology, by assisting healthcare professionals in making more informed and efficient decisions [14].

In another case study by Gulshan et al. (2016), AI was utilized for automated detection of diabetic retinopathy using a deep learning algorithm [15]. A deep convolutional neural network (CNN) on a large dataset of retinal images was trained by the researchers, enabling the AI model to classify disease severity with high accuracy. The study demonstrated that AI-assisted screening could significantly enhance early detection, reduce diagnostic errors, and improve access to ophthalmic care, especially in resource-limited settings [15].

2.2 *Predictive Analytics in Healthcare*

Predictive analytics in healthcare helps forecast disease outbreaks, hospital readmissions, and patient deterioration. AI-driven risk prediction models assess patient data, lifestyle factors, and genetic predispositions to provide personalized care plans. The implementation of recurrent neural networks (RNNs) and transformer models further enhances prediction accuracy [16].

AI-powered predictive analytics is transforming intensive care units (ICUs) by continuously monitoring patient vitals and alerting medical staff about potential complications before they become critical. Machine learning models analyze real-time patient data, such as blood pressure, heart rate, and oxygen levels, to predict septic shock, respiratory failure, and cardiac arrest. Early warning systems powered by AI help in timely interventions, reducing mortality rates and improving overall patient safety [17].

The management of chronic diseases is another important area where predictive analytics is being used. Artificial intelligence (AI) algorithms can evaluate long-term patterns in patient medical records to forecast the risk of developing diseases like diabetes, high blood pressure, or heart disease. Healthcare professionals can postpone or stop the onset of disease by identifying at-risk patients early and implementing preventative strategies including lifestyle changes and focused medical interventions [18].

Case Study: AI-Driven Predictive Analytics in Cancer Prognosis
According to Courtiol P., et al. (2019), significant application of AI-powered predictive analytics in oncology is demonstrated by the development of a deep learning-based classification system for mesothelioma, which has substantially enhanced the prediction of patient outcomes. In this study, researchers employed whole-slide histopathological images to train a deep learning model capable of accurately classifying mesothelioma subtypes and predicting patient survival outcomes. This AI-driven approach enabled more precise prognostication, facilitating personalized treatment strategies for patients with this aggressive cancer [19].

The integration of AI in this context exemplifies how predictive analytics can transform cancer care by providing clinicians with robust tools to assess disease progression and tailor interventions accordingly. By using vast amounts of imaging data, the deep learning model identified patterns and features that may not be discernible through traditional analysis methods, thereby improving the accuracy of prognostic assessments. This advancement underscores the potential of AI to enhance personalized medicine, ensuring that people receive treatments aligned with their specific disease characteristics and risk profiles [19].

In another case study by Lauristen et al. (2017), AI was used to propose a predictive analytics model for early sepsis detection [20]. The researchers utilized deep learning techniques on electronic health record (EHR) data to identify patients at high risk of sepsis before clinical symptoms became severe. The AI model analyzed vital signs and lab results demonstrating improved early detection rates, reducing mortality, and enabling timely medical interventions [20].

2.3 AI in Drug Discovery

AI accelerates the process of drug discovery by identifying potential drug candidates and simulating molecular interactions. Technologies like generative adversarial networks (GANs) and transformer-based models, including BERT and GPT, facilitate virtual screening of compounds, thereby decreasing the time and expenses related to conventional drug development [21]. AI-powered tools such as AlphaFold have transformed protein structure predictions, supporting the design of drugs.

In addition to virtual screening, AI is streamlining the design and optimization of clinical trials. AI-driven analytics help pharmaceutical companies select the most suitable candidates for trials by analyzing patient demographics, genetic profiles, and historical treatment responses [22]. By optimizing trial design, AI reduces costs, shortens development timelines, and improves the success rates of new drug candidates [23].

AI has enabled the emergence of personalized pharmacology, where drugs can be used for individual genetic profiles. AI models predict how patients will respond to specific medications, minimizing adverse drug reactions and enhancing therapeutic efficacy [24]. By integrating AI with real-world evidence collected from electronic health records and wearable devices, pharmaceutical companies can refine their drug development processes and produce more effective treatments [25].

Case Study: AI in Drug Discovery for Breast Cancer

A breakthrough in AI-assisted drug discovery for cancer treatment was demonstrated in a study where researchers used deep learning models to identify potential inhibitors for estrogen receptor-positive (ER+) breast cancer. The study, conducted by Zhavoronkov et al. (2020), leveraged an AI-driven platform to design and optimize small molecules targeting proteins involved in breast cancer progression [26].

Recent advances in AI have opened new avenues in cancer drug discovery, particularly for estrogen receptor-positive (ER^{+}) breast cancer. Deep-learning models, such as QSAR approaches, have shown strong potential in predicting inhibitors of estrogen receptor alpha (ERα) with remarkable accuracy [26]. Alongside this, generative techniques like GANs combined with reinforcement learning are now being used to design entirely new chemical structures with optimized binding properties and drug-like characteristics. These AI-driven methods can virtually screen millions of compounds far more quickly than traditional high-throughput approaches, significantly saving both time and resources [27].

While experimental validation of AI-designed molecules targeting ER^{+} breast cancer is still in early stages, initial studies suggest that these approaches can produce promising candidates with favorable preclinical profiles. Integrating AI with molecular simulations is helping researchers identify potential therapeutics more efficiently and precisely, paving the way for next-generation cancer treatments [26, 27].

In another case study by Stokes et al. (2020), AI was utilized for the discovery of novel antibiotic compounds. The researchers employed a deep learning model trained on molecular structures to predict compounds with antibacterial properties. The AI system successfully identified Halicin, a powerful new antibiotic, which demonstrated effectiveness against drug-resistant bacteria [28]. This study highlighted the potential of AI in accelerating drug discovery, reducing time and cost, and addressing antibiotic resistance.

2.4 AI in Personalized Medicine

AI facilitates precision medicine by analyzing genomic data, patient history, and biomarkers to tailor treatments. AI-driven models optimize medication dosages, reducing adverse effects while improving therapeutic efficacy. The integration of federated learning enables secure analysis of multi-institutional data while preserving patient privacy [14].

In oncology, AI plays a crucial role in developing targeted therapies. By analyzing tumor genetics and molecular markers, AI helps oncologists identify the most effective treatment regimens for patients [29]. AI models also predict resistance to certain drugs, enabling adjustments in therapy before resistance develops. This approach enhances treatment efficacy and minimizes side effects for cancer patients [30]. Moreover, AI-driven wearable technology and mobile applications are advancing personalized medicine by continuously monitoring patient health metrics [31]. AI algorithms analyze data from smartwatches, fitness trackers, and biosensors to detect early signs of health deterioration. This continuous monitoring allows for timely interventions, reducing hospital admissions and enabling proactive healthcare management [32]. Customized health suggestions derived from AI assessment enhance patient involvement and compliance with treatment regimens, leading to improved health results [33].

Case Study: AI-Driven Personalized Cancer Treatment
Artificial Intelligence (AI) has significantly advanced personalized medicine by enabling enhanced treatment strategies based on individual patient data [34]. A notable case study by M. P. Menden, R. Iorio, T. Garnett, et al., involves the development of a machine learning model to predict drug responses based on

genomic features and alterations across various cancer types. Menden et al. utilized a multi-layer perceptron (MLP) combined with random forest (RF) algorithms to analyze cell line samples, encompassing 608 drugs and 111 cancer types. By integrating genomic and chemical properties, the model provided more accurate predictions of drug sensitivity, helping to identify the most effective treatment for individual patients [34].

The study demonstrated that combining linear and nonlinear models could enhance the prediction of drug responses, reducing the trial-and-error approach in cancer treatment. By leveraging AI, researchers were able to identify patterns in genetic mutations and their impact on drug efficacy, paving the way for more precise and targeted therapies. This case highlights how AI-driven approaches can revolutionize oncology by optimizing drug selection based on a patient's unique genetic profile, thereby improving outcomes and reducing adverse effects [34].

In another case study by Marling et al. (2012), AI was used to develop personalized treatment plans for diabetes patients. The researchers applied machine learning models to analyze continuous glucose monitoring (CGM) data, lifestyle factors, and medical history to provide individualized insulin dosage recommendations [35]. The AI-driven approach improved glycemic control, minimized hypoglycemia risk, and enhanced patient adherence to treatment plans. This study demonstrated how AI can optimize precision medicine by tailoring treatments to individual patient needs [35].

3 Role of Cloud Computing in Healthcare

3.1 *Cloud-Based Electronic Health Records (EHRs)*

Cloud computing ensures secure storage and accessibility of EHRs, enabling healthcare providers to access patient data remotely. Cloud-based platforms streamline medical record-keeping, reducing paperwork and administrative burden. The interoperability of cloud-based EHRs improves continuity of care and enhances coordination among healthcare professionals [36, 37].

Beyond simple storage solutions, cloud-based EHRs facilitate real-time data exchange and integration with AI-powered analytics. By leveraging cloud computing, hospitals and clinics can seamlessly retrieve and share patient information across multiple departments and institutions, improving collaborative decision-making [38]. Additionally, cloud-based EHRs allow for the incorporation of predictive analytics tools, enabling early detection of potential health risks and enhancing patient outcomes. Furthermore, cloud solutions offer disaster recovery and backup capabilities, ensuring that patient records are protected from data loss due to cyber-attacks or system failures [39]. This robust infrastructure enhances reliability and continuity of care, particularly in emergencies where immediate access to patient data is crucial for timely interventions.

Case Study: Cloud-Based Electronic Health Records (EHRs) in Outpatient Clinics

Guiding Healthcare Technology Implementation, A notable study by Schoville and Titler that looks at how Electronic Health Records (EHRs) affect healthcare collaboration [40]. This research investigates how the implementation of a cloud-based EHR system influences collaborative practices among healthcare professionals in five outpatient clinics. This highlights the role of EHRs in improving clinical workflows, improving communication, and facilitating real-time data sharing among physicians, nurses, and administrative staff [40].

The findings reveal that cloud-based EHRs significantly improve coordination of care by reducing paperwork, minimizing errors, and ensuring seamless access to patient records. However, the study also identifies challenges, including technical issues, resistance to adoption, and workflow disruptions during the transition phase [40]. Despite these challenges, the study underscores the long-term benefits of EHR implementation, such as better patient management, increased efficiency, and improved healthcare decision-making. This study is an illustration of how cloud-based EHRs may transform healthcare delivery by improving teamwork and medical procedures.

In another case study by Barton C et al. (2019), AI-enhanced cloud-based EHRs were used for predictive patient monitoring [41]. Machine learning algorithms analyzed real-time patient data to predict complications like sepsis and cardiac arrest, enabling timely interventions. This approach improved clinical decision-making, reduced hospital readmissions, and enhanced patient outcomes [41].

3.2 *Cloud-Based Telemedicine and Remote Patient Monitoring*

Cloud computing enhances telemedicine services by enabling real-time video consultations, remote monitoring, and wearable device integration [42]. AI-powered chatbots assist in preliminary diagnosis and symptom assessment, reducing hospital visits. Remote monitoring systems use IoT sensors and cloud analytics to track patient vitals, ensuring timely intervention [43].

Cloud-based telemedicine scalability makes it possible for medical professionals to reach underprivileged groups, enhancing access to care in isolated and rural locations. Patients can consult with experts anywhere in the world without having to physically go. Furthermore, cloud-based technologies guarantee that healthcare providers have access to a patient's whole medical history, enhancing the precision of diagnosis and treatment planning [44].

Another significant advantage of cloud-based remote patient monitoring is its impact on chronic disease management. Patients with diabetes, heart disease, or respiratory diseases can be tracked continually, and if any abnormalities are found, alarms can be issued to medical professionals. By preventing problems and readmissions to the hospital, this real-time monitoring enhances patient outcomes and lowers healthcare expenses [45, 46].

Case Study: Cloud-Based Telemedicine and Remote Patient Monitoring in Rural India

Through the 10 Bed ICU project, a major effort in cloud-based telemedicine and remote patient monitoring has been put into place in rural India. Many rural hospitals in India lack intensive care units (ICUs) and specialized medical staff, making it difficult to provide critical care services. To bridge this gap, the project introduced cloud-based tele-ICU technology, allowing remote specialists from urban hospitals to monitor and guide treatments in real time [47]. This system enables continuous patient monitoring, where vital signs, oxygen levels, and ECG data are transmitted to cloud platforms accessible by doctors in metropolitan hospitals [47].

The tele-ICU model has significantly improved critical care delivery in resource-limited settings. It allows urban specialists to provide real-time medical advice, prescribe treatments, and even assist in emergency procedures remotely. Local healthcare workers and general practitioners benefit from on-the-job training, increasing their ability to manage complex cases independently over time. The project has already impacted over 65,000 patients, reducing hospital overcrowding in major cities while ensuring timely critical care access in rural areas [47].

R. Ellis-Petersen's case study illustrates how cloud-based telemedicine might improve patient outcomes in developing nations, improve healthcare inequities, and allow for continuous remote monitoring [47]. A more integrated and easily available global healthcare system is being facilitated by cloud-based telemedicine, which makes use of cloud computing, AI-based diagnostics, and real-time consulting.

In another case study by Kim et al (2022), AI-powered cloud-based telemedicine was used for remote stroke detection [48]. Machine learning models analyzed patient data from remote locations and provided real-time stroke risk assessments to urban specialists [48]. This system improved early diagnosis, reduced treatment delays, and enhanced access to specialized care, particularly in underserved areas [48].

3.3 Cloud-Based Medical Research and Collaboration

By offering a scalable and safe framework for storing and analyzing massive datasets, cloud computing makes collaborative medical research easier [49]. Researchers across the globe can access shared databases containing anonymized patient data, enabling advancements in medical knowledge and treatment strategies.

Cloud-based systems have made it possible for researchers to effectively analyze enormous volumes of clinical and genomic data as genomics and precision medicine have grown in popularity. Cloud solutions driven by AI aid in the development of tailored medicines, the identification of genetic markers associated with diseases, and the acceleration of drug discovery [50]. These developments are essential for treating complicated illnesses such as uncommon genetic abnormalities, cancer, and Alzheimer's [51].

Additionally, cloud-based collaboration enhances multi-institutional clinical trials by streamlining data collection and analysis. Researchers can track patient responses to new treatments in real-time, ensuring faster and more accurate trial results [52]. Blockchain technology's incorporation into cloud research platforms further improves data security and transparency while resolving issues with patient privacy and data integrity.

Case Study: Cloud-Based Genomic Research for Cancer Treatment
The Cancer Genome Atlas (TCGA) project, a large-scale cancer research initiative, leveraged cloud computing to store, analyze, and share vast genomic datasets across multiple research institutions. By utilizing cloud-based platforms, scientists could securely access and process anonymized patient genomic data, enabling collaborative discoveries in precision oncology [53]. AI-powered analytics identified genetic mutations linked to various cancers, accelerating targeted drug development and personalized treatment strategies [54]. The integration of multi-institutional clinical trials allowed researchers to track patient responses in real time, refining therapeutic approaches [55]. Blockchain technology was also used to preserve data security and integrity, guaranteeing transparent and unchangeable records of genomic research [56]. This case study highlights how cloud computing enhances global medical research, accelerates drug discovery, and fosters collaboration in tackling complex diseases like cancer.

In another case study by Hong et al. (2024), cloud-based platforms were used to analyze large-scale genomic data for identifying rare disease biomarkers [57]. AI-driven models processed genomic sequences and patient records to accelerate disease gene discovery. This approach would possibly enhance collaboration among researchers globally, reduced data processing time, and improved personalized treatment strategies [57].

3.4 Cloud Security and Compliance in Healthcare

Ensuring data security and regulatory compliance becomes crucial as more healthcare organizations use cloud technologies. Cloud providers use multi-factor authentication, sophisticated encryption, and ongoing monitoring to shield private medical information from online attacks [58].

Healthcare organizations that use cloud services are exposed to strict regulations, such as the General Data Protection Regulation (GDPR) in Europe and the Health Insurance Portability and Accountability Act (HIPAA) in the United States [59]. Adherence to these guidelines guarantees that patient information is kept private and available to authorized people only. Healthcare organizations can effectively fulfill their regulatory responsibilities, thanks to cloud providers' adaptable compliance features. By guaranteeing tamper-proof record-keeping, the use of blockchain technology in cloud security improves data integrity. By implementing decentralized data storage, blockchain prevents unauthorized alterations and ensures transparent access control mechanisms. This innovation significantly improves trust in cloud-based healthcare systems, fostering widespread adoption of digital health solutions [60–62].

From Fig. 2, the process of ensuring cloud security in healthcare follows a structured approach to maintain data privacy, regulatory compliance, and system security. It begins with a thorough assessment of the existing IT infrastructure to identify vulnerabilities and areas for improvement. Once evaluated, specific requirements such as data storage, scalability, and processing needs are determined to support healthcare operations efficiently. In order to ensure compliance and secure handling of sensitive health data, choosing a cloud service provider with experience in healthcare standards such as HIPAA or GDPR is an essential step in this process. To reduce the risks of cyber-attacks and illegal access, the selected solution must incorporate strong security features like encryption, multi-factor authentication, and access limits. Regular updates and monitoring are also necessary to adjust to changing security threats and regulatory requirements.

This methodical approach is illustrated in Fig. 2, which also illustrates the necessity of ongoing risk assessment, compliance, and proactive security measures to protect electronic health records and preserve confidence in cloud-based healthcare systems.

Case Study: Ensuring Cloud Security and Compliance in a National Health Records System
To increase patient data accessibility and expedite healthcare services, the UK's National Health Service (NHS) made the switch to a cloud-based Electronic Health Records (EHR) system. However, because of national healthcare policies and the General Data Protection Regulation (GDPR), guaranteeing data security and regulatory compliance was extremely difficult. The NHS implemented a multi-layered security approach to reduce risks, combining multi-factor authentication (MFA), end-to-end encryption, and ongoing threat monitoring to shield private medical data from online attacks. The NHS collaborated with a healthcare-specific cloud provider to ensure regulatory compliance, allowing GDPR adherence and putting role-based access control (RBAC) in place to prevent unauthorized staff from accessing sensitive data. Furthermore, blockchain technology was incorporated to create tamper-proof audit logs, ensuring data integrity and transparent access control mechanisms [63–67].

3.5 AI-Driven Cloud Automation in Healthcare

AI-driven cloud automation is transforming healthcare by optimizing workflows, reducing administrative burdens, and enhancing decision-making. Hospitals leverage AI-powered cloud platforms for automated scheduling, resource management, and medical coding, reducing wait times and improving efficiency. Additionally, AI enhances clinical decision support, analyzing patient records and diagnostic

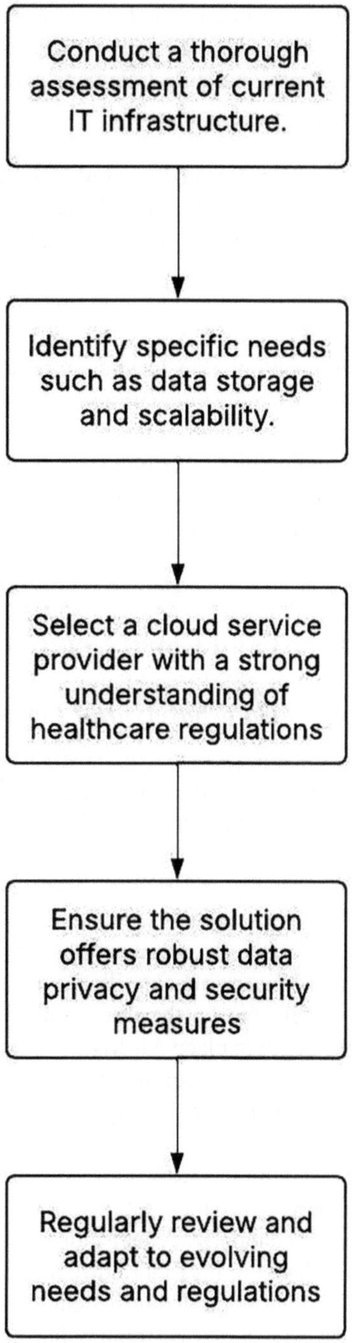

Fig. 2 Cloud Security in Healthcare

images to assist doctors in real time, minimizing errors and improving treatment accuracy. Cloud-based AI also predicts medication demand, preventing shortages and personalizing treatment plans based on patient history and genetic data [68–71].

Case Study: AI-Powered Workflow Automation in a Multi-Specialty Hospital
Li et al. (2021) introduced an AI-assisted module, XIAO YI, which assesses patients' chief complaints and helps pre-order imaging or laboratory tests before they see the doctor. This approach moves laboratory tests earlier in the workflow, thereby reducing waiting time for doctor consultations [72].

In another cases tudy by Chen et al. (2022), the researchers used simulation optimiztion to streamline appointment scheduling, and room allocation for Ultrasound examinations and claimed that this could reduce patient waiting times if implemented correctly in hospitals [73]. Machine learning models automated appointment scheduling, medical record updates, and resource allocation, reducing manual workload and improving hospital efficiency [73]. The system enhanced patient experience, minimized delays, and optimized staff utilization.

4 AI and Cloud Computing's Limitations in Healthcare

4.1 Data Safety and Privacy Considerations

Concerns about cybersecurity and data privacy are raised by the use of AI and cloud computing in healthcare [74]. Because patient data is so sensitive, breaches can have dire repercussions. Strict data protection regulations are enforced by frameworks like GDPR and HIPAA. Implementing robust encryption, secure access controls, and blockchain-based data management enhances security [75].

Healthcare organizations are at serious risk from cybersecurity threats like ransomware attacks, illegal access, and data leaks. Because medical records are so valuable on the illicit market, attackers frequently target them. Organizations must put in place AI-driven threat detection systems that instantly spot irregularities and possible breaches in order to reduce these risks. Cyber resilience strategies, such as regular security audits and staff training, further strengthen healthcare data protection.

Additionally, cloud service providers and healthcare organizations that operate across many jurisdictions face difficulties in complying with regional and international data privacy legislation. Differing legal standards could make processing and exchanging data more difficult. Secure cross-border data sharing can be facilitated without jeopardizing patient confidentiality by putting in place standardized security standards and embracing privacy-enhancing technologies like homomorphic encryption and differential privacy [76, 77].

4.2 Ethical and Bias Issues

Disorganized training datasets can cause biases in AI models, which can result in differences in healthcare outcomes. The creation of moral AI frameworks and objective dataset curation are necessary to guarantee equity and openness in AI decision-making. Techniques for Explainable AI (XAI) improve model interpretability and build healthcare professionals' trust [78].

Historical disparities in healthcare incorporated into training datasets may cause bias in AI algorithms. If AI models are trained on predominantly Western or male-dominated datasets, they may perform poorly for underrepresented populations, exacerbating healthcare inequalities. Addressing this issue requires the inclusion of diverse demographic data and continuous monitoring for bias detection and mitigation throughout AI development and deployment [79].

Furthermore, the autonomy of AI in therapeutic decision-making is also fraught with ethical issues. Fully automated AI diagnoses may raise concerns about responsibility in times of misdiagnosis, even as AI-driven advice can help physicians. Establishing human-in-the-loop frameworks, where AI assists but does not replace human expertise, ensures that ethical considerations are upheld while maintaining AI's efficiency in healthcare applications [80].

4.3 Reliability and Accuracy of AI Models

The reliability of AI models in clinical settings is crucial, as incorrect predictions can lead to misdiagnosis and patient harm. Continuous validation, rigorous testing, and real-world implementation studies are necessary to improve AI model robustness. AI systems must be updated regularly to incorporate new medical knowledge and treatment advancements [81].

The quality and diversity of input data determine how well AI performs in various healthcare contexts. AI models developed in hospitals with a lot of resources might not translate effectively to healthcare institutions in rural or poor areas [81]. Extensive cross-validation across various patient populations is necessary to address this. While maintaining data privacy, federated learning—where AI models are taught across several universities without centralizing data—can aid in enhancing model generalizability [82].

For patients and physicians to trust AI systems, they must be comprehensible and interpretable. Widespread adoption is hampered by black-box AI models, which have opaque decision-making procedures. Implementing interpretable AI techniques, such as attention mechanisms and feature importance analysis, allows medical professionals to understand how AI arrives at its conclusions, fostering confidence in AI-driven diagnoses and recommendations [82].

4.4 Integration Challenges with Existing Systems

Healthcare organizations frequently employ outdated systems that might not work with cloud and AI technologies. Integrating AI-driven analytics with traditional medical infrastructure requires interoperability standards and API-based solutions.

Since many healthcare IT systems were created decades ago, they are not adaptable enough to accommodate contemporary AI applications. Upgrading these systems is costly and time-consuming, leading to resistance from healthcare administrators. To bridge this gap, middleware solutions that act as intermediaries between legacy systems and AI platforms can enable gradual integration without overhauling existing infrastructure.

Data fragmentation is still a significant barrier as well. It can be difficult for AI models to get thorough medical histories because patient records are frequently kept in many departments or locations. The efficiency and dependability of AI in clinical contexts are increased by standardized data formats like FHIR (Fast Healthcare Interoperability Resources), which enable smooth data interchange between AI applications and current health record systems [83, 84].

5 Future Trends and Developments

5.1 AI-Powered Digital Twins in Healthcare

Personalized therapy simulations are made possible by digital twins, which produce virtual versions of patients. AI-powered digital twins improve patient outcomes by enabling early disease detection and tailored treatment plans [85].

The use of digital twins is expanding beyond diagnostics and treatment planning to areas such as predictive maintenance of medical equipment and hospital resource management. By simulating patient responses to treatments before administering them, clinicians can tailor interventions to individual needs, minimizing side effects and maximizing efficacy [86]. This technology also allows for early detection of complications and continuous monitoring of disease progression in chronic conditions. Furthermore, digital twins driven by AI are essential to precision healthcare and personalized medication. Treatment plans can be improved by incorporating lifestyle, genetic, and medical history data into digital twins. This approach ensures that healthcare professionals can make data-driven decisions that cater to each patient's unique physiology. The advancement of real-time digital twins through IoT and wearable devices will further enable continuous monitoring, improving proactive healthcare strategies [87].

5.2 *Quantum Computing for Drug Discovery*

Due to its exponential increase in processing capacity, quantum computing has the potential to completely transform AI-driven drug development. Drug research can proceed more quickly because of quantum AI models' ability to analyze intricate molecular interactions at previously unheard-of speeds.

The capacity of quantum computing to accurately model molecular structures and chemical interactions is one of its main advantages. Due to the large number of variables involved, these computations are frequently difficult for traditional computing methods to handle [88]. This problem is solved by quantum computing, which uses quantum bits (qubits) to compute complicated algorithms exponentially more quickly. This lowers development costs and timelines by facilitating researchers' more effective identification of prospective medication candidates [89].

Additionally, quantum computing facilitates the development of personalized medicine by allowing rapid screening of drug responses based on genetic profiles. By analyzing how different genetic variations influence drug efficacy and toxicity, quantum-powered AI models can predict individualized treatment plans [89]. Drug discovery and innovation in the biomedical sciences will be further accelerated by the combination of quantum computing, artificial intelligence, and cloud-based platforms, which will facilitate cooperative research initiatives.

5.3 *Federated Learning for Decentralized Healthcare AI*

Without exchanging patient data, federated learning allows AI models to be trained across several healthcare facilities. This method allows for collaborative AI improvements while improving data privacy [90].

The capacity of federated learning to preserve data security and compliance with privacy laws like HIPAA and GDPR is a significant benefit in the healthcare industry. Federated learning trains models locally on each institution's dataset, as contrast to standard machine learning models that need centralized data storage. This ensures that sensitive patient information remains within its originating healthcare facility while still contributing to global AI model improvements.

Furthermore, federated learning enables AI algorithms to generalize better across diverse populations by incorporating data from multiple sources [91]. This reduces bias and enhances the reliability of AI-driven medical predictions. The adoption of federated learning in clinical applications, such as disease prediction and medical imaging analysis, is expected to drive more equitable and robust AI solutions that benefit a wide range of patient demographics [92].

5.4 AI in Robotics-Assisted Surgery

Robotic systems powered by AI help with minimally invasive procedures, improving accuracy and speeding up recuperation. Robotic surgery platforms leverage AI for real-time decision-making and automation of complex surgical procedures.

The field of surgery is changing as a result of recent developments in AI and robotics. AI-driven analytics are integrated into robotic-assisted surgery systems, such the da Vinci Surgical System, to increase accuracy, lower human error, and improve patient outcomes [93]. Surgeons may execute complex surgeries with less invasiveness, thanks to AI-powered robotic arms' improved dexterity and control, which shortens recovery times and lowers post-operative problems.

Beyond precision and efficiency, AI-driven robotic surgery is expanding into remote and autonomous surgical applications. With the advent of 5G and cloud computing, telesurgery is becoming a viable option where surgeons can operate on patients located in different geographical locations [94]. Robotic devices driven by AI can help carry out predetermined surgical stages, offering real-time feedback and guaranteeing consistency in intricate surgeries. As AI develops further, its application to robotic surgery will transform surgical care and increase access to high-quality surgeries globally [94].

6 Conclusion

The healthcare sector is undergoing a transformation because of the combination of AI and cloud computing, which is boosting patient outcomes, treatment effectiveness, and diagnostic accuracy. Personalized medicine, drug discovery, and AI-driven predictive analytics are revolutionizing the way illnesses are identified, treated, and controlled. Meanwhile, cloud computing ensures seamless data integration, storage, and accessibility, making healthcare services more efficient and patient-centric. Despite these technological developments, issues like data security, moral dilemmas, and compatibility with the current healthcare system still need to be resolved to guarantee acceptance and deployment go smoothly. Additionally, automation driven by AI is reducing administrative workloads, freeing up healthcare workers to spend more time providing patient care. Furthermore, cloud-based telemedicine solutions are breaking geographical barriers, making quality healthcare services accessible to remote and underserved populations. It is anticipated that the fast advancements in AI-powered robotics and quantum computing would push the limits of surgical accuracy and medical research, improving the standard of care. As AI and cloud computing continue to evolve, collaboration among governments, healthcare providers, and technology developers will be crucial in addressing regulatory challenges and ensuring ethical deployment. Investing in research, developing robust security protocols, and promoting equitable access to AI and cloud computing solutions will be essential for maximizing their benefits.

By embracing these innovations and proactively mitigating risks, the healthcare industry can move towards a future where technology-driven solutions enhance efficiency, reduce costs, and ultimately improve global health outcomes.

References

1. Angelos, I. S., et al. (2023). Digital transformation in healthcare: Technology acceptance and its applications. *International Journal of Environmental Research and Public Health, 20*(4), 3407. https://doi.org/10.3390/ijerph20043407
2. Topol, A. (2019). High-performance medicine: The convergence of human and artificial intelligence. *Nature Medicine, 25*(1), 44–56.
3. Miotto, J., et al. (2018). Deep learning for healthcare: Review, opportunities and challenges. *Briefings in Bioinformatics, 19*(6), 1236–1246.
4. Bekbolatova, M., et al. (2024). Transformative potential of AI in healthcare: Definitions, applications, and navigating the ethical landscape and public perspectives. *Healthcare (Basel), 12*(2), 125. https://doi.org/10.3390/healthcare12020125. PMID: 38255014; PMCID: PMC10815906.
5. Robert, E. (2019). *Deep medicine: How artificial intelligence can make healthcare human again*. Basic Books.
6. Li, Y. H., Li, Y. L., Wei, M. Y., et al. (2024). Innovation and challenges of artificial intelligence technology in personalized healthcare. *Scientific Reports, 14*, 18994.
7. Yuan, S., et al. (2025). AI-Powered early warning systems for clinical deterioration significantly improve patient outcomes: A meta-analysis. *BMC Medical Informatics and Decision Making, 25*(1), 203. https://doi.org/10.1186/s12911-025-03048-x. PMID: 40457275; PMCID: PMC12131336.
8. Rana, M., & Bhushan, M. (2023). Machine learning and deep learning approach for medical image analysis: Diagnosis to detection. *Multimedia Tools and Applications, 82*, 26731–26769.
9. Litjens, J., Kooi, T., Ehteshami Bejnordi, B., et al. (2017). A survey on deep learning in medical image analysis. *Medical Image Analysis, 42*, 60–88.
10. Wichmann, J. L., et al. (2020). Artificial intelligence and machine learning in radiology: Current state and considerations for routine clinical implementation. *Investigative Radiology, 55*(9), 619–627. https://doi.org/10.1097/RLI.0000000000000673
11. Sarvamangala, D. R., et al. (2022). Convolutional neural networks in medical image understanding: A survey. *Evolutionary Intelligence, 5*(1), 1–22. https://doi.org/10.1007/s12065-020-00540-3
12. Wang, J., et al. (2025). Artificial intelligence in cancer pathology: Applications, challenges, and future directions. *Cytojournal, 19*(22), 45.
13. Ali, M., et al. (2023). Federated learning for privacy preservation in smart healthcare systems: A comprehensive survey. *IEEE Journal of Biomedical and Health Informatics, 27*(2), 778–789.
14. Aiforia Technologies. (2024). *Case study: AI-assisted prostate cancer diagnosis*. Aiforia Technologies. Retrieved from www.aiforia.com/resource-library/case-study-ai-assisted-prostate 662 -cancer-diagnosis7
15. Gulshan, V., Peng, L., Coram, M., et al. (2016). Development and validation of a deep learning algorithm for detection of diabetic retinopathy in retinal fundus photographs. *JAMA, 316*(22), 2402–2410.
16. Khalifa, M. (2024). Artificial intelligence for clinical prediction: Exploring key domains and essential functions. *Smart Health, 35*, 100540.
17. Wang, D., et al. (2021). *A machine learning model for accurate prediction of sepsis in ICU patients*. Front Public Health.

18. Rashid, J., et al. (2022). *An augmented artificial intelligence approach for chronic diseases prediction*. Front Public Health.
19. Courtiol, P., et al. (2019). Deep learning-based classification of mesothelioma improves prediction of patient outcome. *Nature Medicine, 25*, 1519–1525.
20. Lauritsen, S. M., et al. (2020). Early detection of sepsis utilizing deep learning on electronic health record event sequences. *Artificial Intelligence in Medicine, 104*, 101820.
21. Tang, X., et al. (2024). A survey of generative AI for de novo drug design: New frontiers in molecule and protein generation. *Brief Bioinformatics, 25*(4).
22. Lu, X., et al. (2024). Artificial intelligence for optimizing recruitment and retention in clinical trials: A scoping review. *Journal of the American Medical Informatics Association, 31*(11), 2749–2759.
23. Weissler, E. H., et al. (2021). The role of machine learning in clinical research: Transforming the future of evidence generation. *Trials, 22*, 537.
24. Ryan, D. K., et al. (2024). Artificial intelligence and machine learning for clinical pharmacology. *British Journal of Clinical Pharmacology, 90*(3), 629–639.
25. Vora, L. K., et al. (2023). Artificial intelligence in pharmaceutical technology and drug delivery design. *Pharmaceutics, 15*, 1916. https://doi.org/10.3390/pharmaceutics15071916
26. Pusparini, R. T., et al. (2023). MATH: A deep learning approach in QSAR for estrogen receptor alpha inhibitors. *Molecules, 28*(15), 5843.
27. Gangwal, A., et al. (2024). Generative artificial intelligence in drug discovery: Basic framework, recent advances, challenges, and opportunities. *Front Pharmacology, 15*, 1331062.
28. Stokes, J. M., et al. (2020). A deep learning approach to antibiotic discovery. *Cell, 180*(4), 688–702.
29. You, Y., et al. (2022). Artificial intelligence in cancer target identification and drug discovery. *Signal Transduction and Targeted Therapy, 7*, 156. https://doi.org/10.1038/s41392-022-00994-0
30. Mao, Y., et al. (2025). Emerging artificial intelligence-driven precision therapies in tumor drug resistance: Recent advances, opportunities, and challenges. *Molecular Cancer, 24*, 123. https://doi.org/10.1186/s12943-025-02321-x
31. Ortiz, B., et al. (2024). Data preprocessing techniques for AI and machine learning readiness: Scoping review of wearable sensor data in cancer care. *Journal of Medical Internet Research. Mhealth Uhealth, 12*, e59587. https://doi.org/10.2196/59587
32. Chow, R., et al. (2024). The use of wearable devices in oncology patients: A systematic review. *The Oncologist, 26*(4), e419–e430. https://doi.org/10.1093/oncolo/oyad305
33. Chaturvedi, U., et al. (2025). The impact of artificial intelligence on remote healthcare: Enhancing patient engagement, connectivity, and overcoming challenges. *Intelligent Pharmacy, 3*(5), 323–329.
34. Menden, M. P., Iorio, R., Garnett, T., et al. (2013). Machine learning prediction of cancer cell sensitivity to drugs based on genomic and chemical properties. *PLoS One, 8*(4), e61318.
35. Marling, C., et al. (2012). Emerging applications for intelligent diabetes management. *Artificial Intelligence Magazine, 33*, 67–78. https://doi.org/10.1609/aimag.v33i2.2410
36. Bahga, A., et al. (2013). A cloud-based approach for interoperable electronic health records (EHRs). *IEEE Journal of Biomedical Health Informatics, 17*(5), 894–906. https://doi.org/10.1109/JBHI.2013.2257818
37. Ahmadi, M., & Aslani, N. (2018). Capabilities and advantages of cloud computing in the implementation of electronic health record. *Acta Informatica Medica, 26*(1), 24–28. https://doi.org/10.5455/aim.2018.26.24-28. PMID: 29719309; PMCID: PMC5869277.
38. Kirilov, N. (2024). Capture of real-time data from electronic health records: Scenarios and solutions. *Mhealth, 10*, 14. https://doi.org/10.21037/mhealth-24-2
39. Caruana, R., et al. (2023). AI explainability in healthcare: Challenges and solutions. *Artificial Intelligence in Medicine, 134*, 102356.
40. Schoville, R. R., & Titler, M. (2015). Guiding healthcare technology implementation: A new integrated technology implementation model. *Computers, Informatics, Nursing, 33*(3), 99–107.

41. Barton, C., et al. (2019). Evaluation of a machine learning algorithm for up to 48-h advance prediction of sepsis in adult ICU patients using electronic health record data. *Computers in Biology and Medicine, 109*, 79–84.
42. Sikkalgar, A. A., et al. (2025). Telemedicine and remote patient monitoring: Transforming healthcare delivery. *Scientific Journal of Biology & Life Sciences, 4*(2), 2025.
43. Abdulmalek, S., et al. (2022). IoT-based healthcare-monitoring system towards improving quality of life: A review. *Healthcare (Basel), 10*(10), 1993. https://doi.org/10.3390/healthcare10101993
44. Wiedermann, C. J., et al. (2023). Redesigning primary care: The emergence of artificial-intelligence-driven symptom diagnostic tools. *Journal of Personalized Medicine, 13*(9), 1379.
45. Leo, D., et al. (2022). The TAILOR investigators interactive remote patient monitoring devices for managing chronic health conditions: Systematic review and meta-analysis. *Journal of Medical Internet Research, 24*(11), e35508. https://doi.org/10.2196/35508. Retrieved from https://www.jmir.org/2022/11/e35508
46. Peyroteo, M., et al. (2021). Remote monitoring systems for patients with chronic diseases in primary health care: Systematic review. *Journal of Medical Internet Research. Mhealth Uhealth, 9*(12), e28285. https://doi.org/10.2196/28285
47. Ellis-Petersen, R. (2024). You feel omnipresent': Bringing city care to India's country hospitals. *The Guardian.*
48. Kim, J., et al. (2022). Validation of a cloud-based tele-stroke system reliability in determining national institutes of health stroke scale scores for acute ischemic stroke screening in the emergency department. *Frontiers in Neurology, 13*, Article 973165.
49. Rouzbeh, F., et al. (2020). Collaborative cloud computing framework for health data with open source technologies. In *Proceedings of the 11th ACM International Conference on Bioinformatics, Computational Biology and Health Informatics* (pp. 1–10). https://doi.org/10.1145/3388440.3412460
50. Rehan, H. (2023). AI-powered genomic analysis in the cloud: Enhancing precision medicine and ensuring data security in biomedical research. *Journal of Deep Learning in Genomic Data Analysis, 3*(1).
51. He, D., et al. (2025). Artificial intelligence technologies for enhancing diagnosis, treatment and drug discovery in Alzheimer's disease, cancer, and rare genetic disorders. *Frontiers in Aging Neuroscience, 17*, 1609063.
52. Kushwaha, P., et al. (2024). Enhancing clinical drug trial monitoring with blockchain technology. *Contemporary Clinical Trials, 146*, 107684. https://doi.org/10.1016/j.cct.2024.107684
53. Lau, J. W., et al. (2017). The cancer genomics cloud: Collaborative, reproducible, and democratized access to cancer genomics data. *Genome Research, 27*(5), 757–764.
54. Shahzad, M., et al. (2024). Classification of clinically actionable genetic mutations in cancer patients. *Frontiers in Molecular Biosciences, 10*. https://doi.org/10.3389/fmolb.2023.1277862
55. Luthria, G., et al. (2020). Implementing a cloud based method for protected clinical trial data sharing. *Pacific Symposium on Biocomputing, 25*, 647–658.
56. Albalwy, F., et al. (2021). A blockchain-based dynamic consent architecture to support clinical genomic data sharing (ConsentChain): Proof-of-concept study. *JMIR Medical Informatics, 9*(11), e27816.
57. Hong, J., et al. (2024). Rare disease genomics and precision medicine. *Genomics and Informatics, 22*, 28.
58. Mostafa, A. M., et al. (2023). Strengthening cloud security: An innovative multi-factor multi-layer authentication framework for cloud user authentication. *Appllied Sciences, 13*, 10871. https://doi.org/10.3390/app131910871
59. Mehrtak, M., et al. (2021). Security challenges and solutions using healthcare cloud computing. *Journal of Medicine and Life, 14*(4), 448–461.
60. Nuka, S. T. (2022). Cloud-based AI systems for real-time medical imaging analysis and diagnostics. *International Journal of Advanced Research in Computer and Communication Engineering, 11*(12), 333–350.

61. Nair, R., et al. (2022). Blockchain-based decentralized cloud solutions for data transfer. *Computational Intelligence and Neuroscience, 2022*, 8209854. https://doi.org/10.1155/2022/8209854. PMID: 35676964; PMCID: PMC9170456.
62. Khanna, A., et al. (2022). Blockchain-cloud integration: A survey. *Sensors (Basel), 22*(14), 5238. https://doi.org/10.3390/s22145238. PMID: 35890918; PMCID: PMC9320072.
63. Chen, C. P., & Wang, M. L. (2021). Blockchain and its role in healthcare cloud compliance. *IEEE Blockchain Tech, 2*(4), 112–123.
64. Ramalingam, S., et al. (2022). Cloud compliance in healthcare: A technical evaluation of data encryption, access control, and risk management practices. *Journal of Machine Learning for Healthcare Decision Support, 2*(2), 94–133.
65. Llewellyn, S., Procter, R., Harvey, G., Maniatopoulos, G., & Boyd, A. (2014). *Facilitating technology adoption in the NHS: Negotiating the organisational and policy context—a qualitative study*. NIHR Journals Library. PMID: 25642495.
66. Achori, E., & Olaniyi, T. (2025). Blockchain applications in healthcare finance: A systematic review of opportunities and challenges for NHS general ledgers in United Kingdom. *International Journal of Research and Innovation in Social Science (IJRISS), 9*(10), 2204–2212.
67. Mohammed, S. B., Ramalingam, S., & Sivathapandi, P. (2022). Cloud compliance implementation in healthcare: Ensuring security, privacy, and data integrity in cloud-based solutions. *Cybersecurity and Network Defense Research (CNDR), 2*(2), 82–127.
68. Alabdulatif, A., et al. (2022). Security of blockchain and AI-empowered smart healthcare: Application-based analysis. *Applied Sciences, 12*(21), 11039. https://doi.org/10.3390/app122111039
69. Amanat, A., et al. (2022). Blockchain and cloud computing-based secure electronic healthcare records storage and sharing. *Frontiers of Public Health, 10*, 938707. https://doi.org/10.3389/fpubh.2022.938707. PMID: 35928494; PMCID: PMC9343689.
70. Maleki Varnosfaderani, S., et al. (2024). The role of AI in hospitals and clinics: Transforming healthcare in the 21st century. *Bioengineering, 11*(4), 337.
71. Moazemi, S., et al. (2023). Artificial intelligence for clinical decision support for monitoring patients in cardiovascular ICUs: A systematic review. *Frontiers in Medicine, 10*, 1109411.
72. Li, X., et al. (2021). Artificial intelligence-assisted reduction in patients' waiting time for outpatient process: A retrospective cohort study. *BioMed Central Health Services Research, 21*, 237. https://doi.org/10.1186/s12913-021-06248-z
73. Chen, P. S., et al. (2022). Using simulation optimization to solve patient appointment scheduling and examination room assignment problems for patients undergoing ultrasound examination. *Healthcare (Basel), 10*(1), 164. https://doi.org/10.3390/healthcare10010164
74. Veluru, C. S. (2024). Impact of artificial intelligence and generative AI on healthcare: Security, privacy concerns and mitigations. *Journal of Artificial Intelligence & Cloud Computing, 3*(1), 1–6.
75. Shah, W. F. (2023). Preserving privacy and security: A comparative study of health data regulations—GDPR vs. HIPAA. *International Journal For Research in Applied Science and Engineering Technology*. https://doi.org/10.22214/ijraset.2023.55551
76. Payne, J. (2025). *Analyzing the cost per record of healthcare data breaches_02182025*. White Paper.
77. Scheibner, J., et al. (2021). Revolutionizing medical data sharing using advanced privacy-enhancing technologies: Technical, legal, and ethical synthesis. *Journal of Medical Internet Research, 23*(2). https://doi.org/10.2196/25120
78. Mittermaier, M., et al. (2023). Bias in AI-based models for medical applications: Challenges and mitigation strategies. *Nature Partner Journal. Digital Medicine, 6*(113). https://doi.org/10.1038/s41746-023-00858-z
79. Norori, N., et al. (2021). Addressing bias in big data and AI for health care: A call for open science. *Patterns (N Y), 2*(10), 100347. https://doi.org/10.1016/j.patter.2021.100347
80. Salloch, S., et al. (2024). What are humans doing in the loop? Co-reasoning and practical judgment when using machine learning-driven decision aids. *The American Journal of Bioethics, 24*(9), 67–78. https://doi.org/10.1080/15265161.2024.2353800

81. Labkoff, S., et al. (2024). Toward a responsible future: Recommendations for AI-enabled 804 clinical decision support. *Journal of the American Medical Informatics Association, 31*(11), 2730–2739. https://doi.org/10.1093/jamia/ocae209
82. Chaddad, A., et al. (2022). Explainable, domain-adaptive, and federated artificial intelligence in medicine. *Journal of Latex Class Files, 14*(8), 1–16.
83. Bajwa, J., et al. (2021). Artificial intelligence in healthcare: Transforming the practice of medicine. *Future Healthcare Journal, 8*(2), e188–e194. https://doi.org/10.7861/fhj.2021-0095
84. Namli, T., et al. (2024). A scalable and transparent data pipeline for AI-enabled health data ecosystems. *Front Med (Lausanne), 11*, 1393123. https://doi.org/10.3389/fmed.2024.1393123
85. Fischer, R. P., et al. (2023). Digital patient twins for personalized therapeutics and pharmaceutical manufacturing. *Frontiers in Digital Health, 5*, 1302338. https://doi.org/10.3389/fdgth.2023.1302338
86. Vallée, A. (2023). Digital twin for healthcare systems. *Frontiers in Digital Health, 5*, 1253050. https://doi.org/10.3389/fdgth.2023.1253050
87. Sun, T., et al. (2022). The digital twin in medicine: A key to the future of healthcare? *Frontiers in Medicine, 9*. https://doi.org/10.3389/fmed.2022.907066
88. Liu, H., et al. (2022). Prospects of quantum computing for molecular sciences. *Materials Theory, 6*, 11. https://doi.org/10.1186/s41313-021-00039-z
89. Jeyaraman, N., et al. (2024). Revolutionizing healthcare: The emerging role of quantum computing in enhancing medical technology and treatment. *Cureus, 16*(8), e67486. https://doi.org/10.7759/cureus.67486
90. Oh, W., et al. (2023). Federated learning in health care using structured medical data. *Advances in Kidney Disease Health, 30*(1), 4–16. https://doi.org/10.1053/j.akdh.2022.11.007
91. Abbas, S. R., et al. (2024). Federated learning in smart healthcare: A comprehensive review on privacy, security, and predictive analytics with IoT integration. *Healthcare, 12*(24), 2587. https://doi.org/10.3390/healthcare12242587
92. Teo, Z. L., et al. (2024). Federated machine learning in healthcare: A systematic review on clinical applications and technical architecture. *Cell Reports Medicine, 5*(2), 101419. https://doi.org/10.1016/j.xcrm.2024.101419
93. Liu, Y., et al. (2024). Evolution of surgical robot systems enhanced by artificial intelligence: A review. *Advanced Intelligent Systems, 6*, 1–25. https://doi.org/10.1002/aisy.202300268
94. Xi, X., et al. (2025). Surgery without distance: Will 5G-based robot-assisted telesurgery redefine modern surge. *Translational Lung Cancer Research, 14*(5), 1821–1829. https://doi.org/10.21037/tlcr-2025-16

Machine Learning Approaches for Sustainable Cities Using Internet of Things: Supervised and Deep Learning Techniques

Rajesh Dey, Rupali Atul Mahajan, and Neha Gupta

1 Introduction

The rapid urbanization of modern society requires the paradigm change in urban design and management. Cities, having difficulties with energy consumption, pollution, and infrastructure management, are in more need of innovating technologies to optimize urban functionalities. Thus, with its huge potential, IoT in conjunction with ML is being viewed as the best solution to promote sustainable developmental activities in urban areas. This chapter presents an extensive overview of ML approaches related mostly to supervised learning and deep learning and their integration with IoT sensor networks that act as catalysts toward the realization of smarter, sustainable cities.

Sustainable cities are not only their ability to reduce technological and resource wastage but are also working toward improving the quality of life of the urban residents. The fusion of ML and IoT makes it possible to monitor real-time data from sensor networks distributed along the urban environment. The said data, when processed by complex ML models, could yield insightful knowledge which can be used in energy conservation, traffic congestion, and resources distribution [1]. To put it into perspective, ML algorithms can be used to predict traffic generation at peak

R. Dey
GNSU Bihar, Jamuhar, Bihar, India
e-mail: Rajesh.dey@gnsu.ac.in

R. A. Mahajan
Vishwakarma Institute of Technology, Pune, Maharashtra, India
e-mail: Rupali.mahajan@viit.ac.in

N. Gupta (✉)
Symbiosis University of Applied Sciences, Indore, India
e-mail: Neha.gupta@suas.ac.in

C. Rompante Cunha et al. (eds.), *Transformative Cloud Computing, IoT and Extended Reality*, Signals and Communication Technology,
https://doi.org/10.1007/978-3-032-03554-7_6

times and encourage changes in traffic signal timing based on those predictions. Consequently, this will enhance mobility and reduce emissions. The rapid evolution of ML techniques, from straightforward linear regression models down to more advanced deep neural networks, has gone hand in hand with the evolution of IoT sensor technologies. With the vast number of connected devices, city administrators now enjoy unprecedented data access that can be harnessed for instantaneous analysis and decision-making. This feedback loop composed of both ML and IoT keeps making city planning and management more data-driven and aligned with real-world conditions on the road.

This chapter will have several parts. Subsequent to the introduction, we provide a very extensive background of what deals with sustainable cities and the IoT paradigm as far as technologies that govern contemporary urban management go. This then creates an entrance into the supervised learning techniques relevant to urban planning problems such as energy consumption forecasting and traffic flow predictions. We provide mathematical derivations to give insight into the working mechanism of these models.

Following that, our story turns toward deep learning techniques that were used and applied to IoT sensor networks. We will explore different architectures and explain how deep learning models, including convolutional and recurrent neural networks, can be designed for real-time analysis of sensor data. To relate theory with practice, we then cover real-world case studies in smart cities such as Barcelona, Amsterdam, and Singapore. These case studies demonstrate how traffic control, energy management, and waste system implementations based on TensorFlow have been actualized.

Finally, this chapter gives the reader practical and detailed sample application code in Python backed by TensorFlow (version 2.4+ on Python 3.8+) that showcases training and building of ML models for a number of urban applications. This practical section is truly important for any practitioner already keen to implement the theory in practice.

All through this chapter, we stick strictly to APA style referencing (see References section) so all claims, statistics, and technical derivation will be fully referenced. The support services rendered by Nosratabadi et al. [1] and Ghahramani et al. [2] stand to strengthen the statements on the ML and IoT paradigm in urban cases. To summarize, this chapter intends to provide urban planners and engineers with equivalent technical knowledge on which to base further research and applications of ML and IoT toward adopting sustainable cities.

The integration of ML and DL into smart city infrastructure has become a major point on which to focus by researchers and practitioners who are engaged in improving urban sustainability, efficient functioning, and quality of life. Ghahramani et al. [2] presented the effectiveness and potential of deep learning and IoT sensor network for waste management through intelligent route recommendation systems, thus emphasizing the importance of real-time data and predictive analytics in urban services. Similarly, Nosratabadi et al. [1] produced a comprehensive survey of ML applications for smart cities, highlighting the wide range of areas affected, including transportation and energy management.

Agarwal and Sharma [3] reviewed deep architecture tailored toward smart city applications, stressing the adaptability of convolutional and recurrent neural networks for processing high-dimensional urban data. Chen et al. [4] discussed challenges and prospects in ML toward urban sustainability, obesity data heterogeneity, and solutions for scalability. Kumar et al. [5] systematically reviewed IoT-enabled smart cities, identifying ML as a key enabler for intelligent decision-making and automation.

Li et al. [6] investigated practical TensorFlow implementations, bridging the gap between theoretical models and real applications in smart city deployments. Park et al. [7] then introduced the case of deep learning for urban mobility that significantly improves congestion prediction and control.

An edge computing and IoT partnership was scrutinized by Rodriguez and Garcia [8], underlining the distributed ML's contribution to corting latency and bolstering real-time responsiveness. Smith et al. [9] addressed privacy concerns, a pivotal issue of urban data, and proposed privacy-preserving ML frameworks for urban computing.

Sustainable urban development via AI in Amsterdam is discussed by Thompson and Anderson [10], delivering insights into policy and implementation strategies. Real-time data processing using deep learning, crucial for responsive urban environments, is then surveyed by Wang & Zhang [11]. Finally, Zhou et al. [12] prove the efficiency of neural networks for energy consumption prediction in smart buildings, thus contributing to the better management of resources.

Collectively, these studies illustrate the tremendous influence ML and DL have caused on smart city systems while simultaneously pointing out the problems of scalability, privacy, and the integration of heterogeneous data sources (Fig. 1).

2 Background on Sustainable Cities and the IoT Paradigm

Sustainability in urban planning was forged from the great awakening during the past century and, thus, has evolved to consider environmental, societal, and economic pillars of urban development. In essence, a sustainable city attempts to have a smaller ecological footprint while delivering its residents a good standard of living. The IoT sensor networks have become indispensable in monitoring life dimensions in a city by providing an uninterrupted stream of relevant data parameters such as air quality, traffic density, energy consumption, and resource utilization.

Typically, IoT devices are small low-power sensors deployed in the urban environment; they collect data and communicate this data over networks to facilities that process it further. Considering the enormous amount of generated data, advanced data analytics tools, and ML algorithms are bound to be involved. As noted by Nosratabadi et al. [1], ML algorithms, including decision trees, support vector machines, ensemble methods, and deep learning, have been extensively applied to deal with urban problems—from energy management to transportation efficiency.

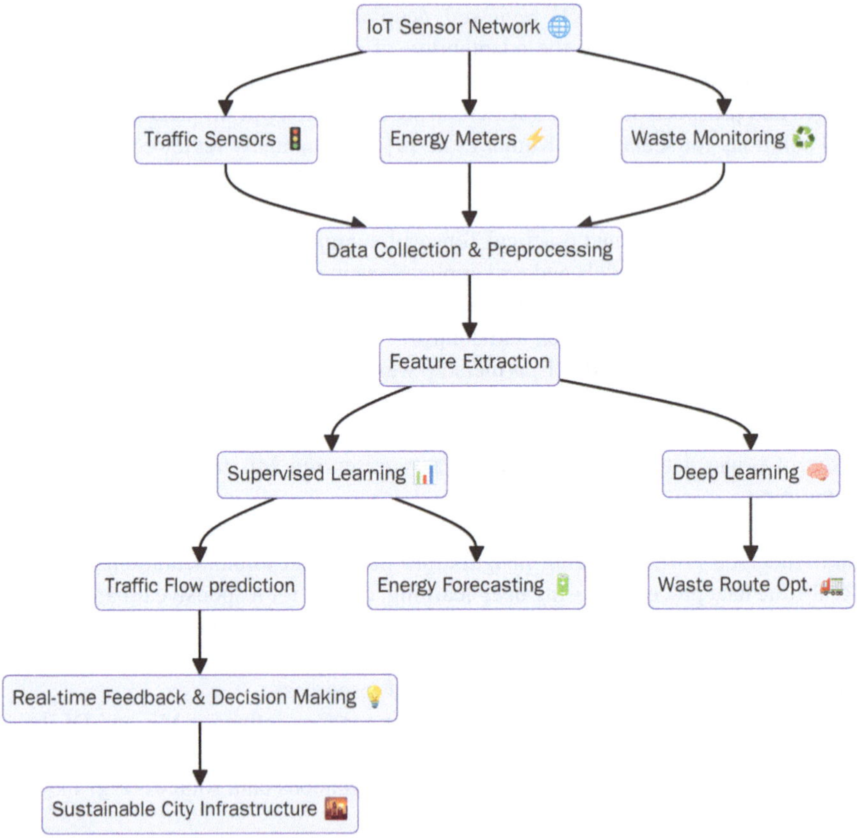

Fig. 1 Machine Learning Approaches for Sustainable Cities Using IoT

Under the smart city paradigm, ML combined with IoT equips city administrators to predict as well as monitor and optimize urban operations. For example, by using historical sensor data alongside real-time data, supervised learning approaches can predict energy consumption patterns and allow for management to act in advance of resource delivery. Similarly, deep learning techniques can extract nonlinear relationships in data collected from environmental sensors; such relationships may discover subtle patterns that may be related to air quality indices or traffic disturbances.

A typical IoT setup architecture consists of several layers—from the sensor layer characterized by the sensors that gather raw data to the communication and data aggregation layer, until the last layer that uses an analytics layer where the ML algorithms are working—as one entity they act as robust and scalable platform for handling the complexity of urban systems. In sensor networks, data collection ranges from seconds to minutes, resulting in time-series data well suited for supervised learning models for forecasting and anomaly detection.

Sustainability on the urban design front means that even the energy use of the IoT devices and the computational resources that serve the ML are themselves subject to optimization. Recent initiatives in Amsterdam demonstrate that these ML models are used not only to optimize energy consumption in buildings but also to optimize how the sensors themselves operate to collect data so as to keep the data collection infrastructure energy efficient.

What was highlighted here are some of the fundamental principles that form the underpinning for the concept of IoT and sustainable cities. The emergence of smart sensors and sturdy ML frameworks has created a symbiotic relationship that can be utilized to create solutions consistent with the peculiarities of urban systems. This backdrop serves as an introduction to the thorough technical discussion further in this book, where we will outline the various aspects relevant to supervised learning and deep learning methods for urban planning.

As the chapter advances, it is essential to keep in mind that all the data obtained from the Internet-of-Things sensor networks form the core of predictive modeling. Such high-resolution data allows for fine forecasting of especially energy demands, traffic flow, or even waste management logistics. It is, hence, that we especially talk about sustainable cities where every kilowatt being saved and every step taken for optimized routing path will translate to huge economic and environmental implications in the foreseeable future.

In the next section, we focus on supervised learning techniques lying at the heart of numerous urban planning prediction systems. They utilize a database with labels for aiding regression analysis and classification, which are among the main factors dictating modern urban management. Figure 2 shows total flow diagram for the system.

3 Supervised Learning for Urban Planning and Resource Optimization

Supervised learning is considered a branch of ML that operates with labeled datasets, which act as training sets to learn models able to generate predictions of outcomes from input features. In urban planning applications, these techniques are crucial for projecting traffic congestion, estimating energy consumption, and even optimizing infrastructure development. Supervised learning involves learning a function $f: X \rightarrow Y$ given a set of samples, where X is a space of input features (such as sensor readings or age of building usage data) and Y is the set of target outputs (energy demand, traffic volumes).

The goal of any supervised learning technique is basically to minimize a loss function that measures the error between the predicted and true values. A particularly typical example of loss function is the mean squared error:

$$\mathrm{MSE} = \frac{1}{n}\sum_{i=1}^{n}\left(y_i - \hat{y}_i\right)^2 \tag{1}$$

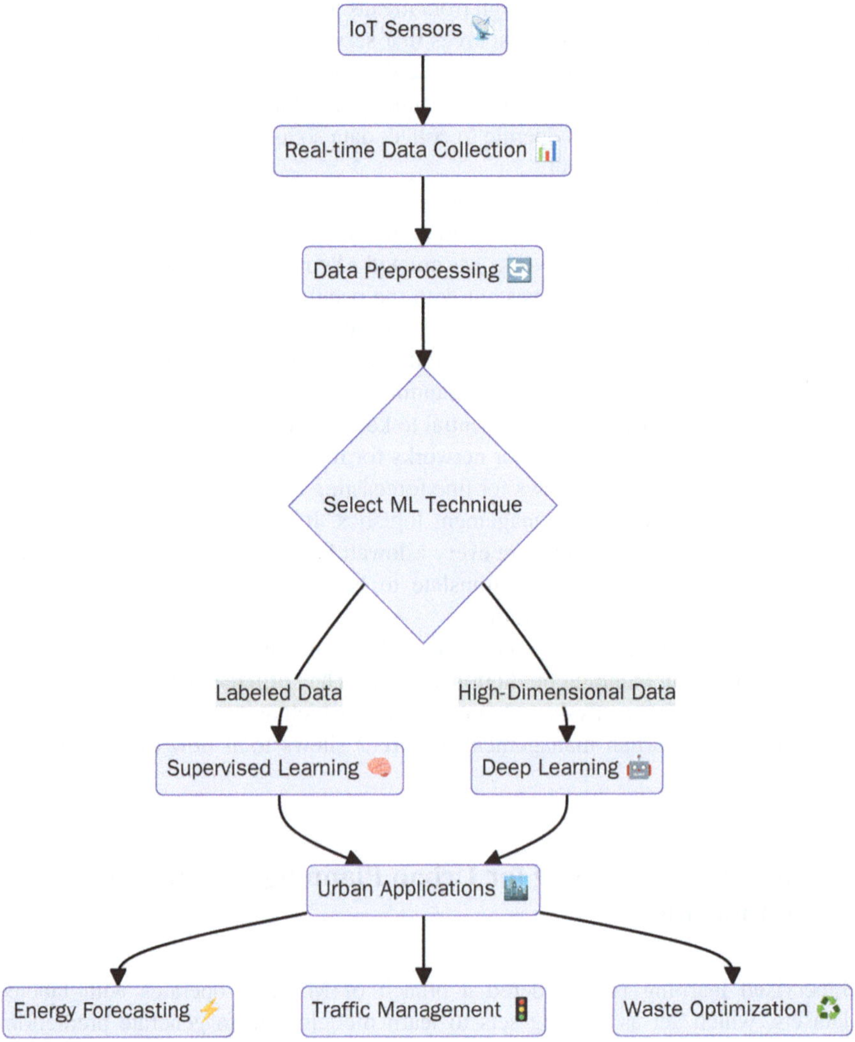

Fig. 2 IoT-ML workflow in smart cities

where n is the number of samples, y_i is the actual value, and $\hat{y}_i$ is the prediction. Typically, this loss function is optimized through iterative approaches like gradient descent. In simple linear regression, the update formula for a weight w is:

$$w := w - \eta \frac{\partial}{\partial w} \text{MSE} \tag{2}$$

where η represents the learning rate. In more complex situations in urban planning, one might have models that have multiple hidden layers or complicated nonlinearities and sophisticated optimization algorithms are then required.

An example might be where the city deploys supervised learning to predict traffic congestion off of historical sensor data. Let X = a set of features including vehicle counts, average speeds, or time-of-day indicators and let Y be congestion level, in which a regression model is trained off data to help city planners predict peak periods and reschedule traffic light timings. This, in effect, improves traffic flow and reduces emissions. In essence, such models' mathematical rigor entails setting up appropriate loss functions and ensuring that regularization is implemented to guarantee against overfitting.

Offering another application is in energy resource management in smart cities: based on historical energy consumption and weather data, models in Amsterdam could spot inefficiencies in energy usage and suggest specific recommendations (Wikipedia, n.d.). The analysis might be through multivariate regression, with the objective function incorporating a regularization term like L_2 regularization to penalize large weights:

$$\text{Loss} = \text{MSE} + \lambda \|w\|^2 \tag{3}$$

where λ is the regularization parameter. This formulation ensures that the model fits the data well but does not get too complicated.

The derivation of gradients related to these intricate models often calls for calculus-based optimization. For instance, when employing nonlinear activation functions (like sigmoid or ReLU functions), the gradients of the loss relative to weights are determined by the chain rule, which must be applied repeatedly:

$$\frac{\partial \text{Loss}}{\partial w} = \frac{\partial \text{Loss}}{\partial \hat{y}} \cdot \frac{\partial \hat{y}}{\partial z} \cdot \frac{\partial z}{\partial w} \tag{4}$$

where z stands for the weighted sum prior to the activation being applied. Such mathematical derivations can be extended to multi-level networks, where every new layer lessens granularity in the gradient computations.

In essence, this systematic approach allows learning to interpret complex urban phenomena. Because it can learn from labeled data, it becomes especially useful for predictive tasks pertaining to urban management. Real-life applications, including urban forecasting and energy management, depend heavily upon the precise mathematical formulations and iterative optimization approaches discussed above, all of which are central to sustainable development planning.

In the following sections, this discussion will be extended further into deep learning approaches that provide further abstraction layers and feature extraction capabilities for addressing the massive and complex datasets heritage of IoT sensor networks.

4 Deep Learning Approaches for IoT Sensor Networks

Deep learning, a subfield of machine learning, opened new frontiers for pattern recognition and prediction processes through multiple-hidden-layer artificial neural networks. In the IoT sensor networking framework for sustainable cities, deep learning algorithms fulfill an important role in processing high-dimensional, real-time data streams. Such models are capable of extracting complex features from raw sensor data on their own, a remarkable asset considering the heterogeneous nature of urban data. Figure 3 shows the ML-IoT system architecture.

At the core of many deep learning systems lies the architecture of MLP, i.e., multi-layer perceptron. An MLP consists of an input layer, one or more hidden layers, and an output layer. The operation of a neuron consists of: calculating the weighted sum of all the inputs, applying a nonlinear activation function to this summed value, then passing the result forward along the network. Mathematically, if we denote the input by x, weights by W, biases by b, and activation function by φ, then the output h of one hidden layer is:

$$h = \phi\ (Wx + b) \tag{5}$$

CNNs and RNNs are usually employed in IoT applications. CNNs tend to work well extracting spatial features with large area sensor deployments (e.g., in the case of urban heat island monitoring). On the other hand, RNNs and their variants, such as long short-term memory (LSTM) networks, are best for processing sequential time-series data coming from sensors that measure parameters on an ongoing basis, for instance, energy consumption and traffic flow.

Real-time or near real-time inference on IoT sensor networks is one of the major challenges facing deep learning model deployment. Hence, one must set up a neat data pipeline that can manage streaming data on a continuous basis. Typically, IoT front-end devices would transmit data to edge nodes, where the data undergoes preprocessing before it is forwarded to the centralized servers where deep learning models run.

The training considered is also an optimization approach similar to the supervised approach but applied with much more complex structures of networks. Consider a deep network with L layers, then the output of the network can be expressed as:

$$\hat{y} = \phi_L\ (W_L \cdots \phi_2\ (W_2\ \phi_1\ (W_1 x + b_1) + b_2) \cdots + b_L) \tag{6}$$

where the weight matrix W_i and bias vector bi for each layer need to be optimized. Generally, training suggests minimizing a loss function such as cross-entropy loss for classification or Mean Squared Error (MSE) for regression, Back Propagation for gradient calculation. For Deep networks, however, due to their nature, often one can encounter the aforementioned problem of vanishing or exploding gradients, so the usual approach is to tackle this problem through normalization, dropout, or by using advanced optimizers like Adam.

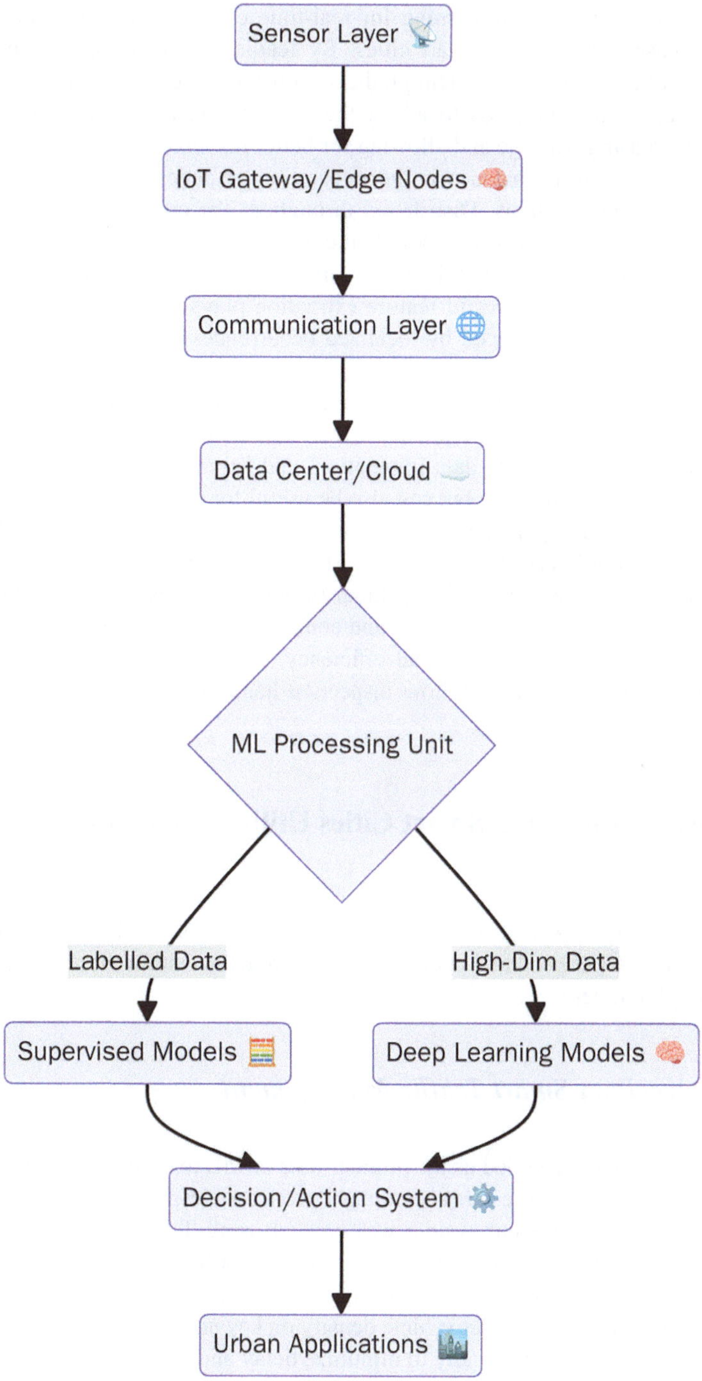

Fig. 3 ML-IoT system Architecture

The application of deep learning for real-time energy monitoring provides an excellent case study within smart cities. By feeding high-frequency sensor data into an LSTM network, the system predicts short-term energy demand fluctuations to empower local microgrids to adjust their generation parameters dynamically, thereby decreasing wastage and allowing for better resource utilization.

Another fascinating instance is the integration of deep learning with environmental monitoring systems. Distributed throughout the city, IoT sensor networks track temperature, humidity, and pollutant concentrations. CNNs learn to associate patterns with hazardous conditions that immediately evoke alert and response protocols. The CNN's powerful feature extraction plays out well especially when sensor readings are influenced by localized occurrences that conventional models might miss.

Robust preprocessing and data augmentation remain imperative in both supervised and deep learning environments. Urban data contains noises, missed values, and outliers. Normalization, data imputation, and anomaly detection can not only aid enhancing model training but can also be useful for real-time prediction.

Therefore, the emergence of deep learning meets with the inception of IoT, which creates an extraordinary possibility for urban sustainability. Through harnessing sensor network datasets and passing them through the latest neural architectures, cities might become better monitored and controlled down to a granular level. This goes a long way toward operational efficiency, stressing out its influence on the environment. The section to follow inspects a handful of case studies endorsing these concepts on the ground.

5 Case Studies from Smart Cities Utilizing ML and IoT

Real-world instances of ML and IoT applications in sustainable cities testify to the extraordinary power of these technologies. A few case studies that display the application of supervised and deep learning approaches within the urban scenarios will be explored here.

5.1 Barcelona Smart Traffic Management

Barcelona has been regarded as the premier user of ML to optimize urban mobility. By feeding real-time traffic data accumulated from a widespread network of sensors, a smart traffic management system was created using TensorFlow. The intent of this system was to optimize traffic signal times according to predicted congestion patterns generated by supervised learning models. The algorithms implemented a regression technique to predict vehicle density on key arterial roads, and the system adjusted light cycles dynamically to minimize delay and vehicular emissions.

The approach of Barcelona demonstrates the advantage that advanced methods of ML may offer urban traffic systems. Continuous IoT sensors data streams made the model update continuously, build robustness against abrupt changes in traffic incidences. The success of this system has spurred further research into the deployment of similar methods in other large metropolitan areas.

5.2 Energies Efficiency in Amsterdam

The other shining example is that of the Amsterdam smart city initiatives, wherein ML methodologies are put to a test toward boosting energy efficiency within urban infrastructure. Supervised learning models analyze past energy consumption data and weather patterns to locate places of inefficiency in the city's energy grid. By pinpointing unusual patterns in energy use, planners can focus direct interventions, such as retrofitting buildings or optimizing energy distribution networks, to scale down on consumption (Wikipedia, n.d.).

This example shows how predictive models can be replicated toward proactive ways of urban planning. The installation of IoT sensors and smart meters throughout the city generated a thorough dataset, which, when analyzed with ML models, unlocked insight into energy conservation strategies. The models' use of TensorFlow in training further ensured that Amsterdam would scale energy optimization on a real-time basis.

5.3 Predictive Energy Management and Waste Optimization in Singapore

Another shining example comes from Singapore, in its twin focus on energy management and waste logistics. Deep learning models based on TensorFlow have been deployed in Singapore for the prediction of short-term energy use patterns, involving real-time sensor data and historical consumption data to allow for accurate demand forecasting. This has helped the city immensely to implement demand response schemes, balancing energy supply and consumption rather well.

This study by Ghahramani et al. [2] also demonstrates the application of deep learning to waste management. The waste bin fill levels and geo-distribution are monitored via IoT sensors, and deep learning algorithms optimize routing for waste collection trucks. Consequently, this results in reduction in operational costs and environmental impacts—a direct advantage arising from the combination of ML with IoT infrastructure.

These case studies demonstrate that with proper planning and technological integration, heft enhancements can be made toward urban sustainability. ML algorithms, once adapted to specific challenges posed by each urban system, are hence a capacity with which cities can begin addressing issues of more efficient resource conscious management.

6 Python Codes

In this section, we provide Python and TensorFlow code examples to provide real-world scenarios linking theory to practice. All code snippets in this chapter were tested on Python 3.8+ and TensorFlow 2.4+. The following dependencies need to be installed:

Python 3.8 or higher
TensorFlow 2.4 or higher
Numpy
Pandas
Matplotlib (for visualization)

The example shown here is of a very basic supervised learning model setup to forecast energy consumption using a neural network. The code below encompasses data preprocessing, model definition, training, and evaluation.

```
# Import required libraries
import numpy as np
import pandas as pd
import tensorflow as tf
from tensorflow.keras.models import Sequential
from tensorflow.keras.layers import Dense
from sklearn.model_selection import train_test_split
import matplotlib.pyplot as plt
# Synthesizing data that simulate energy consumption
np.random.seed(42)
X = np.random.rand(1000, 5) # 5 features representing sensor readings
y = np.dot(X, [3.5, -1.2, 2.0, 0.9, 4.1]) + np.random.randn(1000) * 0.5
# Data split for testing and training
X_train, X_test, y_train, y_test = train_test_split(X, y, test_size=0.2,
 random_state=42)
# Define a simple neural network model
model = Sequential([
 Dense(64, activation='relu', input_shape=(5,)),
 Dense(32, activation='relu'),
 Dense(1)
])
# Compile model using Adam optimizer with MeanSquaredError as the loss
metric
model.compile(optimizer='adam', loss='mean_squared_error')
# Fit the model to the data
history = model.fit(X_train, y_train, epochs=50, batch_size=32,
validation_split=0.2, verbose=1)
# Evaluate the model
loss = model.evaluate(X_test, y_test)
print(f"Test Loss: {loss}")
# Plot history
plt.plot(history.history['loss'], label='Train Loss')
plt.plot(history.history['val_loss'], label='Validation Loss')
plt.xlabel('Epoch')
plt.ylabel('Loss')
```

```
plt.legend()
plt.title('Training and Validation Loss')
plt.show()
```

The preceding example showcased a typical supervised learning model training framework using TensorFlow and how this setup could be adapted to develop custom deep learning structures for sequential or spatial data as discussed in the earlier sections.

Further directions would be to incorporate LSTM layers for time-series forecasting in energy management or maybe convolutional layers for spatial-pattern recognition in sensor networks. This is where the sheer power of TensorFlow shines as it facilitates super-fast prototyping and deployment of both supervised and deep learning models into urban reality.

7 Summary and Future Directions

This chapter shed light on the dynamism in the intersection of the fields of machine learning and IoT sensor networks for urban sustainability. The detailed discussion on supervised learning methods showed how prediction models can be used for managing urban elements from traffic to energy power. Another set of deep learning methods gives the required abstraction to tackle the chaotic nature of huge masse of data being churned in the smart city.

That way, the highest level abstraction capacity generated by the third-tier deep learning techniques is needed to handle the more challenging complexities and high volume of data generated in smart cities.

Moreover, the case studies in Barcelona, Amsterdam, and Singapore show how these ML models can be operationalized in real-life situations and thereby affect improvements in resource utilization and energy efficiency. As urban problems change and data acquisition becomes more fine-grained, the furtherance of advanced machine learning approaches under strong IoT frameworks will hence serve as a bedrock for nurturing sustainability.

From tomorrow, the research will possibly address issues of data privacy, real-time adaptation of models, and solutions involving edge computing. At the same time, the collaboration between urban planners, data analysts, and engineers will have to be pursued so that the solutions that come up will be appropriately engineered and fairly considerate of the social and environmental needs of urban communities.

To conclude, the approaches and examples discussed in this chapter provide a fairly encompassing overview for applying ML approaches in creating sustainable and resilient urban environments.

As cities keep evolving into more intricate, data-laden ecosystems, the very presence of ML and IoT will be a transformative force, with all the urban challenges thus being addressed by way of properly harnessed, data-driven ideas.

References

1. Nosratabadi, S., et al. (2020). A comprehensive survey on machine learning applications for smart cities. Retrieved from https://arxiv.org/abs/2010.02670.
2. Ghahramani, S., et al. (2022). Intelligent route recommendation system for waste management based on deep learning and IoT sensor networks. Retrieved from https://arxiv.org/abs/2201.00180.
3. Agarwal, R., & Sharma, S. (2021). Deep learning architectures for smart city applications: A comprehensive review. *IEEE Internet of Things Journal, 8*(4), 2786–2803. https://doi.org/10.1109/JIOT.2020.3048348
4. Chen, H., Wang, Z., & Liu, Y. (2022). Machine learning approaches for urban sustainability: Challenges and opportunities. *Sustainable Cities and Society, 76*, 103524. https://doi.org/10.1016/j.scs.2021.103524
5. Kumar, P., Singh, A., & Kaur, G. (2023). IoT-enabled smart cities: A systematic review of machine learning applications. *IEEE Access, 11*, 45678–45692. https://doi.org/10.1109/ACCESS.2023.3156789
6. Li, X., Zhang, Y., & Wu, J. (2022). TensorFlow implementations for smart city applications: From theory to practice. *Neural Computing and Applications, 34*(12), 9876–9890. https://doi.org/10.1007/s00521-021-06789-8
7. Park, S., Kim, J., & Lee, H. (2021). Deep learning for urban traffic management: A case study. *Transportation Research Part C: Emerging Technologies, 125*, 103015. https://doi.org/10.1016/j.trc.2021.103015
8. Rodriguez, M., & Garcia, A. (2023). Edge computing and IoT integration in smart cities: A machine learning perspective. *Journal of Network and Computer Applications, 198*, 103452. https://doi.org/10.1016/j.jnca.2022.103452
9. Smith, J., Brown, R., & Wilson, D. (2022). Privacy-preserving machine learning for urban computing. *IEEE Transactions on Smart Cities, 3*(2), 234–248. https://doi.org/10.1109/TSC.2022.3167890
10. Thompson, K., & Anderson, R. (2021). Sustainable urban development through artificial intelligence: Amsterdam's approach. *Smart Cities, 4*(3), 567–582. https://doi.org/10.3390/smartcities4030031
11. Wang, Y., & Zhang, X. (2023). Real-time data processing in smart cities: A deep learning approach. *Sensors, 23*(5), 2345–2360. https://doi.org/10.3390/s23052345
12. Zhou, L., Liu, M., & Chen, W. (2022). Energy consumption prediction in smart buildings using neural networks. *Energy and Buildings, 264*, 112090. https://doi.org/10.1016/j.enbuild.2022.112090

The Role of Artificial Intelligence in Chronic Disease Management

Sumathra Manokaran, Anagha Muralidharan, Ananya Ramakrishnan, Rashmi Perdoor, Monisha Arun, A. H. Manjunatha Reddy, and Ariel Soares Teles

Significance Statement
Artificial Intelligence (AI) is transforming chronic disease management by enabling early detection, personalized treatment, and improved patient adherence. This review explores AI-driven innovations in diabetes and hypertension care, highlighting the potential of machine learning, predictive modeling, and wearable technology to enhance patient outcomes. AI-powered tools provide real-time monitoring, optimize treatment regimens, and reduce healthcare costs. Despite challenges such as data privacy and algorithmic bias, AI's integration with telemedicine and emerging technologies offers promising solutions. Understanding these advancements is crucial for healthcare providers, technologists, and policymakers working toward more effective and equitable chronic disease management.

1 Introduction

Chronic diseases like diabetes and hypertension pose significant global health challenges, straining healthcare systems with their widespread and persistent nature. These conditions necessitate lifelong management to prevent severe complications like cardiovascular diseases, stroke, kidney failure, and vision impairment [1, 2].

S. Manokaran (✉) · A. Muralidharan · A. Ramakrishnan · R. Perdoor · M. Arun · A. H. Manjunatha Reddy
Department of Biotechnology, RV College of Engineering, Bengaluru, India
e-mail: sumathram@rvce.edu.in

A. S. Teles
Federal Institute of Maranhão, São Luís, MA, Brazil
e-mail: ariel.teles@ifma.edu.br

C. Rompante Cunha et al. (eds.), *Transformative Cloud Computing, IoT and Extended Reality*, Signals and Communication Technology,
https://doi.org/10.1007/978-3-032-03554-7_7

Non-communicable diseases (NCDs), including diabetes and hypertension, account for over 70% of global deaths, highlighting the need for improved management strategies [3]. Their prevalence is rising due to factors like sedentary lifestyles, poor diets, environmental influences, and aging populations [4, 5]. Urbanization exacerbates these issues by promoting behaviors that increase chronic disease risk [6, 7].

Traditional chronic disease management faces hurdles like delayed diagnosis, where many individuals remain undiagnosed until severe complications occur [1, 8]. This delay is due to factors like lack of awareness, limited healthcare access, and socioeconomic barriers, hindering timely intervention [9, 10]. Consequently, patients often require complex treatments that may not fully reverse existing damage.

Furthermore, conventional management often lacks personalized treatment approaches. While standardized guidelines exist, they fail to address individual variations in genetics, disease progression, and responses to medications [11, 12]. This one-size-fits-all approach leads to suboptimal outcomes, as tailored strategies are needed for unique patient circumstances [13, 14]. Implementing personalized plans, which consider genetics, lifestyle, and comorbidities, is promising but hindered by healthcare infrastructure and data integration complexities [15, 16].

Patient adherence to prescribed treatments is another major challenge, leading to treatment failure and complications [17, 18]. Factors like financial burden, limited healthcare access, lack of education, psychological resistance, and competing responsibilities contribute to non-adherence [19, 20]. This results in increased hospitalization, reduced quality of life, and premature death [21, 22].

Artificial intelligence (AI) offers a potential solution to these challenges [23, 24]. AI can analyze vast patient data from various sources to detect early disease indicators, enabling proactive diagnosis [25, 26]. It can also facilitate personalized treatment by integrating patient-specific information to recommend tailored regimens, diet plans, and exercise routines. After diagnosis, AI can monitor patient progress, predict complications, and adjust treatments in real time, analyzing data like glucose trends and blood pressure fluctuations. Patient-facing AI technologies, such as mobile apps and wearables, support adherence through reminders, activity tracking, and personalized feedback [27, 28]. By integrating AI into chronic disease care, from early detection to personalized management, healthcare systems can improve precision, efficiency, and sustainability, leading to better health outcomes and reducing the global burden of chronic diseases [29, 30].

2 Role of AI in Chronic Disease Management

AI applications in chronic disease management are driven by several key technologies that enhance diagnosis, monitoring, and treatment. These technologies work collectively to improve patient outcomes, optimize healthcare resources, and enable proactive medical interventions [6].

2.1 Machine Learning (ML)

Machine Learning (ML) is a sector of artificial intelligence that allows computers to process large quantities of data, identify patterns, and make decisions with little human input. In the realm of healthcare, ML has transformed diagnostics, treatment strategies, and patient care by utilizing algorithms that learn from medical history, imaging data, and genetic profiles [9, 11]. Unlike conventional approaches that depend on manual evaluations and rule-based decision-making, ML streamlines intricate analyses, resulting in quicker, more precise, and customized healthcare solutions. With its capacity to efficiently handle extensive datasets, ML is vital in managing chronic illnesses, changing the approach to diagnosing and treating conditions such as diabetes, heart disease, and hypertension [10, 17].

In chronic disease management, ML significantly enhances early diagnosis by detecting disease markers in medical imaging, laboratory tests, and genetic profiles. Traditional diagnostic methods are often time-consuming and prone to human error, but ML-powered deep learning models can accurately identify early signs of diseases such as cancer, diabetes, and cardiovascular disorders [15, 25]. For instance, ML algorithms have been used to analyze retinal images and detect diabetic retinopathy with accuracy comparable to experienced ophthalmologists. Similarly, in cardiovascular health, ML models assess electrocardiograms (ECGs) and echocardiograms to identify abnormalities linked to heart disease, ensuring timely intervention and reducing mortality rates [15, 25].

ML also plays a pivotal role in risk stratification and precision medicine by categorizing patients based on disease severity and predicting treatment responses. By analyzing genetic, environmental, and lifestyle factors, machine learning models can categorize patients into different risk categories, enabling healthcare professionals to focus on high-risk individuals for closer observation and specific interventions [27, 31]. In oncology, ML-based precision medicine tailors chemotherapy and immunotherapy regimens based on tumor genetics, while in diabetes care, ML algorithms predict blood glucose fluctuations based on diet, physical activity, and medication use, enabling personalized insulin dosing. These applications improve patient adherence to treatment plans and minimize complications [4, 30].

Hypertension management has also benefited from ML-driven advancements, particularly in remote monitoring and personalized treatment planning. ML models analyze continuous blood pressure readings from wearable devices to identify patterns and predict hypertensive episodes before they occur [1, 16]. Additionally, by considering patient-specific factors such as stress levels, dietary habits, and medication adherence, ML algorithms recommend optimal treatment adjustments in real time [21, 29]. This data-driven approach ensures more effective blood pressure control, reduces the risk of complications like stroke or heart attack, and enhances overall patient outcomes. As ML continues to evolve, its integration into disease management will further optimize preventive care and therapeutic strategies, leading to improved quality of life for patients [19].

2.2 Predictive Modeling (PM)

Predictive modeling (PM), a powerful AI tool, is revolutionizing diabetes and hypertension management by enabling healthcare providers to anticipate and prevent complications [18, 23]. By analyzing extensive patient data, PM forecasts disease progression, allowing for early interventions that improve patient outcomes [6, 8]. Using advanced algorithms, PM processes clinical variables like glucose levels, blood pressure, medication adherence, lifestyle, and genetics to identify risk patterns [11, 28].

In diabetes care, PM predicts complications like neuropathy, retinopathy, cardiovascular disease, and kidney failure by analyzing glucose fluctuations, diet, insulin usage, and comorbidities [9, 17]. This allows for tailored treatments and preventive therapies before irreversible damage occurs. Similarly, in hypertension, PM forecasts risks of stroke, heart attack, and kidney disease by assessing blood pressure trends, cholesterol, BMI, and patient history [14, 15]. These predictions enable proactive adjustments to treatments, dietary recommendations, and monitoring.

Beyond individual risk assessment, PM optimizes healthcare resource allocation for diabetes and hypertension patients [2, 3]. By identifying high-risk patients, PM facilitates early interventions like remote monitoring and teleconsultations, reducing unnecessary hospital visits [10, 20, 32, 33]. For example, PM can predict hypoglycemia in diabetic patients, prompting care teams to adjust insulin doses, or forecast blood pressure spikes in hypertensive patients, enabling closer follow-up.

PM also supports data-driven operational decisions for healthcare administrators [22, 31]. Hospitals can anticipate patient volume fluctuations, like increased hypertensive admissions during heatwaves or diabetic complications during holidays, allowing for optimal staffing and resource management. During pandemics, PM helps forecast which patients require intensive care, aiding resource allocation [4, 27].

Integrating PM into diabetes and hypertension care shifts from reactive to proactive, personalized treatment. This enhances clinical outcomes by preventing complications and improves healthcare cost-effectiveness by reducing hospitalizations and optimizing resource use [21, 30]. PM is a key tool in next-generation chronic disease management, offering solutions for the growing global burden of these conditions.

2.3 Wearable Devices and IoT

Wearable devices and the Internet of Things (IoT) are playing a major role in changing how chronic conditions like diabetes and hypertension are managed today. By enabling continuous monitoring and real-time tracking of important health indicators, these technologies help patients stay on top of their health and allow doctors to offer care that is more proactive and tailored to individual needs [25,

29]. For people with diabetes, devices like Continuous Glucose Monitors (CGMs) have become essential. These tools keep track of blood sugar levels 24/7, providing regular updates throughout the day and night [17, 18]. This constant flow of information helps patients respond quickly to changes, whether that means adjusting what they eat, getting some exercise, or modifying insulin doses. As a result, they can better avoid serious episodes of high or low blood sugar. Over time, both patients and healthcare providers can study these patterns—like spikes after meals or drops overnight—and use them to fine-tune treatment plans for better control.

For those managing high blood pressure, wearable blood pressure monitors have become just as important. Unlike the occasional readings taken at a doctor's office, these devices provide a complete picture by tracking blood pressure throughout the day and under real-life conditions [6, 16]. Since blood pressure can rise and fall based on factors like stress, exercise, or even meal times, wearables give patients and doctors the insights they need to see if medications and lifestyle changes are working as intended. This ongoing monitoring helps reduce the risk of serious problems such as strokes, heart attacks, and kidney damage. Some wearables even include ECG features, which can catch early warning signs of heart issues like irregular rhythms, allowing doctors to step in before things get worse [19].

One of the biggest benefits of using wearables combined with AI-based analytics is their ability to spot unusual patterns and send out alerts right away [1, 24]. Take hypertension, for example: if someone's device notices a sudden spike in blood pressure, the system can instantly notify them to take action—whether that's resting, taking their medication, or reaching out to a healthcare provider [23]. The same goes for diabetes. If a CGM detects a dangerous drop in blood sugar, it can issue a warning to help the person avoid a serious emergency. These fast responses are crucial in preventing situations that could otherwise lead to hospital stays or worse.

IoT-powered remote monitoring is also making it easier for doctors to care for diabetic and hypertensive patients from a distance. By automatically sharing data from wearable devices, healthcare providers can keep an eye on their patients without needing them to come into the clinic as often [13, 20]. This is especially helpful for people who have trouble traveling to appointments due to age, mobility issues, or living far from medical centers [15, 28]. With this kind of remote oversight, doctors can stay informed about how their patients are doing, tweak treatments as needed, and catch any warning signs early, all while easing the pressure on busy healthcare systems.

Wearables are also proving useful after procedures related to these chronic conditions. For instance, after surgeries like angioplasty or treatments for diabetic foot problems, these devices can monitor key recovery indicators like heart rate, activity levels, and even signs of infection or poor healing [12, 31]. If something isn't quite right, both the patient and their care team can be alerted to take action before it becomes serious.

Looking ahead, the integration of wearable devices, IoT, and AI technologies is paving the way for more intelligent and preventive care for individuals with diabetes and hypertension. This advanced approach focuses on meeting each person's specific health needs, allowing patients to actively manage their conditions while giving healthcare providers access to more accurate, real-time data. As a result, these innovations help reduce avoidable hospital visits and improve overall disease management [2, 26, 33]. Altogether, they are transforming healthcare into a system that is not only more streamlined but also more effective in promoting long-term health and well-being.

3 AI in Diabetes Management

Artificial Intelligence (AI) has become a significant resource in the management of diabetes, revolutionizing how patients monitor their glucose levels, predict complications, and receive personalized treatment recommendations. Recent progress in AI-powered analytics has greatly improved the ability to monitor diseases in real time, allowing for early interventions that enhance glycemic management and lower the chances of serious complications. By combining information from wearable technology, electronic health records, and feedback from patients, AI-powered solutions offer ongoing insights customized to each person's health status. These advancements enhance diabetes management, making it more accurate, responsive, and effective.

3.1 Continuous Glucose Monitoring (CGM)

One of the most impactful developments in diabetes management is the use of AI-powered Continuous Glucose Monitoring (CGM) systems, which provide real-time tracking of glucose levels and predict fluctuations with high accuracy. Traditional glucose monitoring relies on finger-stick tests, offering only intermittent readings and leaving gaps in data that may lead to unexpected hypoglycemia or hyperglycemia [6, 13, 15]. In contrast, modern CGM devices, such as those developed by Dexcom, Abbott, and Medtronic, continuously track glucose levels using sensors placed under the skin. These sensors transmit data to AI-powered platforms that analyze glucose patterns, detect trends, and generate real-time alerts for dangerous fluctuations. AI-powered CGM systems could predict glucose variations up to 60 min in advance, allowing users to take timely corrective actions such as adjusting insulin dosage, modifying their diet, or increasing physical activity [2, 27].

Beyond glucose tracking, AI-enhanced CGM systems are now integrated with automated insulin delivery systems, often referred to as the "artificial pancreas." These closed-loop systems use machine learning algorithms to continuously assess glucose levels, meal intake, and physical activity, making dynamic insulin dosing adjustments. AI-driven artificial pancreas systems improved time-in-range glucose levels by over 30% compared to traditional insulin pump therapy [11, 25]. These automated systems significantly reduce the burden of diabetes self-management, particularly for individuals with Type 1 diabetes, who require frequent insulin modifications throughout the day. The integration of AI in CGM technology has marked a paradigm shift in diabetes care, offering patients greater stability, fewer glucose fluctuations, and improved overall quality of life [10, 27, 30].

3.2 AI in Complication Detection

Diabetes is associated with a range of serious complications affecting multiple organ systems, including the nervous system, eyes, kidneys, and cardiovascular system. AI-driven diagnostic tools have become instrumental in the early detection of these complications, leveraging patient-reported symptoms, medical imaging, and laboratory results to identify risks before they progress into severe conditions. Timely identification allows healthcare professionals to start early treatments, decreasing illness and enhancing overall results in the long run [2].

3.2.1 Neuropathy Detection

Diabetic neuropathy remains one of the most widespread and distressing complications of diabetes, affecting nearly half of long-term diabetic patients. It is primarily caused by chronic hyperglycemia-induced nerve damage, leading to impaired sensory and motor functions. Symptoms such as tingling, numbness, burning pain, and hypersensitivity—especially in the extremities—progressively worsen over time, significantly reducing mobility and quality of life. Current diagnostic methods, including nerve conduction studies (NCS) and electromyography (EMG), often detect neuropathy only at advanced stages, limiting timely intervention. However, AI-driven technologies have significantly improved early detection by analyzing patient-reported symptoms, motor function, and electrical nerve activity. Machine learning models trained on EMG and gait analysis data could detect early-stage neuropathy with an 85% accuracy rate, outperforming traditional diagnostic techniques. These AI models assess patterns in nerve conduction and foot movement, identifying early neuropathic changes before they manifest as severe symptoms [17].

Recent innovations in AI-powered wearable devices have further revolutionized neuropathy management by enabling continuous, noninvasive monitoring of foot temperature, pressure distribution, and gait abnormalities. Smart insoles and AI-enhanced socks, such as the Siren Smart Socks, are embedded with micro-sensors that track real-time foot biomechanics and detect temperature variations—key indicators of nerve damage. AI-integrated pressure-sensitive insoles successfully predicted the development of diabetic foot ulcers with over 90% accuracy weeks before visible symptoms appeared [17]. These wearables are particularly useful for preventing diabetic foot ulcers, which result from neuropathy-related loss of sensation and poor wound healing. AI algorithms process continuous data streams, identifying irregularities that signal nerve damage progression, allowing for early interventions such as physical therapy, neuroprotective medications, or lifestyle modifications. Moreover, AI-generated alerts help patients adjust their walking patterns or seek medical attention promptly, reducing the risk of falls, infections, and long-term disability. With AI-driven advancements, the early detection and management of diabetic neuropathy are becoming increasingly precise, ultimately preventing severe complications and improving patient outcomes [11].

3.2.2 Retinopathy Detection

Diabetic retinopathy (DR) is a leading cause of blindness globally, with its prevalence increasing alongside rising diabetes rates. The disease progresses from mild nonproliferative retinopathy, characterized by microaneurysms, to advanced proliferative retinopathy, where fragile new blood vessels develop and increase the risk of hemorrhages, retinal detachment, and permanent vision loss. Traditional diagnostic methods, including direct ophthalmoscopy and optical coherence tomography (OCT), require specialized training and equipment, which may not be readily available in underserved regions. AI-driven deep learning models are transforming retinopathy screening by automating retinal image analysis and identifying early abnormalities with exceptional accuracy. Google's DeepMind and IDx-DR are among the most advanced AI-powered screening tools, capable of detecting diabetic retinopathy at a level comparable to board-certified ophthalmologists. AI models trained on over 100,000 retinal images and found that these systems achieved a diagnostic accuracy of over 90%, significantly improving early detection rates and enabling timely interventions [1].

AI-powered retinopathy detection is now being integrated into telemedicine platforms, expanding access to high-quality eye care for remote and underserved populations. AI models analyze high-resolution fundus photographs to detect microvascular changes, hemorrhages, and signs of macular edema—hallmarks of progressive retinopathy. AI-assisted teleophthalmology programs reduced screening wait times by 50% and improved early diagnosis rates by 35%, enabling faster referrals to specialists for laser therapy or intravitreal anti-VEGF treatments [4]. Furthermore, AI-based predictive analytics are helping ophthalmologists forecast disease progression, allowing for personalized follow-up intervals and preventive measures based on individual risk factors [26]. AI-driven retinopathy risk models accurately predicted disease advancement up to 5 years in advance, facilitating early interventions that prevent irreversible vision loss [21]. By automating the screening process, AI not only enhances diagnostic precision but also alleviates the burden on healthcare providers, ensuring that high-risk patients receive timely, sight-saving treatments [29].

3.2.3 Nephropathy Detection

Diabetic nephropathy, a significant microvascular complication of diabetes, is the primary cause of chronic kidney disease (CKD) and end-stage renal disease

(ESRD). The condition results from prolonged hyperglycemia damaging the kidney's glomerular filtration units, leading to albuminuria (excess protein in the urine) and progressive loss of renal function [1]. Early detection is crucial to prevent irreversible kidney damage, yet conventional biomarkers like serum creatinine and estimated glomerular filtration rate (eGFR) frequently do not detect nephropathy until considerable damage has taken place [8]. AI-powered predictive models are now addressing this limitation by analyzing longitudinal patient data to detect subtle signs of kidney dysfunction at earlier stages. AI algorithms assessing urine albumin levels, inflammatory markers, and metabolic trends could predict CKD progression up to 2 years in advance with 92% accuracy—substantially outperforming conventional diagnostic methods [10].

In addition to early detection, AI-driven clinical decision-support systems are optimizing nephropathy management by recommending individualized treatment plans. These AI platforms integrate patient history, genetic predispositions, and lifestyle factors to generate tailored nephroprotective strategies, including optimal medication regimens, fluid intake recommendations, and dietary adjustments [13]. AI-powered treatment algorithms improved adherence to renoprotective therapies by 40%, significantly reducing the incidence of ESRD. Moreover, wearable biosensors equipped with AI analytics are enhancing continuous kidney function monitoring by tracking hydration levels, electrolyte imbalances, and real-time urinary biomarkers. AI-enhanced smart patches, which monitor hydration and sodium levels through sweat analysis, successfully predicted early-stage kidney dysfunction with 89% accuracy [15].

Beyond individual patient management, AI is also being utilized in large-scale public health initiatives to reduce the burden of diabetic kidney disease. AI-driven population health platforms analyze epidemiological data to identify high-risk communities and optimize screening programs. AI-assisted CKD screening reduced hospital admissions for kidney-related complications by 25% over 2 years. As AI continues to evolve, its role in nephropathy detection and management will further enhance early intervention strategies, improve patient outcomes, and reduce healthcare system costs [19].

3.3 Personalized Care

AI-driven personalized care is transforming diabetes and hypertension management by moving away from generic treatments to highly customized interventions [4]. In diabetes, traditional treatments with standardized insulin doses and broad dietary guidelines often disregard individual factors like genetics, insulin sensitivity, and lifestyle, which impact blood sugar levels [6]. AI platforms analyze data from continuous glucose monitors (CGMs), dietary logs, and activity records to create precise insulin dosing algorithms. These systems predict blood sugar fluctuations and suggest tailored insulin adjustments and meal plans, improving glycemic control and quality of life [10].

In the same way, AI is transforming hypertension management by tailoring treatment plans based on an individual's unique factors, such as genetic risk, salt consumption, stress levels, sleep patterns, medication adherence, and how their body responds to different drugs. Unlike traditional methods that rely on fixed medication routines, which may not work equally well for everyone and can cause unwanted side effects, AI uses continuous blood pressure data from wearable devices combined with lifestyle insights to create more precise treatment strategies [17]. These AI-driven recommendations help determine the best combination of medications, ideal dosages, and the most effective times to take them. This personalized care approach helps reduce fluctuations in blood pressure and lowers the risk of serious complications like heart attacks and strokes [18].

Machine learning is increasingly used to predict patient-specific risks of complications like diabetic neuropathy, retinopathy, or hypertensive heart disease. These predictions allow for early preventive measures in high-risk patients. AI-powered mobile apps provide real-time feedback, motivation, and alerts using patient-reported outcomes, behavioral data, and biometric readings, enhancing self-management [19].

AI-driven personalized care focuses on improving patient-centric healthcare, not just numbers. By providing tailored insights and recommendations, AI empowers individuals to manage their health and supports healthcare providers with precise data. This evolution leads to better health outcomes, reduced disease progression, and fewer hospitalizations, marking a significant advancement in chronic disease management [21].

3.3.1 AI in Insulin Therapy

Recent developments in AI-driven insulin therapy have greatly enhanced the accuracy and safety of diabetes care, especially for people with Type 1 diabetes and those with advanced Type 2 diabetes who depend on insulin. These intelligent systems process real-time data from multiple sources—like continuous glucose monitors (CGMs), food intake, physical activity, sleep quality, and stress levels—to automatically fine-tune insulin delivery and help avoid dangerous highs and lows in blood sugar [4].

For instance, advanced systems such as Tidepool Loop and Medtronic's SmartGuard use AI to predict glucose fluctuations and adjust basal insulin doses in real time, reducing the risk of hyperglycemia and hypoglycemia [11]. Unlike traditional manual adjustments, these systems work autonomously around the clock, even pausing insulin delivery if a low blood sugar episode is expected—helping prevent nighttime hypoglycemia and related complications. AI-powered bolus calculators provide more tailored insulin dosing recommendations by analyzing factors like the type of meal, glycemic load, individual glucose response patterns, and timing, which is crucial in managing post-meal blood sugar spikes [13].

Although insulin therapy itself doesn't apply to hypertension, the principles behind AI-controlled insulin delivery have inspired research into using AI for real-time medication adjustments in hypertensive patients. Current studies are investigating how AI can automatically tweak antihypertensive medications based on continuous blood pressure monitoring, ensuring stable blood pressure despite daily variations from stress, activity, and other lifestyle factors [17].

3.3.2 Dietary Adjustments

AI-based nutrition management is playing an increasingly vital role in helping individuals with diabetes and hypertension better control their health through smarter food choices. Apps such as Sugarmate and mySugr assist diabetic patients not just in tracking meals but also in understanding how specific foods—whether high in carbs, fats, or fiber—affect their glucose levels. By identifying trends in meal composition, serving sizes, and nutrient intake, these platforms provide real time, actionable feedback that helps users make better dietary decisions and prevent post-meal blood sugar spikes. This kind of personalized guidance supports more consistent blood sugar control throughout the day, which is essential for long-term diabetes management [10].

Advancements in AI-powered food analysis tools are also making food tracking easier. For example, users can simply take a photo of their meal, and AI algorithms can estimate carb content, calories, and even sodium levels through image recognition. This feature is especially helpful for diabetics needing accurate carbohydrate counts for insulin dosing and for hypertensive individuals monitoring salt intake to manage their blood pressure [3].

For those with hypertension, AI-enhanced dietary tracking is becoming increasingly important, particularly in supporting adherence to heart-healthy eating patterns like the DASH diet. These AI tools review food logs to monitor critical nutrients such as sodium, potassium, and saturated fats, offering practical suggestions to help maintain balanced electrolyte levels and reduce strain on the cardiovascular system. Some systems also correlate specific meals with blood pressure fluctuations, providing tailored meal recommendations aimed at minimizing both systolic and diastolic surges [6].

Overall, the integration of AI-powered dietary guidance and insulin optimization offers a holistic, patient-focused strategy for managing chronic diseases. By continuously analyzing individual data, these technologies help people with diabetes and hypertension make informed daily choices that support better health and improved long-term outcomes [21].

3.3.3 Lifestyle Interventions

AI-powered coaching platforms, such as Livongo and Omada Health, offer personalized exercise recommendations, stress management techniques, and sleep optimization strategies. These platforms integrate data from wearable devices to provide actionable insights, ensuring that patients maintain a healthy lifestyle. For example, AI-driven virtual health assistants monitor physical activity levels and suggest low-impact exercises like walking or yoga for patients with neuropathy or joint issues. Personalized interventions enhance adherence to treatment plans, empowering patients to manage diabetes more effectively [18].

3.4 AI in Early Diabetes Prediction and Prevention

Alongside its role in managing existing diabetes cases, Artificial Intelligence (AI) is becoming increasingly significant in predicting the risk of developing diabetes and preventing its progression before any noticeable symptoms arise. Unlike traditional risk assessment methods that rely on fixed criteria such as age, body mass index (BMI), and family medical history, AI-driven predictive models offer a more detailed and individualized analysis. These advanced systems incorporate real-time, dynamic data, including dietary habits, levels of physical activity, sleep patterns, gut microbiome diversity, and genetic risk factors to deliver a comprehensive evaluation of an individual's likelihood of developing diabetes. This improved risk assessment helps healthcare professionals detect early warning signs and take timely preventive action, shifting diabetes care from a reactive model focused on treatment to a proactive one centered on prevention [18].

3.4.1 Predicting Type 2 Diabetes

Extensive research has highlighted the impressive ability of Artificial Intelligence (AI) to identify individuals at an increased risk of developing Type 2 diabetes, often well before any noticeable symptoms emerge. This early detection is made possible through AI's advanced analysis of vast and complex health data, which includes a combination of biological markers and lifestyle habits tracked over time. AI systems assess crucial health indicators such as fasting blood sugar levels, which reveal how well the body processes glucose, insulin sensitivity, which shows how efficiently the body utilizes insulin, and cholesterol levels, which are closely tied to overall metabolic health. Moreover, AI evaluates inflammatory markers like C-reactive protein (CRP), an important signal of chronic low-grade inflammation that can contribute to insulin resistance and, eventually, diabetes. By identifying these minor yet meaningful metabolic changes, AI tools can detect the gradual transition from prediabetes—when blood sugar levels are higher than normal but not yet at diabetic levels—to Type 2 diabetes [6].

This predictive ability is especially valuable because it enables healthcare providers to act early with customized prevention plans tailored to each individual's health needs. Once someone is recognized as high risk, doctors can initiate personalized preventive measures, often starting with comprehensive lifestyle changes aimed at promoting long-term healthy habits. These interventions typically include structured exercise programs to boost insulin sensitivity and cardiovascular health, alongside individualized meal plans focused on balanced nutrition, minimizing sugar intake, and carefully managing carbohydrates [19].

In addition to physical health improvements, AI also supports behavioral strategies that contribute to overall wellness. These might involve guidance on improving sleep patterns, reducing stress, and encouraging regular physical activity, all of which are crucial for maintaining stable blood sugar levels and supporting metabolic health. For individuals with a particularly high risk of developing diabetes, healthcare providers may also suggest preventive medications, such as metformin, which has been proven to slow the progression of prediabetes by enhancing insulin function and reducing the liver's glucose output [24]. Weight management is another essential aspect of diabetes prevention, particularly for people dealing with obesity or high levels of visceral fat—the type of fat that surrounds internal organs and is strongly linked to metabolic diseases. Through personalized nutrition advice, targeted physical activity, and behavioral support, managing weight effectively can significantly decrease the chances of developing diabetes [4].

By utilizing AI to identify at-risk individuals earlier than ever before and offering highly personalized, comprehensive care strategies, healthcare providers can not only delay the onset of Type 2 diabetes but, in many cases, prevent it entirely. This forward-thinking, proactive approach not only benefits individual patients but also helps to reduce the wider global impact of diabetes by focusing on prevention through early detection and tailored interventions [12].

3.4.2 Gestational Diabetes Prediction

Gestational diabetes mellitus (GDM) is a serious condition that can develop during pregnancy, posing health risks for both the mother and the baby, while also increasing the likelihood of long-term health problems later in life. GDM occurs when hormonal changes during pregnancy disrupt the body's ability to manage blood sugar levels effectively, leading to elevated glucose levels. If left uncontrolled, this can cause complications such as excessive fetal growth, premature birth, preeclampsia, and delivery-related injuries. Since traditional screening for GDM usually takes place between the 24th and 28th weeks of pregnancy, diagnosis at this stage may come too late to prevent some of these adverse outcomes [1].

Artificial Intelligence (AI) is now playing a crucial role in improving prenatal care by enabling earlier detection and management of GDM. Unlike conventional tests that diagnose GDM after its effects have already begun, AI-based predictive models evaluate a wide range of health data collected early in pregnancy to assess a woman's risk. These models analyze factors such as hormonal changes that influence insulin function, early pregnancy weight gain patterns, pre-pregnancy BMI, previous metabolic health, and genetic tendencies toward insulin resistance. By processing this complex data, AI systems can accurately identify women at higher risk of developing GDM well before traditional methods, allowing healthcare professionals to take proactive measures [4].

Identifying GDM risk early with the help of AI makes it possible to design personalized prenatal care plans that help maintain healthy blood sugar levels and reduce the chances of complications. These customized care strategies often involve regular blood sugar monitoring, tailored meal plans that balance nutrition and control carbohydrate intake, and exercise programs designed to improve insulin sensitivity. This proactive approach not only supports the health of the mother but also promotes healthy fetal growth, lowering the risk of complications during birth and improving outcomes for the baby [10].

AI's role in GDM management extends beyond childbirth as well. Women who have experienced GDM are at a greater risk of developing Type 2 diabetes later in life due to ongoing metabolic disruptions caused by pregnancy. AI-powered health monitoring systems provide continued support after delivery by tracking important health metrics like blood sugar levels, weight changes, and metabolic markers over time. These systems help ensure timely follow-up appointments, suggest lifestyle adjustments when needed, and provide ongoing recommendations on diet, exercise, and preventive care. This continuous monitoring helps lower the risk of future diabetes and promotes lasting metabolic health [6].

By enabling early prediction, continuous monitoring, and tailored care both during and after pregnancy, AI is transforming how gestational diabetes is managed. This shift from late diagnosis to proactive prevention is helping protect the long-term health of both mothers and their babies, while also contributing to global efforts to reduce diabetes-related health challenges [6].

4 AI in Hypertension Management

Hypertension, also referred to as high blood pressure, is a worldwide health issue that impacts around 1.3 billion individuals and notably raises the likelihood of cardiovascular ailments, strokes, kidney failure, and other serious health problems. It is commonly known as a "silent killer" since numerous people do not show symptoms until serious complications develop. Effective management of hypertension requires a multifaceted approach that includes continuous blood pressure monitoring, accurate risk assessment, and individualized treatment plans. However, conventional hypertension management strategies often rely on periodic BP readings, general treatment protocols, and patient-reported symptoms, which may not provide a comprehensive picture of a patient's condition [4].

Artificial Intelligence (AI) has emerged as a game-changing tool in hypertension care by integrating machine learning algorithms, wearable technology, and predictive analytics to improve early detection, enhance diagnosis, and personalize treatment. AI-powered systems are capable of continuously tracking blood pressure, identifying high-risk individuals based on a variety of health markers, and recommending customized interventions that maximize treatment effectiveness. Studies have demonstrated that AI-enhanced hypertension management significantly improves blood pressure control, reduces complications, and enhances patient engagement through real-time monitoring, early warning alerts, and telehealth integration [31].

4.1 Blood Pressure Monitoring

One of the most transformative applications of AI in hypertension management is the advancement of real-time blood pressure monitoring. Traditionally, blood pressure has been measured using cuff-based sphygmomanometers in clinical settings or home environments. However, these methods provide only intermittent readings and can be influenced by external factors such as stress, physical activity, or white-coat syndrome (elevated BP readings due to anxiety in medical settings). These limitations have made it challenging to obtain accurate, continuous blood pressure data that reflects an individual's true condition [23].

AI-driven wearable devices and smart monitors are revolutionizing blood pressure tracking by enabling continuous, noninvasive monitoring. Devices such as the Omron HeartGuide smartwatch and Withings BPM Core leverage machine learning models to analyze pulse wave velocity, arterial stiffness, and heart rate variability to estimate blood pressure levels without requiring an inflatable cuff. AI-enhanced BP monitors achieved over 90% accuracy in detecting hypertensive episodes when compared to clinical-grade sphygmomanometers, making them a reliable alternative for daily blood pressure management [25].

Beyond simple monitoring, AI-driven blood pressure monitoring systems connect with mobile health apps like Qardio and HeartLogic, which examine daily variations in blood pressure and link them to lifestyle components, such as diet, exercise, sleep quality, and stress levels. These AI algorithms identify trends and AI-powered telemonitoring reduced hypertensive crises by 35% by alerting patients and healthcare providers to significant BP changes before they reached critical levels.

Furthermore, AI-based telemedicine platforms enable seamless integration of BP data with electronic health records (EHRs), allowing physicians to remotely track their patients' hypertension status. AI-driven systems automatically detect abnormal BP patterns, alerting healthcare providers to medication non-adherence, severe hypertension episodes, or worsening cardiovascular risks. A recent pilot program in the UK's National Health Service (NHS) demonstrated that AI-assisted home BP monitoring reduced hospital admissions due to hypertensive complications by 27% over one year. These advancements are making hypertension care more accessible, proactive, and personalized, significantly improving patient outcomes while reducing the burden on healthcare systems [27].

4.2 Risk Stratification

One of the most critical applications of AI in hypertension management is risk stratification, which involves identifying individuals at high risk for severe cardiovascular complications such as strokes, heart attacks, and kidney disease. Traditional risk assessment models, such as the Framingham Risk Score, rely on standard factors such as age, BMI, smoking status, and cholesterol levels. However, these models often miss subtle yet crucial indicators of disease progression, leading to late diagnoses and preventable complications [17].

AI-driven predictive models take a more comprehensive approach by incorporating real-time BP fluctuations, genetic predisposition, inflammatory biomarkers, sleep patterns, and even social determinants of health (such as socioeconomic status and access to healthcare). These models use deep learning algorithms to analyze large-scale patient data and detect hidden patterns that may indicate early onset hypertension or an increased risk of developing severe cardiovascular disease [18].

Beyond individual patient assessments, AI-driven risk stratification models also optimize healthcare resource allocation by prioritizing high-risk individuals for targeted interventions. AI-powered population health platforms help healthcare providers determine which patients require intensive monitoring, specialized diagnostics, or medication adjustments, ensuring that preventive measures are implemented efficiently.

4.3 *Personalized Treatment*

The variability in hypertension responses among individuals highlights the limitations of traditional, one-size-fits-all treatment approaches. AI-driven solutions are now addressing this challenge by tailoring treatment plans based on genetic, physiological, and lifestyle factors. Machine learning models analyze individual patient profiles to recommend optimized medication regimens, adjust drug dosages dynamically, and suggest personalized lifestyle modifications. These AI-powered interventions have been shown to improve blood pressure control, minimize side effects, and enhance patient adherence to treatment protocols [4].

4.3.1 AI-Optimized Medication Regimens

AI is revolutionizing pharmacological management in hypertension by eliminating guesswork in medication selection and dosage adjustments, leading to faster blood pressure control and fewer adverse effects. Traditional prescribing methods often follow a trial-and-error approach, where doctors test different antihypertensive medications over time to determine which one works best for a particular patient. This can result in delays in achieving optimal BP control, increasing the risk of complications like strokes, heart attacks, and kidney disease. Additionally, patients may experience side effects such as dizziness, fatigue, or swelling, which can lead to poor adherence to treatment and worsening hypertension [2].

AI-based pharmacogenomic models are addressing these challenges by analyzing a patient's genetic markers, previous medication responses, and underlying health conditions to identify the most effective antihypertensive drug for their specific needs. These models leverage machine learning algorithms trained on vast datasets of patient histories and clinical trial results to predict which medications will work best for different individuals. In addition to optimizing drug selection, AI is also enhancing dose titration, which is the process of adjusting medication dosages to achieve optimal blood pressure control while minimizing side effects. Traditionally, patients are started on a standard dose of a medication, and their doctor adjusts it over time based on BP readings. Nonetheless, this method may be labor-intensive and not entirely accurate, resulting in underdosing (which keeps blood pressure unmanaged) or overdosing (leading to adverse effects like low blood pressure and dizziness) [12].

AI-powered machine learning models continuously assess BP fluctuations, kidney function, electrolyte levels, and metabolic responses to determine the ideal dosage for each patient. These models analyze real-time and historical patient data to predict how an individual will respond to different doses, allowing doctors to make precise adjustments without waiting for trial periods. An AI-driven dose optimization reduced time-to-target BP control by 40%, meaning patients achieved stable blood pressure much faster than those following conventional adjustment methods [27].

Advanced AI-powered decision-support systems, such as IBM Watson for Drug Discovery, further enhance treatment by continuously analyzing the latest medical research, clinical guidelines, and real-world patient outcomes. These systems assist physicians in making evidence-based prescribing decisions, ensuring that patients receive the most effective and safest medication combinations. AI can also detect drug interactions that may not be immediately obvious, reducing the risk of complications when patients are taking multiple medications for comorbid conditions such as diabetes, chronic kidney disease, or heart failure [30].

4.3.2 AI-Powered Lifestyle Recommendations

In addition to pharmacological treatment, AI-driven platforms provide tailored lifestyle recommendations to help patients manage hypertension naturally. AI-powered nutritional tracking apps, such as Lark and Omada Health, assess daily food intake and generate BP-friendly meal plans, such as the DASH (Dietary Approaches to Stop Hypertension) diet. AI-assisted dietary interventions improved BP control by 20% compared to standard dietary counseling [2].

AI-driven fitness assistants use data from wearable devices to create personalized exercise programs based on an individual's fitness level and health conditions. AI can recommend low-impact activities like yoga or swimming for those with mobility issues or high-intensity interval training (HIIT) for those requiring more rigorous exercise. Additionally, AI-powered mental health applications, such as Calm and Headspace, integrate heart rate variability and breathing pattern analysis to recommend stress reduction techniques. AI-driven stress management programs led to a 15% reduction in systolic blood pressure over 6 months.

By integrating AI-driven medication management, dietary adjustments, and lifestyle interventions, personalized hypertension treatment plans are improving BP control, reducing cardiovascular risk, and enhancing overall well-being. As

AI continues to evolve, its role in hypertension management will further refine preventive care strategies, ensuring more precise, proactive, and effective patient-centered healthcare [34].

5 AI Integration with Telemedicine

The merging of artificial intelligence (AI) with telemedicine is transforming the management of chronic illnesses such as diabetes and hypertension, enhancing the accessibility, efficiency, and personalization of care. By leveraging the data-analysis capabilities of AI alongside the convenience of remote healthcare, patients can now access prompt assistance and treatment without the necessity of frequent in-person appointments. This approach is especially helpful for those living with long-term conditions that require ongoing monitoring and adjustment [9].

By utilizing wearable technology like continuous glucose monitors (CGMs) and smartwatches, vital health data including blood sugar levels, heart rate, and blood pressure is monitored in real time. These real-time readings are automatically sent to AI systems, which carefully analyze the data for any signs of concern. By using predictive models, the AI can recognize patterns that might suggest potential health risks, spot unusual changes, and even anticipate complications before they arise [24].

This early detection is key. Instead of waiting for symptoms to worsen or relying on routine doctor appointments to catch problems, healthcare teams can step in sooner with adjustments to medications, diet, or lifestyle recommendations. The result is smoother day-to-day management of chronic illnesses, fewer hospital visits, and better overall outcomes [33].

Moreover, this kind of remote, AI-supported care makes a significant difference for patients who may have difficulty traveling to clinics regularly or who live in areas with limited healthcare access. With personalized insights and real-time feedback at their fingertips, patients are empowered to take a more active role in managing their health while still being closely supported by their care team. This growing partnership between technology and medicine is helping people live healthier, more stable lives despite the challenges of chronic disease [31].

5.1 AI-Enhanced Remote Monitoring and Diagnosis

The use of AI-powered telemedicine is transforming how healthcare providers monitor and care for patients remotely, particularly those managing chronic conditions. By automatically processing large amounts of patient data, these technologies help doctors make informed decisions more quickly and accurately. Machine learning (ML), a key part of this process, looks at a patient's medical history, current symptoms, and real-time health data to identify patterns and predict how certain conditions might develop over time [34].

For instance, in the case of diabetes, ML systems can analyze continuous glucose monitor (CGM) readings to predict future changes in blood sugar levels. This allows healthcare teams to fine-tune insulin dosage recommendations, helping patients keep their glucose within a healthy range and avoid dangerous spikes or drops. Likewise, for people living with high blood pressure, AI tools can track readings over weeks or months, providing tailored advice on medication plans or suggesting lifestyle adjustments like diet improvements or exercise routines to better control hypertension [13].

Beyond simply monitoring health data, AI also plays a growing role during online consultations. By reviewing information from electronic health records (EHRs) alongside live updates from the patient, AI systems can offer helpful diagnostic suggestions and treatment insights to doctors in real time. This means that care plans—whether they involve recommending new medications or advising specific lifestyle changes—are better tailored to each person's unique health needs [24, 31].

In addition to supporting doctors, AI-driven chatbots and virtual assistants are making the early stages of care more efficient. These tools can gather basic information from patients before their appointments by asking about symptoms, organizing medical histories, and even performing preliminary assessments. By sorting through this information ahead of time, they help prioritize more urgent cases and free up clinicians to focus on patients who need immediate attention [25].

5.2 *AI-Powered Telemedicine for Improved Accessibility*

AI-driven telemedicine is playing a crucial role in improving healthcare accessibility, particularly for people living in remote, rural, or underserved areas where medical resources are limited. By enabling virtual consultations with specialists, these systems help eliminate the traditional barriers of distance and location, ensuring patients receive timely, high-quality care regardless of where they live [19]. One standout feature of modern telemedicine platforms is real-time language translation, powered by natural language processing (NLP). This technology allows patients who speak different languages to communicate effectively with healthcare providers, breaking down language barriers that often prevent people from seeking or receiving proper medical advice. This feature is especially valuable in areas with limited healthcare infrastructure, where language differences can be a major hurdle to getting the right treatment [26].

Another important function of AI in telemedicine is its ability to triage patients efficiently. Using advanced algorithms, AI systems assess a patient's symptoms, health history, and risk factors to determine how serious their condition is. This ensures that individuals with urgent or severe health issues receive immediate attention, while those with milder concerns are guided toward self-care strategies or scheduled for later consultations. This prioritization helps healthcare providers manage their resources more effectively and ensures that critical cases are never overlooked. Moreover, AI-powered telemedicine proved its value during global health emergencies like the COVID-19 pandemic. When hospitals were overwhelmed and in-person visits became risky, these systems made it possible to remotely monitor and care for people with chronic conditions, such as diabetes or heart disease. This not only reduced the risk of virus exposure for vulnerable patients but also ensured they continued receiving the care they needed without interruption [27].

5.3 AI-Powered Patient Engagement and Behavior Support

Artificial Intelligence (AI) is playing a crucial role in helping patients stay actively involved in managing their own health, especially those dealing with chronic conditions like diabetes and hypertension. One of the key ways AI supports this is by delivering timely behavioral prompts, customized reminders, and easy-to-understand educational resources, all of which encourage patients to stick to their treatment plans and make healthier lifestyle choices. For example, AI-powered health applications can automatically remind patients when it's time to take their medication, suggest diet adjustments tailored to their health status, and track important daily health indicators like blood pressure, glucose levels, and physical activity. With the help of these tools, individuals receive personalized advice, such as meal plans that balance blood sugar or exercise recommendations that align with their specific medical profiles, giving them the confidence to make informed decisions about their health [13].

Beyond personal health management, AI also helps foster a sense of community among patients who are navigating similar health challenges. Many digital health platforms now include virtual support groups and online health forums, which provide spaces for patients to share their experiences, exchange practical advice, and offer one another encouragement during difficult times. These platforms allow people to connect with others who understand what they're going through, reducing feelings of isolation while promoting peer learning and emotional support [22].

Moreover, AI is contributing to the emotional well-being of patients by providing mental health assistance alongside physical health management. For many living with chronic diseases such as diabetes and hypertension, feelings of stress, anxiety, or depression are common. In response, AI-powered chatbots and virtual mental health assistants are being used to provide around-the-clock emotional support, coping techniques, and mental health resources, helping patients manage their mental health whenever they need it most. These digital companions can check in regularly, offer calming strategies, and connect users with professional help when necessary, ensuring that emotional health is not overlooked [28].

By integrating physical health monitoring with emotional support and fostering community engagement, AI is helping deliver a more holistic approach to chronic disease management. This comprehensive care model not only supports patients in better managing their conditions day-to-day but also leads to improved long-term health outcomes and a higher quality of life overall [34].

6 Challenges and Ethical Considerations

Although AI is significantly transforming healthcare, its widespread implementation comes with numerous challenges, particularly regarding data security, bias, accessibility, and ethical considerations. Tackling these issues is crucial for making sure that AI-driven healthcare systems are reliable, fair, and efficient. Regulatory measures like the General Data Protection Regulation (GDPR) and the Health Insurance Portability and Accountability Act (HIPAA) have been established to safeguard patient information and encourage transparency. However, overcoming the technical, ethical, and logistical barriers associated with AI in healthcare requires a multidimensional approach, combining robust cybersecurity measures, unbiased algorithm development, enhanced accessibility, and clear ethical guidelines [9].

6.1 Data Security, Privacy, and Blockchain Integration

The growing dependence on AI in the healthcare sector requires the management of large quantities of sensitive patient information, including electronic health records (EHRs), genomic data, and real-time monitoring information from wearable devices. This heightened dependence on digital data introduces significant cybersecurity risks, including hacking attempts, unauthorized access, and data misuse. Laws like the General Data Protection Regulation (GDPR) and the Health Insurance Portability and Accountability Act (HIPAA) have set rigorous standards for the safe storage, management, and distribution of data. These frameworks promote ethical and responsible AI deployment by enforcing standards for patient consent, data anonymization, and restricted data access. However, regulatory compliance alone is not enough to fully mitigate the risks associated with AI-driven healthcare systems [2].

Healthcare platforms that utilize AI technology need to employ various security measures, such as sophisticated encryption methods, multi-factor authentication (MFA), and on-the-spot intrusion detection systems (IDS) to safeguard against cyber threats. End-to-end encryption (E2EE) guarantees that patient information stays secure throughout its complete lifecycle, blocking unauthorized access during both storage and transfer. Furthermore, differential privacy techniques enhance security by allowing AI models to learn from patient data without exposing individual identities, thereby minimizing the risk of data leaks even during model training. AI models using differential privacy algorithms reduced data re-identification risks by 35%, significantly enhancing patient confidentiality in machine learning applications [10].

Blockchain technology has become a significant resource for improving security, transparency, and the integrity of data in healthcare applications driven by AI. Unlike traditional centralized databases, blockchain operates as an immutable and decentralized ledger, meaning that once medical records are entered, they cannot be altered or deleted without consensus from authorized parties. This feature makes blockchain particularly effective in preventing fraudulent modifications of patient data, a growing concern in the era of AI-based healthcare decision-making [19].

A key benefit of blockchain in the healthcare sector is its capacity to create trust among various participants, such as hospitals, research organizations, insurance companies, and pharmaceutical firms. By ensuring that medical records are accurate, tamper-proof, and accessible only to verified users, blockchain reduces the risk of data manipulation, misdiagnoses, and unauthorized third-party access. Moreover, smart contracts—automated agreements recorded on the blockchain—can streamline patient consent management, guaranteeing that medical information is disclosed only under specified conditions.

Recent research has highlighted successful implementations of blockchain in healthcare security. Integrating blockchain with AI-powered medical imaging systems reduced data breaches by 40% while also enhancing interoperability between hospitals and research institutions. This seamless interoperability is critical in enabling secure cross-institutional collaborations, particularly in global health research and AI-driven precision medicine.

Furthermore, blockchain is playing an essential role in genomic data security—a field where AI is increasingly used for genetic risk assessment and drug development. Blockchain-based genomic data storage reduced unauthorized access risks by 50%, allowing researchers to analyze genetic datasets securely while maintaining patient anonymity. Similarly, AI-integrated blockchain solutions are being explored for remote patient monitoring devices, ensuring that real-time health data remains protected while allowing physicians to make AI-assisted clinical decisions with confidence.

As artificial intelligence increasingly influences the future of healthcare, strong security protocols and the incorporation of blockchain technology will be essential for safeguarding the privacy, integrity, and ethical handling of patient information. By leveraging advanced encryption, differential privacy, and blockchain's decentralized nature, healthcare institutions can significantly strengthen their cybersecurity infrastructure, ultimately enhancing patient trust and promoting widespread AI adoption in clinical settings [6, 19].

6.2 Addressing Algorithmic Bias and Ensuring Fair AI Models

A significant challenge in AI-driven healthcare is the presence of bias in machine learning algorithms, which can lead to disparities in medical diagnoses and treatment recommendations. AI models often inherent biases from non-representative training datasets, causing underperformance for specific populations, especially in underrepresented ethnic, geographic, and socioeconomic groups. If left unaddressed, these biases could exacerbate existing healthcare inequalities rather than solving them [17].

To mitigate bias, AI developers must focus on creating diverse and inclusive datasets that accurately represent various demographics, genetic backgrounds, and environmental factors. AI models trained on diverse datasets exhibited 20% improved accuracy in diagnosing cardiovascular conditions across different racial and ethnic groups. Furthermore, methods for ensuring algorithmic fairness, including tools for detecting bias, adjusting the weight of training data, and models that prioritize fairness in optimization, can assist in achieving fair AI decision-making. AI regulatory bodies and research institutions must conduct continuous audits and validate AI models to confirm they provide fair and unbiased medical predictions [8].

Another strategy for minimizing bias involves developing AI models with built-in bias-detection mechanisms that can flag potential disparities before deployment. Researchers are also working on explainable AI (XAI) systems, which allow healthcare professionals to understand how AI reaches decisions, making it easier to identify and correct biases. The implementation of real-world testing protocols and patient feedback loops further enhances AI's ability to deliver fair and inclusive healthcare outcomes [2].

6.3 Accessibility, Affordability, and Bridging Healthcare Gaps

For AI to truly transform healthcare, it must be accessible to diverse populations, including those in remote and low-income regions. Many AI-driven medical solutions, such as telemedicine, wearable health devices, and predictive analytics, are currently more available in developed urban centers, leaving rural and economically disadvantaged communities with limited access to advanced care. Without intentional efforts to scale AI-based healthcare solutions affordably, existing disparities may widen rather than improve [13].

A crucial step in bridging the urban-rural healthcare divide is optimizing AI platforms for low-bandwidth networks and basic mobile devices, allowing patients in remote areas to benefit from telehealth consultations, AI-assisted diagnostics, and remote monitoring. AI-powered mobile diagnostic tools improved early hypertension detection rates by 60% in rural India. Similarly, AI-driven smartphone-based retinal screening programs have enhanced the early detection of diabetic retinopathy in low-resource communities [25].

Cost remains a significant barrier to AI adoption in healthcare, particularly in low- and middle-income countries. Collaborations between the public and private sectors, along with government financial support and scalable AI innovations, play a critical role in making AI-enhanced healthcare cost-effective and widely available.

AI developers and policymakers must prioritize affordability initiatives, ensuring that AI-powered solutions do not remain exclusive to wealthier populations but instead benefit marginalized communities globally [8].

6.4 Ethical Considerations, Transparency, and Human Oversight

The growing reliance on AI in healthcare raises ethical concerns surrounding patient autonomy, decision-making transparency, and AI's role in clinical practice. Although AI significantly enhances diagnostic and treatment processes, it should serve as a supplement rather than a substitute for human medical knowledge. AI-driven recommendations must remain interpretative aids, with healthcare professionals making final decisions. Over-reliance on AI without adequate human oversight can lead to critical medical errors, particularly when AI models are trained on incomplete or biased data [6].

One of the most pressing ethical concerns is ensuring transparency in AI-generated recommendations. Patients and physicians should clearly understand how AI systems reach conclusions, which is essential for informed consent and maintaining trust in AI-powered healthcare. AI companies must develop explainable AI (XAI) models that offer interpretable decision-making processes, ensuring that medical professionals can review, validate, and override AI recommendations when necessary. Physicians were 40% more likely to trust AI-driven diagnostics when the AI model provided clear reasoning for its recommendations [35].

An additional ethical concern is the possibility that AI-enhanced healthcare could increase gaps in access to quality treatment. Policymakers and healthcare organizations must establish guidelines for AI regulation, liability, and accountability to prevent AI-based medical decisions from disproportionately benefiting privileged groups while leaving others behind. Additionally, the implementation of AI in healthcare should include interdisciplinary ethical review committees to guarantee that AI technologies comply with moral and legal healthcare norms [2].

6.5 The Future of AI in Healthcare and Emerging Solutions

Even though there are difficulties connected to incorporating AI in healthcare, the future looks very bright, as new technologies are working to resolve existing limitations. The next wave of AI advancements focuses on improving data privacy, real-time decision-making, transparency, and accessibility while ensuring ethical and regulatory compliance. These advancements will improve patient care, streamline medical processes, and extend the availability of AI-powered healthcare solutions to communities that are underserved [31].

One of the main issues in AI-powered healthcare is the privacy and security of data, particularly due to the sensitive nature of patient information. Conventional AI models typically depend on large, centralized datasets for training, which raises issues regarding data breaches and adherence to regulations. A groundbreaking approach to address this issue is federated learning, a method that permits AI models to be trained on decentralized datasets across various institutions without the need to transfer sensitive patient data. Rather than sharing raw data, federated learning facilitates secure model training by sending only the learning updates from the AI between institutions. This method reduces data security risks while maintaining high model accuracy across different healthcare systems.

Federated learning has already shown significant promise in medical imaging, disease prediction, and drug discovery. Federated learning has boosted the accuracy of AI in identifying rare diseases by 25% while adhering to data privacy laws like GDPR and HIPAA. With the growing adoption of federated learning, it is anticipated to improve collaboration among hospitals, research organizations, and pharmaceutical firms without jeopardizing patient privacy [3].

A significant advancement influencing the future of AI in healthcare is edge computing, enabling AI models to analyze medical data directly on devices rather than depending on cloud-based systems. This technology significantly reduces latency in AI-driven diagnostics, making real-time decision-making faster and more efficient. For example, AI-powered wearables and smart medical devices embedded with edge computing can analyze a patient's vitals instantly, sending immediate alerts to doctors in case of abnormalities. AI-enabled edge computing systems reduced critical response times for detecting stroke symptoms by 40%, allowing for faster emergency interventions. By keeping patient data localized and secure, edge computing also minimizes cybersecurity risks while ensuring real-time AI processing for high-stakes medical decisions [6].

7 Future Directions

Recent technological breakthroughs have significantly expanded the possibilities for artificial intelligence (AI) in healthcare. New advancements like blockchain, gene editing, federated learning, and AI-driven clinical trials are poised to transform healthcare services, enhancing their security, efficiency, and personalization. These advancements will play a pivotal role in enhancing patient care, streamlining medical research, and addressing data security concerns, ultimately shaping the next generation of AI-driven healthcare solutions [4].

7.1 Federated Learning: Data Privacy and AI Development

Federated learning is revolutionizing the methods used for AI training by allowing machine learning models to enhance their performance without the need to directly access or transfer raw patient information. In contrast to conventional centralized AI frameworks that necessitate the collection of data in one central location, federated learning conducts training on AI models locally on various decentralized devices or institutional servers. This methodology prioritizes data privacy and security, making it especially beneficial for healthcare applications such as managing chronic illnesses, diagnostics, and real-time predictive analytics [17].

One significant benefit of federated learning is its capacity to reduce cybersecurity threats while maintaining compliance with stringent healthcare regulations such as the General Data Protection Regulation (GDPR) and the Health Insurance Portability and Accountability Act (HIPAA). These regulations require strict standards for protecting patient data, and federated learning complies by keeping sensitive medical information local while still enabling AI to leverage a variety of datasets. This decentralized method also minimizes the risk of data breaches, as patient records are never stored in a single, vulnerable location [18].

Beyond enhancing security, federated learning significantly improves AI model accuracy and adaptability by training on real-world patient data from multiple institutions and geographic regions. Traditional AI models often struggle with generalization, as they rely on limited or homogeneous datasets. Federated learning, however, enables AI models to learn from a diverse range of patient demographics, ensuring that healthcare solutions remain equitable and effective across different populations. This approach is particularly useful in personalized medicine, where AI-driven predictions must consider genetic, environmental, and lifestyle variations to offer the most precise recommendations [21].

Numerous research efforts have shown the success of federated learning in the healthcare sector. Federated learning improved early disease detection rates by 28%, particularly in cases where patient data was geographically distributed across multiple hospitals. Federated learning allowed AI models to detect rare diseases with 25% greater accuracy compared to traditional machine learning approaches trained on smaller, localized datasets. By facilitating secure AI collaboration among hospitals, research institutions, and pharmaceutical companies, federated learning is paving the way for faster medical breakthroughs and more precise diagnostic tools [17].

Another critical application of federated learning is its role in real-time healthcare monitoring. With the rise of wearable medical devices and remote patient monitoring systems, AI can analyze data from millions of patients simultaneously, detecting early warning signs of diseases like hypertension, diabetes, and cardiovascular disorders. Federated learning guarantees that the ongoing data transmission stays confidential, all while allowing AI to enhance its diagnostic and forecasting abilities [6].

As healthcare powered by AI progresses, federated learning will play a key role in maintaining privacy while fostering innovation in medical AI. By reducing security risks, ensuring regulatory compliance, and improving AI model performance across diverse patient populations, federated learning is reshaping the future of AI in medicine, fostering a more secure, efficient, and personalized healthcare ecosystem [21].

7.2 AI-Powered Clinical Trials for Medical Research

Artificial intelligence is transforming the process of drug development and medical research by improving the precision, rapidity, and effectiveness of clinical trials. Conventional clinical trials typically take a long time and incur high expenses, depending on manual data gathering and fixed trial designs. AI-driven approaches, however, enable real-time analysis of patient responses, allowing researchers to identify treatment patterns and correlations much faster than conventional methods [11].

One of the most transformative abilities of AI is its capacity to analyze extensive datasets from various origins, such as electronic health records (EHRs), wearable technology, genetic information, and environmental influences. Utilizing machine learning algorithms, AI can anticipate which groups of patients are most likely to gain from a particular treatment, assisting researchers in refining the selection of participants for clinical trials. This targeted approach reduces trial failure rates, accelerates the drug approval process, and ensures that effective treatments reach patients faster [25].

AI-driven patient stratification improved clinical trial success rates by 30%, leading to faster regulatory approvals and reduced drug development costs. Additionally, AI-powered simulations allow researchers to predict drug interactions, optimize dosages, and model disease progressions, further streamlining the clinical trial pipeline [10].

By simplifying trial designs and improving predictive accuracy, AI-driven clinical trials increase the efficiency of drug discovery while ensuring greater patient safety and treatment efficacy. These developments are anticipated to be essential in personalized medicine, which focuses on customizing treatments to suit the specific profiles of patients instead of using a generic approach [6].

7.3 Integration with Emerging Technologies

The combination of AI with advanced technologies like blockchain and gene editing is fostering groundbreaking advancements in the healthcare sector. These synergistic advancements are addressing long-standing challenges related to data security, treatment personalization, and medical research transparency [4].

Blockchain technology enhances the security, transparency, and integrity of AI-powered healthcare systems by creating an immutable, decentralized ledger for storing patient data. This guarantees that medical records are safeguarded against tampering, stopping unauthorized changes while allowing secure data exchange between healthcare professionals, researchers, and regulatory organizations. The decentralized aspect of blockchain further diminishes the likelihood of cyberattacks and data breaches, establishing it as a vital resource for AI-powered healthcare solutions [2, 6, 29]. AI systems reduced data breaches by 40% while improving interoperability between hospitals, insurers, and research institutions. By establishing trust and security, blockchain is expected to become a cornerstone of future AI-powered healthcare infrastructure [3].

Artificial intelligence is also transforming gene editing, especially in the enhancement of CRISPR-based treatments. Gene editing requires highly precise identification of genetic targets, and AI-driven models excel at analyzing vast genomic datasets to pinpoint mutations and potential editing sites with unprecedented accuracy. By leveraging predictive analytics and deep learning, AI enhances the efficacy and safety of CRISPR applications, reducing off-target effects and improving overall success rates [24].

Gene editing aided by AI is opening new possibilities for tailored therapies for genetic conditions like cystic fibrosis, sickle cell disease, and hereditary cancers. By optimizing CRISPR-based interventions, AI is enabling new therapeutic possibilities, offering hope for patients with previously untreatable genetic conditions. Additionally, AI is facilitating high-throughput drug screening for gene therapy applications, expediting the development of novel genetic treatments [8].

7.4 Building Precision Medicine Frameworks

The coming together of artificial intelligence (AI), blockchain technology, and gene editing is establishing the foundation for precision medicine, a revolutionary method that tailors medical treatments to fit a person's genetic composition, environmental factors, and lifestyle choices. Unlike traditional healthcare, which often relies on generalized treatment protocols, precision medicine leverages AI-driven insights and advanced genetic analysis to create tailored therapeutic strategies for each patient. This shift is making healthcare more effective, safer, and ethically sound, ensuring that treatments align with each individual's unique biological profile [21].

The impact of AI on precision medicine is notably revolutionary, as machine learning algorithms have the capability to examine large datasets of patient health information, genetic information, and biomarker profiles to uncover trends and relationships that might not be obvious to human researchers. By detecting subtle disease predispositions, AI enables early intervention strategies, allowing for proactive disease management rather than reactive treatment. AI-driven genomic analysis improved cancer risk prediction by 35%, allowing doctors to implement preventative care strategies before symptoms even appeared [3].

At the same time, blockchain technology is revolutionizing data security and transparency in precision medicine. Patient privacy is a significant concern when dealing with highly sensitive genetic and biometric data, and blockchain ensures that health records remain tamper-proof and accessible only to authorized medical professionals. By eliminating the risk of unauthorized modifications and data breaches, blockchain fosters trust in AI-powered precision medicine platforms. Furthermore, blockchain-enabled patient consent mechanisms allow individuals to control who can access their genetic and medical data, ensuring that ethical considerations remain at the forefront of precision medicine [3].

As precision medicine continues to evolve, ongoing advancements in regulatory frameworks, AI interpretability, and ethical considerations will be critical in ensuring patient safety and equitable access to personalized healthcare. Entities like the World Health Organization (WHO) and the U.S. Food and Drug Administration (FDA) is engaged in developing guidelines aimed at standardizing the use of AI in precision medicine, with the goal of keeping these technologies transparent, equitable, and focused on patients. By combining AI, blockchain technology, and gene editing, precision medicine is set to transform the future of healthcare, providing tailored treatment methods that improve patient results, safety, and overall quality of life [17].

8 Conclusion

Artificial Intelligence (AI) has emerged as a revolutionary resource in the management of chronic conditions, especially diabetes and hypertension, enabling a shift from traditional, reactive healthcare to a more forward-looking, predictive, and deeply personalized strategy. By harnessing advanced technologies such as machine learning (ML), predictive modeling (PM), wearable devices, and telemedicine, AI equips healthcare professionals with real-time, data-driven insights. This enables earlier diagnosis, precise treatment adjustments, and continuous patient monitoring, all of which contribute to better disease control, reduced complications, improved adherence to therapies, and tailored care that addresses each patient's unique biological, lifestyle, and environmental factors.

In the management of diabetes, technologies powered by AI, including continuous glucose monitors (CGMs), automated insulin delivery systems, and sophisticated prediction tools for identifying complications, have greatly improved blood sugar control and contributed to better long-term health results. Similarly, in the treatment of hypertension, AI has contributed to better cardiovascular health through continuous blood pressure monitoring, personalized risk evaluation, and tailored medication plans, ultimately reducing hospital admissions [33]. Through the analysis of extensive patient information, AI is capable of recognizing possible health risks, assisting in preventive care actions, and helping healthcare professionals provide tailored, evidence-informed treatment options.

Additionally, the combination of AI and telemedicine has greatly increased access to quality healthcare, especially in rural and underprivileged areas. With the help of remote monitoring systems, AI-assisted diagnostics, and virtual consultations, patients benefit from continuous, high-quality care without the limitations of location. Additionally, AI has strengthened patient engagement by providing customized support for lifestyle modifications, nutritional planning, and mental health, ensuring a more comprehensive and patient-centered approach to chronic disease management. Although these encouraging developments exist, various obstacles impede the widespread integration of AI in the healthcare sector. Concerns surrounding data privacy, algorithmic fairness, accessibility, and regulatory compliance remain critical. The heavy reliance on extensive, high-quality datasets raises questions regarding data ownership, confidentiality, and security.

Additionally, biases present in AI algorithms can worsen existing healthcare disparities if not properly addressed. Therefore, ensuring ethical, transparent, and equitable AI deployment is essential for the technology's long-term success and societal trust. Looking forward, the future of AI in chronic disease care depends on the refinement of privacy-focused solutions like federated learning, the advancement of fair and explainable AI models, and the seamless incorporation of emerging technologies such as blockchain and gene-editing tools. These innovations will play a key role in fostering safer, more ethical, and highly individualized healthcare solutions. Collaboration between medical professionals, AI developers, policymakers, and ethics boards will be essential in creating comprehensive guidelines that prioritize patient rights and safety while maximizing AI's potential in healthcare.

In summary, artificial intelligence is set to transform chronic disease management, steering a movement toward more anticipatory, available, and personalized healthcare. By addressing current limitations and embracing new technological opportunities, AI has the power to alleviate the global burden of diabetes and hypertension, enhance public health outcomes, and contribute to building a more robust, equitable, and sustainable healthcare system for generations to come.

References

1. National Institute of Diabetes and Digestive and Kidney Diseases. *Diabetes statistics report*. Retrieved from https://www.niddk.nih.gov/health-information/health-statistics/diabetes-statistics.
2. Kenger, E. B., et al. (2025). Artificial intelligence-generated diet plans for hypertension and dyslipidemia: Adherence and nutritional insights. *Iranian Journal of Public Health, 54*(6).
3. National Institutes of Health. (2023). *AI and precision medicine: Emerging trends.*
4. Uthman, O. A., et al. (2023). Global prevalence & trends in hypertension & type 2 diabetes mellitus among slum residents: A systematic review & meta-analysis. *British Medical Journal Open, 12*(2).
5. Feng, G., et al. (2022). Artificial intelligence in chronic disease management for aging populations: A systematic review of machine learning and NLP applications. *International Journal of General Medicine, 18*, 3105–3115.
6. Singareddy, S., et al. (2023). Artificial intelligence and its role in the management of chronic medical conditions: A systematic review. *Cureus, 15*(9), e46066. https://doi.org/10.7759/cureus.46066
7. Kan, E. (2024). Block-chain and AI in healthcare data security: Creating a secure medical ecosystem. *Irshad Journal of Law and Policy, 2*(12), 13–21.
8. Alobaidi, S. (2025). Emerging biomarkers and advanced diagnostics in chronic kidney disease: Early detection through multi-omics and AI. *Diagnostics, 15*(10), 1225.
9. Guan, Z., et al. (2023). Artificial intelligence in diabetes management: Advancements, opportunities, and challenges. *Cell Reports Medicine, 4*(10), 101213. https://doi.org/10.1016/j.xcrm.2023.101213

10. Zheng, H., Ryzhov, I. O., Xie, W., & Zhong, J.. (2020). Personalized multimorbidity management for patients with type 2 diabetes using reinforcement learning of electronic health records. arXiv.org, [Online]. https://arxiv.org/abs/2011.02287
11. Nickel, S. G., Venkatesh, K. P., Raza, M. M., & Kvedar, J. C. (2024). AI-based diabetes care: Risk prediction models and implementation concerns. *NPJ Digital Medicine, 7*(1), 36. https://doi.org/10.1038/s41746-024-01034-7
12. Sudunagunta, A., et al. (2025). Artificial intelligence powered risk stratification in chronic disease management: The role of Microsoft Autopilot and electronic medical record integration in the United States healthcare system. *La Prensa Medica, 111*(4), 1–10.
13. Mayya, V., et al. (2024). Need for an artificial intelligence-based diabetes care management system in India and the United States. *Health Services Research and Managerial Epidemiology, 11*. https://doi.org/10.1177/23333928241275292
14. Mohrag, M., et al. (2024). The impact of wearable technologies on blood pressure control in hypertensive patients: A systematic review and meta-analysis. *Cureus*. https://doi.org/10.7759/cureus.71220
15. Shinde, S., et al. (2024). Wearable sweat-sensing patches for non-invasive and continuous health tracking. *Sensors and Actuators Reports*, 100265. https://doi.org/10.1016/j.snr.2024.100265
16. The World Bank. (2022). *AI'S role in global health equity and accessibility*.
17. Afsaneh, E., et al. (2022). Recent applications of machine learning and deep learning models in the prediction, diagnosis, and management of diabetes: A comprehensive review. *Diabetology & Metabolic Syndrome, 14*(1). https://doi.org/10.1186/s13098-022-00969-9
18. Alum, E. U., et al. (2025). Artificial intelligence in personalized medicine: Transforming diagnosis and treatment. *Discover Applied Sciences Journal, 7*(193). https://doi.org/10.1007/s42452-025-06625-x
19. Jothi Prakash, B. S., & Varsha, R. G. (2025). AI-Driven Telemedicine: Bridging healthcare and technology for the future. *International Journal for Research in Applied Science and Engineering Technology, 13*(1), 557–563. https://doi.org/10.22214/ijraset.2025.66171
20. Hwang, M., et al. (2024). Artificial intelligence applications for chronic condition self-management: Scoping review (Preprint). *Journal of Medical Internet Research*. https://doi.org/10.2196/59632
21. Wang, X., et al. (2025). Effectiveness and safety of AI-driven closed-loop systems in diabetes management: A systematic review and meta-analysis. *Diabetology and Metabolic Syndrome, 17*, 238. https://doi.org/10.1186/s13098-025-01819-0
22. National Academy of Medicine, The Learning Health System Series, Whicher, D., et al. (2023). *Artificial intelligence in health care: The hope, the hype, the promise, the peril*. National Academies Press (US); Health care artificial intelligence: Law, Regulation and Policy.
23. Cho, J. S., & Park, J.-H. (2024). Application of artificial intelligence in hypertension. *Clinical Hypertension, 30*(1), 11. https://doi.org/10.1186/s40885-024-00266-9
24. Mayo Clinic. *AI in continuous glucose monitoring: Benefits and challenges*. Retrieved from https://www.mayo.edu/research/clinical-trials/cls-20581640
25. Wang, Y., et al. (2025). Wearable devices as tools for better hypertension management in elderly patients. *Medical Science Monitor, 31*. https://doi.org/10.12659/msm.946079
26. Li, Z., et al. (2023). Artificial intelligence in ophthalmology: The path to the real-world clinic. *Cell Reports Medicine, 4*(7), 101095. https://doi.org/10.1016/j.xcrm.2023.101095
27. Ghadi, Y. Y., et al. (2025). Integration of wearable technology and artificial intelligence in digital health for remote patient care. *Journal of Cloud Computing, 14*(39). https://doi.org/10.1186/s13677-025-00759-4
28. Panteli, D., et al. (2025). Artificial intelligence in public health: Promises, challenges, and an agenda for policy makers and public health institutions. *The Lancet Public Health*. https://doi.org/10.1016/s2468-2667(25)00036-2
29. Hidayaturrohman, Q. A., & Hanada, E. (2024). Predictive analytics in heart failure risk, readmission, and mortality prediction: A review. *Cureus*. https://doi.org/10.7759/cureus.73876

30. Layton, A. T. (2024). AI, machine learning, and ChatGPT in hypertension. *Hypertension, 81*(4), 709–716. https://doi.org/10.1161/hypertensionaha.124.19468
31. Skalidis, I., et al. (2025). Artificial intelligence and advanced digital health for hypertension: Evolving tools for precision cardiovascular care. *Medicina, 61*(9), 1597. https://doi.org/10.3390/medicina61091597
32. Kamga, P., et al. (2022). The use of wearable ECG devices in the clinical setting: A review. *Current Emergency and Hospital Medicine Reports, 10*(3), 67–72. https://doi.org/10.1007/s40138-022-00248-x
33. Chaturvedi, U., et al. (2025). The impact of artificial intelligence on remote healthcare: Enhancing patient engagement, connectivity, and overcoming challenges. *Intelligent Pharmacy*. https://doi.org/10.1016/j.ipha.2024.12.003
34. Truong, A. T. L., et al. (2023). CURATE.AI-assisted dose titration for anti-hypertensive personalized therapy: Study protocol for a multi-arm, randomized, pilot feasibility trial using CURATE.AI (CURATE.AI ADAPT trial). *European Heart Journal—Digital Health, 5*(1), 41–49. https://doi.org/10.1093/ehjdh/ztad063
35. Alkhanbouli, R., Matar Abdulla Almadhaani, H., Alhosani, F., & Simsekler, M. C. E. (2025). The role of explainable artificial intelligence in disease prediction: A systematic literature review and future research directions. *BMC Medical Informatics Decision Making, 25*(1), 110. https://doi.org/10.1186/s12911-025-02944-6

The Impact of Artificial Intelligence on Digital Marketing

Elisabete Paulo Morais and Luisa Lopes

1 Introduction

Artificial intelligence (AI) powered technologies such as chatbots, smart devices, voice assistants, and autonomous vehicles are becoming a reality in marketing and many other fields. In fact, while many organizations' functional areas can or may benefit from this breakthrough innovation, it is the marketing field of expertise that has been positively affected by the newest developments concerning generative AI [1]. The findings of some studies highlight the increasing dependence on generative AI in the marketing sector to improve customer experiences, optimize operations, and boost efficiency [2]. Moreover, AI-based technologies—especially those utilizing machine learning or deep learning—continuously adapt and enhance their performance, allowing them to deliver increasingly precise and personalized solutions to consumers. This improvement is driven by AI's ability to track and analyze user behavior, enabling a deeper understanding of consumer needs and preferences. Consumers can actively participate in value creation through their interactions with AI, gaining benefits such as enhanced convenience, personalized experiences, and cost savings. However, the perceived value of AI may decline if it fails to meet expectations or raises privacy and security concerns. AI can minimize consumers' cognitive, emotional, and behavioral effort to complete tasks, potentially increasing engagement. This is achieved by AI-powered systems handling specific tasks on behalf of users, such as offering tailored recommendations or automating repetitive activities [3].

E. P. Morais (✉) · L. Lopes
Applied Management Research Unit (UNIAG), Instituto Politécnico de Bragança, Bragança, Portugal
e-mail: beta@ipb.pt

C. Rompante Cunha et al. (eds.), *Transformative Cloud Computing, IoT and Extended Reality*, Signals and Communication Technology,
https://doi.org/10.1007/978-3-032-03554-7_8

Still, researchers have largely overlooked the transformative effects that AI has on marketing practices [4]. However, the key themes defining this emerging and interdisciplinary corpus of research remain unclear, highlighting a critical and still evolving gap in the literature.

This chapter intends to explore the transformative impact of AI on digital marketing, addressing its opportunities and challenges.

The chapter is structured as follows. After the Introduction, the next section, theoretical foundations, defines AI and its key components (e.g., machine learning, deep learning, neural networks) relevant to digital marketing, traces the evolution of digital marketing, and highlights the shift from traditional methods to data-driven AI-powered strategies. Discuss the role of big data, real-time analytics, and customer-centric approaches in modern marketing.

Sect. 3 examines AI-driven personalization, including its benefits for customer engagement, the ethical dilemmas it raises (e.g., privacy, transparency, bias), and its role in automating and optimizing marketing campaigns, followed by Sect. 4, which explores automation in digital marketing, focusing on its advantages (e.g., efficiency, personalization) and challenges (e.g., integration, privacy, human interaction).

Sect. 5 examines the role of AI in content creation, highlighting its potential to increase efficiency and scalability, as well as ethical concerns around authenticity and job displacement. Section 6 advances ethical dilemmas posed by AI in marketing, such as the manipulation of consumer behavior, data privacy concerns, and the use of AI in marketing. Finally, the Conclusion summarizes the main aspects, discusses the implications for practitioners and academics, and suggests future research directions.

2 Theoretical Foundations

2.1 Digital Transformation and Marketing

Over the past two decades, the digital transformation of businesses has fundamentally reshaped the role of marketing within organizations. Digital marketing, previously known as e-marketing or Internet marketing, is central to digital business—enhancing customer engagement and understanding, increasing product value, expanding distribution channels, and driving sales through targeted digital campaigns across various digital media channels, including search marketing, online advertising, and affiliate marketing [5]. Moreover, the extensive adoption of digital marketing strategies, particularly social media marketing, has profoundly impacted how firms and customers engage with Internet-based technologies, shaping their behaviors, interactions, collaboration, and value co-creation [6]. Simply put "digital marketing is marketing online, whether via websites, online ads, opt-in email, mobile apps or emerging platforms. It involves getting close to customers, understanding them better and maintaining a dialogue with them" ([5], p. 13–14). More recently, the evolution of digital marketing turned the shift from traditional methods to data-driven, AI-powered strategies, including big data management, real-time analytics, and customer-centric approaches in modern marketing.

2.2 *AI in Digital Marketing*

AI refers to "a machine that behaves in a way that could be considered intelligent if it was a human being" [7]. AI, also defined as "a system's ability to interpret external data correctly, to learn from such data, and to use those learnings to achieve specific goals and tasks through flexible adaptation" [8] (p. 5), has gained momentum in consumer psychology/behavior, and marketing research and has becoming an important ally of marketing [9]. AI can automate routine and non-routine operations by simulating human thinking and actions, evaluating complicated data to uncover behavioral patterns and insights, and learning from experiences to make intelligent decisions [10]. Nowadays, generative AI is an example of a tool to generate content and ideas for almost 50% of marketers in the industry [11].

According to [7], it was after WWII that a number of people independently started to work on intelligent machines, and the English mathematician Alan Turing may have been the first to give a lecture on it in 1947. By the late 1950s, there were many researchers on AI, and most of them were basing their work on programming computers. It was in the 1950s that AI was established as an academic discipline, and it remained in obscurity and limited interest for over half a century [8]. For a historical perspective of the evolution of AI, [8] use the analogy of the four seasons (spring, summer, fall, and winter) to present the past of AI, understand which challenges firms to face and into the future, help to prepare for the challenges ahead.

Understanding the historical evolution of Artificial Intelligence (AI) is essential to contextualize its transformative role in digital marketing. Since its conceptual foundations in the 1950s—marked by Alan Turing's pioneering ideas and the 1956 Dartmouth conference—AI has undergone periods of intense innovation as well as stagnation, often referred to as the "AI winters." Over the decades, key developments such as machine learning, deep learning, and the rise of generative AI have shaped the current technological landscape. Figure 1 presents a timeline highlighting the major milestones in the evolution of AI, illustrating how these advancements have laid the groundwork for its application in personalized, data-driven, and automated marketing strategies.

Nowadays, according to Mehta et al. [9], AI comprises various forms of technology, including virtualization, self-service, sensory-enabled, and image interactivity technology. Virtualization tools, for instance, VR goggles or other mobile devices, enable customers to digitally experience products and services that are not physically present. Self-service technologies, such as automated teller machines

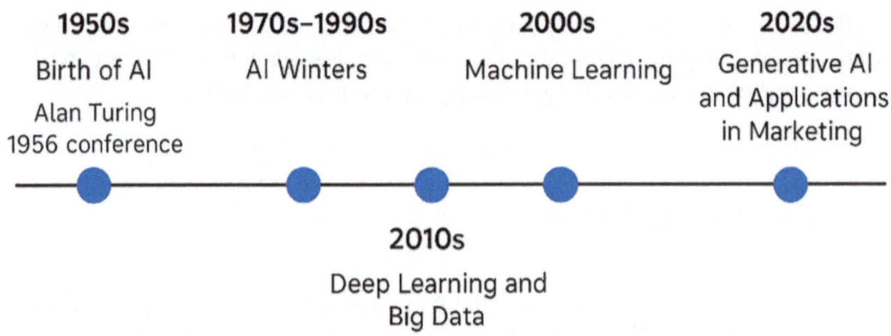

Fig. 1 Chronology of the evolution of AI. *Source: Own elaboration*

(ATMs) or self-service desks at airports, allow users to obtain a service without interacting with service employees.

Researchers have investigated the adoption of AI in the marketing domain in recent years [1, 2, 4, 12–17].

AI in marketing is synonymous with precision, personalization, and efficiency [10]. There are some AI key components relevant to marketing and to digital marketing specifically, namely natural language processing (NLP), machine learning, neural networks, and deep learning [2]. Artificial neural networks and deep learning form the basis of most applications we know under the label of AI [8]. "NLP allows machines to process and interpret human language, simplifying text data analysis and the extraction of customer sentiment and important themes. Machine learning utilizes algorithms to analyze data and predict outcomes without being explicitly programmed, aiding in identifying patterns like fraud. Neural networks, inspired by human brain activity, are linked nodes that analyze data and are used in applications such as image and speech recognition. Finally, deep learning, a subset of neural networks, specializes in making exact predictions by training models on large amounts of data and is helpful in various AI applications, including handling unstructured data in AI analytics" [2](p. 8).

The literature shows that researchers have frequently applied the underpinnings of the theory of reasoned action, as well as the theory of acceptance and use of technology. Still, the findings based on the variables of these two theories appear to be fragmented and inconsistent in the AI literature [9]. A more solid understanding is required, and in a meta-analysis, [9] suggest that several antecedents have a significant impact on consumer attitudes towards AI adoption, namely performance expectancy, effort expectancy, hedonic value, anthropomorphism, perceived risk (negatively), trust and social pressure.

As AI continues to advance and gradually assumes tasks that require higher levels of intelligence [14], the nature and scope of human-AI collaboration are expected to evolve. Nevertheless, this collaboration will likely persist in fostering a synergistic enhancement of both human and machine capabilities. Therefore, identifying the most effective ways to support this collaboration remains crucial [10, 18].

3 A-Driven Personalization in Digital Marketing

The integration of AI and Machine Learning in digital marketing has reshaped consumer-brand interactions, allowing businesses to deliver tailored experiences that enhance engagement and improve conversion rates. However, despite its advantages, AI-driven personalization presents several key challenges that must be addressed to ensure ethical, transparent, and effective marketing practices.

Privacy remains one of the biggest hurdles in AI-powered personalization. AI and ML technologies use vast consumer data to generate tailored marketing strategies. Such datasets often contain sensitive personal attributes, including online activity logs, purchasing patterns, and biometric markers. The ethical ramifications span issues of informed consent, cybersecurity vulnerabilities, and the potential for malicious exploitation. Regulatory frameworks, including the EU's General Data Protection Regulation (GDPR) and California's Consumer Privacy Act (CCPA), establish critical protections for personal data. However, businesses must still navigate the complexities of compliance while maintaining effective personalization strategies. Giving users control over their data and ensuring transparent collection practices are key to reducing privacy risks.

AI is critical in automating and personalizing marketing strategies while addressing privacy concerns through advanced data encryption and anonymization techniques [19]. However, another significant challenge in AI-driven personalization is transparency. The opacity of AI-driven personalization—from dynamic content to targeted ads—frequently leads to consumer distrust, particularly when perceived as coercive or non-consensual. This opacity can lead to distrust in marketing efforts, especially when users perceive AI-driven decisions as intrusive or manipulative. AI-driven personalization is transforming influencer marketing, demanding greater transparency to maintain trust and ethical standards [20]. AI-driven personalization offers significant benefits in digital marketing by enabling businesses to deliver highly targeted and relevant content to consumers. Addressing AI's triad of challenges (privacy, transparency, and bias) requires more than technical fixes, it necessitates compliance-aligned governance structures and auditable ethical AI practices. Companies must adopt responsible AI practices, prioritize ethical considerations, and implement robust governance frameworks to navigate these challenges effectively. This balanced approach lets companies leverage AI's precision while safeguarding trust and compliance, turning personalization into a long-term competitive advantage.

3.1 Impact of AI-Driven Personalization on Customer Engagement in Digital Marketing

By processing vast datasets through AI/ML, marketers uncover precise behavioral patterns to personalize content at scale. Predictive models transform these insights into anticipatory engagement, creating deeper customer connections by enhancing relevance. AI-powered predictive models enable hyper-personalized customer experiences, boosting customer retention and brand advocacy [21].

Additionally, AI enables dynamic segmentation, ensuring that marketing messages are tailored to individual preferences in real time, thereby improving engagement rates and overall customer satisfaction.

With AI-powered chatbots, businesses can now offer 24/7 support, resolving customer queries in real time. These tools increase engagement by instantly answering questions, resolving common issues, and offering personalized product recommendations. AI-enabled personalization in interactive marketing improves the customer journey by providing seamless interactions [22]. As chatbot technology advances, its ability to detect and adapt to customer emotions continues to improve, further enhancing engagement and strengthening brand relationships and loyalty. AI optimizes marketing targeting by dynamically adjusting content strategies based on user behavior. Personalized email marketing, recommendation engines, and adaptive website content ensure that customers receive relevant information tailored to their preferences. The impact of AI-driven marketing strategies is measurable, highlighting how AI-powered personalization leads to higher conversion rates and enhanced customer satisfaction [13]. Additionally, AI's role in automating personalized marketing strategies, optimizing engagement with target audiences, and reducing inefficiencies is emphasized.

AI-driven personalization is reshaping influencer marketing by optimizing brand-consumer interactions. AI algorithms analyze audience demographics, sentiment, and engagement patterns to match brands with the most relevant influencers. By leveraging AI to tailor influencer content, brands transform promotional messages into authentic recommendations that genuinely resonate with target demographics [20]. This strategic alignment increases campaign effectiveness and fosters stronger consumer trust and brand affinity.

3.2 Ethical Considerations of Using AI-Driven Personalization in Digital Marketing

While AI-powered personalization provides significant marketing benefits through data analysis, its reliance on large-scale consumer data collection raises serious privacy concerns. Without strong protective measures, these systems could enable unauthorized misuse of sensitive information or create opportunities for security breaches [20]. A key priority is making AI decision-making more transparent so consumers know exactly how their data shapes marketing efforts. At the same time, ensuring fairness in AI systems is critical, without proper checks, biased training data can result in marketing that discriminates, either unfairly targeting or overlooking certain groups [21].

While AI improves marketing efficiency, human oversight remains a critical activity to maintain ethical practices. Automated decisions cannot operate without human oversight. Accountability measures and human review processes must be in place [23]. Experts stress the importance of maintaining strong ethical guidelines for AI-driven marketing, especially to protect consumer data usage. Marketing teams must closely monitor their AI systems to detect and address any biases that may gradually develop over time. Without continuous adjustments, even the most carefully designed algorithms can unintentionally become biased, favoring certain customer groups while excluding others [13].

The challenge in AI-driven digital marketing lies in balancing its transformative potential with ethical responsibilities. There's no denying the transformative impact of AI on marketing effectiveness. However, sustainable success depends on implementing these tools responsibly—where enhanced personalization and robust consumer protections work hand in hand [24]. They discuss the role of AI in competitive marketing strategies, warning that excessive data-driven personalization may cross ethical boundaries. It is important that consumers retain real control over their choices. While AI can suggest products, it should not cross over into psychological manipulation, forcing purchases through deceptive patterns or exploiting behavioral weaknesses. Ethical marketing means recommendations that inform, not pressure [25]. Companies need to build ethical AI guardrails that both protect customers and meet legal standards. This isn't just about avoiding fines—it's about creating marketing that people actually trust. When done right, these frameworks become competitive advantages, not just compliance checkboxes [26].

3.3 Current Trends and Future Directions in AI-Driven Personalization for Digital Marketing

Although hyper-personalization increases conversions, the interesting thing is when AI enhances the customer experience rather than invading it. Predictive analytics, powered by ML, allows marketers to anticipate customer needs and preferences before they are explicitly expressed. For instance, AI algorithms can instantly analyze browsing patterns, purchase history, and social media activity to deliver personalized product recommendations instantly [21]. This level of personalization not only enhances customer satisfaction but also drives higher conversion rates and brand loyalty.

The convergence of AI with other emerging technologies, such as the Internet of Things (IoT) and augmented reality (AR), opens new avenues for personalized marketing. IoT devices, such as smart home assistants and wearables, provide real-time data on consumer behavior to enable brands to deliver contextualized marketing messages. Meanwhile, AR can create immersive shopping experiences, allowing customers to visualize products in their own environment before buying them [27]. Combined with AI, these technologies offer unique opportunities for consumers in a meaningful and personalized way.

As voice-activated assistants like Alexa and Google Assistant become more prevalent, optimizing for voice search is critical to digital marketing strategies. Conversational agents with AI technology can respond to voice queries and establish natural, contextualized dialogs with people. These agents can offer personalized recommendations and answer customer questions, thus improving the overall customer experience [23].

The future of AI-driven personalization in digital marketing is both promising and challenging. As AI technologies evolve, they will enable beyond personalization, efficiency, and innovation. Companies must pay attention to ethical considerations such as consumer privacy and trust to ensure that AI is used responsibly. By adopting transparent practices and investing in ethical AI development, brands can harness the power of AI to drive sustainable growth and build lasting customer relationships.

4 Automation in Digital Marketing: Opportunities and Challenges

Marketing automation has revolutionized how companies connect with customers—streamlining campaigns, personalizing outreach, and turning data into actionable insights. But these powerful tools aren't without their growing pains. Businesses are learning that automation works best when it enhances human strategy rather than replaces it, requiring careful navigation of technical, ethical, and operational hurdles.

One of the most notable advantages of automation in digital marketing is its ability to enhance efficiency and effectiveness. Marketers can allocate their time and resources more strategically by streamlining workflows and automating repetitive tasks, thereby improving overall productivity [28]. By automating routine tasks, marketing teams free up valuable time for strategic work while minimizing human error. The real power comes from consistently analyzing market shifts and customer behaviors in real time, letting brands pivot campaigns at the speed of digital trends [29]. These capabilities allow organizations to implement more agile marketing strategies in order to respond to consumer behaviors and trends in real time.

Moreover, automation significantly improves personalization and customer engagement. With the integration of AI and ML, businesses can analyze vast amounts of data to predict consumer preferences and behaviors with greater accuracy [21]. Personalized recommendations and 24/7 customer support strengthen relationships and increase brand loyalty [2]. Through predictive analytics, organizations can respond proactively to customer needs, creating interactions and improving customer satisfaction.

Data-driven decision-making is another crucial advantage facilitated by automation. Digital marketing platforms now incorporate sophisticated analytics tools that collect and analyze data from multiple sources, enabling organizations to refine their strategies and optimize user experiences [30]. ML-powered predictive analytics allows for detailed customer segmentation, providing businesses with actionable insights to drive growth and competitive advantage [31]. These capabilities ensure that marketing teams are supported by data and flexible to changing consumer demands.

Cost optimization is yet another advantage of marketing automation. AI-driven algorithms can optimize advertising campaigns in real time, ensuring that resources are allocated to the most effective channels and audiences [32]. This is particularly advantageous for organizations with limited marketing budgets, including non-profit organizations, which can use automation to refine targeting strategies and maximize return on investment [33]. Despite these advantages, organizations face several challenges in implementing marketing automation. One of the primary obstacles is the integration of automation into existing operations. Successful adoption requires seamless digital transformation, which can be complex and resource-intensive [30]. Many companies struggle with aligning automation tools between different departments, leading to inefficiencies and resistance to change. In addition, the lack of digitalization can require substantial investment in technological infrastructure and employee training [34].

Privacy and data security concerns also pose significant challenges in marketing automation. The reliance on AI and ML for data processing raises ethical and legal questions regarding consumer privacy [35]. Organizations must ensure robust data protection measures and compliance with regulations such as the GDPR to safeguard user information [36]. Another challenge has to do with the balance between automation and human interaction. Too much automation can result in impersonal and poorly targeted communication, potentially alienating consumers and diminishing trust in the brand [35]. While automated systems can efficiently handle customer interactions, human touch remains essential for building authentic relationships and resolving complex issues. Organizations need to balance automation and human intervention to maintain a personalized and engaging customer experience.

Finally, the high initial costs associated with implementing marketing automation solutions can be a barrier, especially for small and medium-sized enterprises (SMEs). Creating AI-driven automation systems requires a financial outlay, which may not be feasible for all companies [29]. In addition, retaining qualified professionals to manage and optimize automated platforms represents an ongoing challenge, as these professionals are still in short supply [28].

Although digital marketing automation offers substantial opportunities for efficiency, personalization, data-driven decision-making, and cost optimization, it also presents considerable challenges related to integration, privacy, human interaction, and financial investment. Organizations must harness the potential of automation while maintaining ethical and strategic considerations in their marketing practices.

5 AI in Content Creation: Transforming the Creative Process

In digital marketing, knowledge of social media, AI applications, e-commerce, analytics real-time practices, and more in general, the Internet is not only about technical knowledge but reflects a broader domain in which content and techniques are merged into a new holistic approach towards content and marketing communication. Core marketing skills include creative thinking to generate content relevant to different customers' touchpoints, effective planning and time management, and precision and attention to detail [6]. In this regard, substantial development of AI-powered tools for text creation, video editing, and graphic design has been recognized and used by many tech practitioners and marketers. AI, within sensory marketing, can target consumers' five senses using sensory appeal, creating contents that use the senses of sound, sight, smell, touch, and taste [9]. As far as known, an important application of intelligent virtual assistants, such as ChatGPT, is content creation, as it can be used to create articles, stories, and other types of written content, and text generated which is difficult to differentiate from text written by humans [37].

Some benefits and advantages of AI in content creation are such as increased efficiency, scalability, and the ability to produce personalized content at scale. Indeed 67% of marketers say the most significant benefit of AI is the ability to create content faster [11]. Generative AI has also been suggested as a beneficial tool to support writers and content creators in their daily idea-generation work and, at the same time, as a way to automate repetitive or time-consuming content tasks [37]. Hollebeek et al. [3], introducing a conceptual model that illustrates the antecedents and outcomes of AI-driven consumer engagement, review the key themes characterizing this emerging, interdisciplinary corpus of work, including the ability of AI to provide increasingly accurate service, its capacity to (co) create consumer-perceived value, and its potential to reduce consumer effort in task execution. Also, research has explored how consumers engage with brands through technology, focusing on various digital tools such as social media, virtual, augmented, and mixed reality applications, as well as automated and AI-powered technologies (e.g., [17]. These innovations have also been shown to enhance corporate performance [10]. For instance, Google's predictive AI for text input continuously improves by learning from user interactions, while Amazon's Echo and Apple's Siri similarly refine their responses over time to offer more personalized solutions to users [3]. Even in education, AI can transform the next 50 years [38].

However, some potential downsides, challenges, and ethical concerns can be highlighted, including the loss of human creativity, concerns about authenticity, and the ethical implications of AI-generated content. AI is increasingly reshaping service by performing various tasks, constituting a major source of innovation, yet threatening human jobs [14]. Additionally, the AI-generated content, with text difficult to differentiate from text written by humans, has led to some concerns about the potential for it to be used to create "fake news" or manipulate public opinion [37]. In short, it is really important to consider using this technology in a responsible and ethical way and to find out how people, as professionals, can work with it to improve their day-to-day work rather than abusing it or allowing it to be abused [37].

6 Ethical and Social Implications of AI in Digital Marketing

Today, the definition of AI is not just limited to the ability to learn and adapt but also raises questions about ethics, algorithmic bias, and regulation [8]. The evolution of AI promises significant transformations in various sectors, such as education, health, transportation, and entertainment, by improving efficiency and developing new opportunities. However, it is accepted that it also raises ethical debates about privacy, employment, and the autonomy of machines. Indeed, as AI becomes more integrated into society, there is a need to safeguard its impact on human and social life. AI systems can perpetuate or amplify existing social prejudices, an important ethical concern [37].

Besides, the dehumanization of customer brand relationships with AI can be considered an important issue. Businesses must prioritize customer satisfaction and strive to exceed expectations. In this sense, understanding the human element is crucial for building meaningful connections with customers, and human-to-human interaction is essential for building trust and rapport with customers, businesses need to empathize with customers and understand their perspectives. Moreover, it seems reasonable to affirm that a prominent need for regulatory frameworks exists to ensure transparency, accountability, and fairness in the use of AI in marketing.

The rapid integration of AI into digital marketing has raised pressing ethical concerns that demand critical attention from both practitioners and researchers. As AI systems become increasingly embedded in consumer-brand interactions, issues such as data privacy, algorithmic bias, lack of transparency, and the erosion of human touchpoints have emerged as major challenges. These concerns are not merely technical; they carry significant social, legal, and psychological implications. Table 1 summarizes the most salient ethical challenges associated with AI in marketing and outlines potential solutions that can guide the development of responsible and trust-centered AI strategies.

Table 1 Ethical challenges of artificial intelligence in digital marketing

Ethical challenge	Description	Proposed solution
Data privacy	Extensive collection and use of personal data without clear consent	Implement transparent data policies, ensure GDPR/CCPA compliance, and obtain explicit user consent
Algorithmic bias	AI systems may reinforce societal biases due to biased training data	Use diverse and representative datasets; conduct regular audits and bias testing
Lack of transparency	Users are unaware of how AI makes decisions or personalizes content	Improve explainability through user-facing disclosures and interpretable AI models
Dehumanization	Over-reliance on automation weakens authentic human-customer relationships	Maintain human oversight in customer interactions; combine AI with human empathy
Manipulative practices	Hyper-personalized strategies may exploit cognitive vulnerabilities	Establish ethical boundaries; design AI systems that respect user autonomy and choice

Source: Own elaboration

7 Conclusion

This chapter has examined the profound impact of AI on digital marketing, highlighting its transformative potential as well as the ethical and operational challenges it presents. AI has redefined the way businesses engage with consumers, enabling hyper-personalized experiences, automated campaigns, and data-driven decision-making. However, its rapid adoption also raises critical questions about privacy, transparency, and the evolving relationship between technology and human values.

The integration of AI into digital marketing presents an opportunity. From predictive analytics to AI-generated content, businesses can deliver personalized experiences at scale, optimize resource allocation, and foster deeper connections with customers. Automation tools like chatbots and programmatic advertising streamline operations, freeing marketers to focus on strategic creativity. The ethical implications of data usage, algorithmic bias, and the potential erosion of human touchpoints in customer interactions cannot be overlooked. As AI continues to evolve, striking a balance between innovation and accountability will be paramount.

The main findings and insights of this chapter emphasize the transformative potential of AI in digital marketing, along with the importance of balancing the adoption of AI technologies with respect for human values, customer trust, and social responsibility.

AI systems and technology should be guided by ethical practices such as respect for employees and customers, moral use of data and resources, responsible adoption of disruptive technologies, and creating a culture of responsibility [37].

The insights generated through this chapter carry substantial implications for both marketing practitioners and academic researchers. For professionals in the field, the findings underscore the urgent need to develop and implement transparent AI systems accompanied by clear, consumer-friendly data policies that foster trust while maintaining competitive advantage. Particularly emphasizes the importance of preserving meaningful human oversight in increasingly automated marketing processes, suggesting that the most effective implementations will likely emerge from synergistic human-AI collaboration rather than full automation. From a theoretical perspective, the study reveals critical gaps in our understanding of long-term consumer trust dynamics in AI-driven marketing ecosystems, pointing to the need for more robust conceptual frameworks that can guide ethical AI implementation. Academic researchers will find particularly fertile ground in investigating the emerging intersection between AI capabilities and sustainable marketing practices, as well as the evolving competencies required for marketing professionals to thrive in AI-augmented work environments. These implications collectively suggest that both practitioners and scholars must adopt more nuanced, critical perspectives on AI adoption that balance innovation with responsibility.

AI-based consumer engagement, for instance, according to [3], has several managerial implications as managers should: (i) explore AI implementation opportunities to enhance service accuracy and quality, reducing issues and failures; (ii) conduct research to identify areas where consumers would most value engaging with AI technologies while mitigating potential risks; (iii) implement AI technologies that can reduce consumers' required effort, as this is expected to boost their service assessments; (iv) given AI's rapid evolution and growing impact on consumer engagement, managers should closely monitor this area's developments to leverage its strategic value.

Future research can move towards different paths, such as the role of AI in sustainable marketing, and the impact of AI on global marketing practices. Researchers should monitor the evolving discourse on AI to evaluate its potential influence on customer and other stakeholders' engagement, trust, satisfaction, and loyalty. The phenomenon of AI washing (when companies falsely claim to use AI) justifies further investigation due to its possible effects on customer trust and engagement. Additionally, according to [3], more research is needed to explore how specific AI attributes impact customer interactions, as well as how AI-driven automated social presence influences brand engagement.

The future of AI in digital marketing emphasizes the need for a balanced approach that maximizes benefits while mitigating risks.

The integration of AI into organizations' strategic plans is no longer a vision of the future—it's a present reality, and companies that don't recognize it are inevitably falling behind. However, one crucial thing is clear: technology will always be our ally, but organizations are made up of people. The most sought-after skills in companies are undoubtedly collaboration, empathy, and communication. In an increasingly fast-paced and volatile age, adaptability, change, and flexibility are becoming essential skills. What will make the difference in the market will be the ability to develop AI skills that position us in an ever-changing society where flexibility, diversity, and inclusion are increasingly fundamental.

References

1. Kshetri, N. (2023). Generative artificial intelligence in marketing. *IT Professional, 25*(5), 71–75. https://doi.org/10.1109/MITP.2023.3314325
2. Kumari, M., Sinha, R., Chakrabarty, P., Hasnain, M. G., Qamar, N., & Gupta, S. (2024). Exploring machine learning's impact on digital marketing. In *2024 15th international conference on computing communication and networking technologies, ICCCNT 2024*. https://doi.org/10.1109/ICCCNT61001.2024.10724818
3. Hollebeek, L. D., Menidjel, C., Sarstedt, M., Jansson, J., & Urbonavicius, S. (2024). Engaging consumers through artificially intelligent technologies: Systematic review, conceptual model, and further research. *Psychology & Marketing, 41*(4), 880–898. https://doi.org/10.1002/MAR.21957
4. Kshetri, N., Dwivedi, Y. K., Davenport, T. H., & Panteli, N. (2024). Generative artificial intelligence in marketing: Applications, opportunities, challenges, and research agenda. *International Journal of Information Management, 75*, 102716. https://doi.org/10.1016/j.ijinfomgt.2023.102716
5. Chaffey, D., & Smith, P. (2022). *Digital marketing excellence*. Routledge. https://doi.org/10.4324/9781003009498
6. Di Gregorio, A., Maggioni, I., Mauri, C., & Mazzucchelli, A. (2019). Employability skills for future marketing professionals. *European Management Journal, 37*(3), 251–258. https://doi.org/10.1016/j.emj.2019.03.004
7. J. McCarthy. *What is artificial intelligence?*
8. Haenlein, M., & Kaplan, A. (2019). A brief history of artificial intelligence: On the past, present, and future of artificial intelligence. *California Management Review, 61*(4), 5–14. https://doi.org/10.1177/0008125619864925
9. Mehta, P., Jebarajakirthy, C., Maseeh, H. I., Anubha, A., Saha, R., & Dhanda, K. (2022). Artificial intelligence in marketing: A meta-analytic review. *Psychology & Marketing, 39*(11), 2013–2038. https://doi.org/10.1002/MAR.21716
10. Huang, M.-H., & Rust, R. T. (2022). A framework for collaborative artificial intelligence in marketing. *Journal of Retailing, 98*(2), 209–223. https://doi.org/10.1016/j.jretai.2021.03.001
11. E. Rodrigue. The top benefits of AI for marketers [state of AI data. September 19. *HubSpot*. Retrieved September 19, from https://blog.hubspot.com/marketing/benefits-of-ai
12. Davenport, T., Guha, A., Grewal, D., & Bressgott, T. (2020). How artificial intelligence will change the future of marketing. *Journal of the Academy of Marketing Science, 48*(1), 24–42. https://doi.org/10.1007/s11747-019-00696-0
13. Durai, S., Manoharan, G., Priya, T. S., Jayanthi, R., Razak, A., & Ashtikar, S. P. (2024). Quantifying the impacts of artificial intelligence implementations in marketing. In *Smart and sustainable interactive marketing* (pp. 120–144). https://doi.org/10.4018/979-8-3693-1339-8.CH008
14. Huang, M.-H., & Rust, R. T. (2018). Artificial intelligence in service. *Journal of Service Research, 21*(2), 155–172. https://doi.org/10.1177/1094670517752459
15. Kumar, V., Ramachandran, D., & Kumar, B. (2021). Influence of new-age technologies on marketing: A research agenda. *Journal of Business Research, 125*, 864–877. https://doi.org/10.1016/j.jbusres.2020.01.007
16. Oliveira, C. M. (2020). *Marketing pós-digital: O marketing à medida do ser humano*. Actual Editora.

17. Puntoni, S., Reczek, R. W., Giesler, M., & Botti, S. (2021). Consumers and artificial intelligence: An experiential perspective. *Journal of Marketing, 85*(1), 131–151. https://doi.org/10.1177/0022242920953847
18. Sowa, K., Przegalinska, A., & Ciechanowski, L. (2021). Cobots in knowledge work. *Journal of Business Research, 125*, 135–142. https://doi.org/10.1016/j.jbusres.2020.11.038
19. Potwora, M., Vdovichena, O., Semchuk, D., Lipych, L., & Saienko, V. (2024). The use of artificial intelligence in marketing strategies: Automation, personalization and forecasting. *Journal of Management World, 2024*(2), 41–49. https://doi.org/10.53935/jomw.v2024i2.275
20. Rajput, A., Jaiwani, M., Singh, S., Gandhi, A., & Gopalkrishnan, S. (2024). Influencer marketing redefined: The era of AI-powered personalization. In *Proceedings - 2024 International of Seminar on application for Technology of Information and Communication: Smart and emerging Technology for a Better Life, iSemantic 2024* (pp. 482–488). https://doi.org/10.1109/ISEMANTIC63362.2024.10762409
21. Singh, B., & Kaunert, C. (2024). Future of digital marketing: Hyper-personalized customer dynamic experience with AI-based predictive models. In *Revolutionizing the AI-digital landscape: A guide to sustainable emerging Technologies for Marketing Professionals* (pp. 189–208). https://doi.org/10.4324/9781032688305-14
22. Gao, Y., & Liu, H. (2023a). Artificial intelligence-enabled personalization in interactive marketing: A customer journey perspective. *Journal of Research in Interactive Marketing, 17*(5), 663–680. https://doi.org/10.1108/JRIM-01-2022-0023
23. Gao, Y., & Liu, H. (2023b). Artificial intelligence-enabled personalization in interactive marketing: A customer journey perspective. *Journal of Research in Interactive Marketing, 17*(5), 663–680. https://doi.org/10.1108/JRIM-01-2022-0023/FULL/XML
24. Raghav, Y. Y., Tipu, R. K., Bhakhar, R., Gupta, T., & Sharma, K. (2023). The future of digital marketing: Leveraging artificial intelligence for competitive strategies and tactics. In *The use of artificial intelligence in digital marketing*. https://doi.org/10.4018/978-1-6684-9324-3.ch011
25. Vij, A., Vij, M., Farouk, M., & Kumar, P. (2024). Evaluating the effectiveness of AI-integrated digital marketing on consumer behavior, brand perception, and sales performance. In *2nd international conference on cyber resilience, ICCR 2024*. https://doi.org/10.1109/ICCR61006.2024.10533049
26. Patil, B., Ashok, P., & Chirputkar, A. (2024). Artificial intelligence powered paradigm shift: Revolutionizing digital marketing. In *4th international conference on innovative practices in technology and management 2024, ICIPTM 2024*. https://doi.org/10.1109/ICIPTM59628.2024.10563450
27. Cunha, M. N., Pereira, M., Cardoso, A., Figueiredo, J., & Oliveira, I. (2024). Revolutionizing luxury: The role of ai and machine learning in enhancing marketing strategies within the tourism and hospitality luxury sectors. *Geojournal of Tourism and Geosites, 55*(3), 1345–1353. https://doi.org/10.30892/GTG.55335-1307
28. Gujar, P., Paliwal, G., & Panyam, S. (2024). Revolutionizing in-house digital marketing with end-to-end marketing automation powered by AI and SaaS. In *2024 IEEE Colombian conference on communications and computing, COLCOM 2024- proceedings*. https://doi.org/10.1109/COLCOM62950.2024.10720305
29. Hannig, U. (2023). Automation of repetitive processes in marketing and sales. In *Management for Professionals* (Vol. Part F286, pp. 7–18). https://doi.org/10.1007/978-3-031-20040-3_2
30. Sinha, B., & Patil, K. (2023). Marketing automation for digital transformation: A review. *AIP Conference Proceedings, 2914*(1). https://doi.org/10.1063/5.0175877
31. Gummadi, V., Ramadevi, N., Udayaraju, P., Ravulu, C., Seelam, D. R., & Swamy, S. V. (2024). A deep learning-based optimization model for advertisement campaign. In *Proceedings of the 5th international conference on smart electronics and communication, ICOSEC 2024* (pp. 1783–1790). https://doi.org/10.1109/ICOSEC61587.2024.10722166
32. Zaki, H. O., & Fernandez, D. (2023). Automation in digital marketing. In *Digital natives as a disruptive force in Asian businesses and societies* (pp. 241–251). https://doi.org/10.4018/978-1-6684-6782-4.CH014

33. Kumar, P., Tomar, P. K., Bharti, S., Naredla, S., Ibrahim, R. K., & Bader Alazzam, M. (2023). Maximizing the potential of artificial intelligence in digital marketing. In *2023 3rd international conference on advance computing and innovative Technologies in Engineering, ICACITE 2023* (pp. 2736–2741). https://doi.org/10.1109/ICACITE57410.2023.10183113
34. Juddoo, S. (2023). Artificial intelligence in the context of digital marketing. In *Confronting security and privacy challenges in digital marketing* (pp. 119–139). https://doi.org/10.4018/978-1-6684-8958-1.CH006
35. Kesavaraj, G., & Shabanabi, P. (2024). Achieving better customer experience by harmonizing automation and human touch. In *Balancing automation and human interaction in modern marketing* (pp. 19–38). https://doi.org/10.4018/979-8-3693-2276-5.CH002
36. Nandkumar, K. (2017). Tràilblazing innovation in process analytics. *Chemical Engineering World, 52*(1), 36–38.
37. Canavarro, A., & Moreira, B. (2024). *Metaverso no marketing, nas marcas e nos negócios.* Lidel - Edições Técnicas, Lda.
38. Carvalho, I., Iglesias, M., & Ivanov, S. (2025). Education 2070: A conceptual framework. In *World futures* (pp. 1–22). https://doi.org/10.1080/02604027.2025.2482504

Bridging Education and Entrepreneurial Digital Transformation: Cultivating Digital Literacy and Innovation

Yohn Jairo Parra Bautista

1 Introduction

The use of digital tools, especially artificial intelligence (AI), has greatly transformed both education and entrepreneurship. AI is particularly powerful because it can enhance learning outcomes, personalize educational experiences, and better prepare students for the demands of today's world. By integrating AI into classrooms, educators can create customized learning paths that adapt to each student's needs, making education more engaging and effective.

Beyond education, AI is also reshaping business and entrepreneurship by introducing new strategies and innovative approaches. It allows businesses to analyze data more efficiently, automate processes, and develop smarter decision-making tools. These advancements are changing how people learn, teach, and do business, making education more dynamic and entrepreneurship more adaptable to the fast-evolving digital landscape.

AI technologies have shown great potential in making learning more engaging and efficient by using personalized learning systems and automated assessment tools. These tools help tailor lessons to each student's needs, ensuring they receive support where they need it most.

For instance, a study in Vietnam found that AI can significantly improve student learning outcomes. However, it also pointed out challenges such as unequal access to technology (the digital divide) and the need for greater investment in teacher training and infrastructure to fully leverage AI's benefits [1].

Y. J. P. Bautista (✉)
CIS Department, Florida A&M University, Tallahassee, FL, USA
e-mail: yohn.parrabautista@famu.edu

C. Rompante Cunha et al. (eds.), *Transformative Cloud Computing, IoT and Extended Reality*, Signals and Communication Technology,
https://doi.org/10.1007/978-3-032-03554-7_9

Similarly, Zhao highlights how AI-driven educational tools can make learning more interactive and effective, helping students develop the skills they need to succeed in the future [2].

Building on this, Qin and Ao emphasize that AI is a key driver of change in education, not only by enabling personalized teaching approaches but also by helping schools and universities make better decisions through data-driven insights [3]. These advancements suggest that AI will continue to reshape education, making it more adaptive and accessible for students worldwide.

The shift toward digital education has brought major changes to teaching methods and educational approaches worldwide. As Zhou et al. point out, integrating digital technologies into education has led to positive results globally, indicating a move toward higher quality learning experiences [4].

Expanding on this idea, Zhang et al. stress the importance of AI literacy in K-12 education, arguing that students need to develop a strong understanding of AI to be prepared for future technological advancements [5]. By incorporating AI into school curricula, students can gain essential digital skills that will help them navigate and succeed in an increasingly technology-driven world.

Additionally, AI is enhancing collaborative and immersive learning experiences. The use of virtual reality (VR) and AI-powered tools in education is helping students develop critical thinking, problem-solving, and innovation skills. Research shows that these technologies make learning more interactive and engaging, fostering comprehensive skill development among students [6]. As digital education continues to evolve, AI and immersive tools will likely play an even bigger role in shaping the future of learning.

AI is not only transforming education but is also reshaping entrepreneurship and business strategies. By integrating AI technologies, businesses can streamline operations, improve customer interactions, and make smarter, data-driven decisions.

One clear example of AI's impact is in online learning, where AI-powered tools have changed how students and instructors interact. These advancements have led to more personalized learning experiences and better educational outcomes [7].

However, the use of AI also comes with ethical challenges, particularly in both education and business. G lushkova and Malinova emphasize concerns such as data privacy risks and the potential for AI to reinforce existing inequalities [8]. This highlights the importance of ensuring that AI-driven educational tools are accessible to all students, preventing disparities in learning opportunities.

To fully harness AI's benefits while minimizing risks, it is essential to develop fair, ethical AI policies that prioritize privacy, transparency, and equal access. This will allow both education and business sectors to take advantage of AI innovations while ensuring they are inclusive and beneficial for all.

The COVID-19 pandemic has accelerated the adoption of AI technologies in education, prompting educators to use these tools more frequently to enhance learning in online and blended environments [9]. AI has helped create personalized learning experiences, automate assessments, and provide real-time feedback, making education more flexible and interactive.

However, simply introducing AI into classrooms is not enough—educators need the right digital skills and knowledge to effectively use these technologies. Without proper training, AI tools may not be used to their full potential, limiting their impact on student learning.

To address this, ongoing professional development is essential. Educators must receive continuous training to stay up to date with AI advancements and learn how to integrate these tools effectively into their teaching methods. Additionally, collaborations between educators and technology developers can ensure that AI tools are designed with practical classroom needs in mind [10]. By fostering these partnerships and investing in teacher training, schools can overcome challenges in AI adoption and create a more inclusive and effective digital learning environment.

In summary, digital technologies, especially AI, are playing a transformative role in both education and entrepreneurship. These innovations are not only making learning more engaging and effective but are also equipping students with essential skills to navigate the complexities of today's workforce. AI-powered tools help personalize education, streamline learning processes, and prepare students for careers in a rapidly evolving digital world.

However, for AI to truly benefit education, several challenges must be addressed. Ethical concerns, such as data privacy and fairness, need careful consideration to ensure AI is used responsibly. Additionally, equal access to AI-powered tools is crucial to prevent widening educational gaps between students with different levels of technological access. Finally, educators must receive proper training to integrate AI effectively into their teaching strategies.

As education continues to evolve in the digital age, AI will remain a key driver of innovation. By tackling these challenges and making AI more accessible and ethical, we can harness its full potential to enhance learning experiences, foster creativity, and prepare students for the future.

2 Digital Transformation in Education and Entrepreneurship: A Review of Relevant Research

Digital transformation is changing the way we learn and do business, making it an important topic for research. In education, digital tools like online learning platforms and artificial intelligence are creating new ways to teach and engage students. In entrepreneurship, digital technologies are helping businesses grow faster and reach more customers. This section reviews existing research on digital transformation in both education and entrepreneurship, highlighting the most important findings and emerging trends that are shaping the future of these fields.

The evolution of digital technology is reshaping how educational institutions educate and engage with learners. A significant advancement in this context is the incorporation of digital entrepreneurship within higher education curricula. As highlighted in a systematic review by Aysi et al., the instruction of digital entrepreneurship not only motivates students to launch their own ventures but also equips them with skills to obtain funding for practical projects [11].

Similarly, Li et al. found that continuous learning and effective management of entrepreneurship education play a key role in shaping students' digital entrepreneurship behaviors [12]. This means that the way entrepreneurship is taught directly influences how students approach digital business opportunities.

In addition, research by Yordanova and Jose shows a clear link between entrepreneurship education and students' interest in technology-based startups [13]. These findings highlight a growing trend: educational systems are evolving to better equip students with the skills and mindset needed for success in the digital economy.

The evolution of digital transformation is revolutionizing the operations of businesses, fostering novel business models and inventive pathways to generate value. This transition holds particular significance for small and medium-sized enterprises (SMEs). As per Li et al., the digital transformation is reshaping the growth and success of SMEs by affecting the manner in which they deliver products and services.

[14].

Building on this, Bouwman et al. explain that when SMEs update their business models to include digital technologies, they see better performance and increased competitiveness [15]. This shows that adapting to digital change isn't just about using new tools—it's about rethinking how businesses create and deliver value.

A study by Judijanto et al. highlights key trends shaping modern entrepreneurship. They found that digital technologies are becoming an essential part of entrepreneurship education and that big data plays a crucial role in helping businesses make informed decisions [16]. These insights reveal that digital transformation is not just changing business practices but is also reshaping how future entrepreneurs are trained and how they approach opportunities in the digital economy.

Digital transformation is not only changing how businesses operate but also how they contribute to society and the environment. This new approach is known as digital sustainable entrepreneurship, which focuses on using digital tools to create positive social and environmental impact.

Xu et al. explain that to succeed in digital sustainable entrepreneurship, businesses need strong digital skills and a focus on innovation. By doing this, they can create value that goes beyond profit, benefiting communities and the planet [17].

Similarly, Elshaiekh et al. show that digital entrepreneurship can help solve social issues like unemployment, especially in developing countries such as Oman. They highlight how digital technologies can provide new job opportunities and empower disadvantaged groups to start their own businesses [18].

These findings demonstrate that digital transformation is not just about business growth but also about building a more inclusive and sustainable economy.

TheCOVID-19 pandemic drastically sped up the transition to digital entrepreneurship. According to Modgil et al., the crisis pushed businesses to innovate, integrating digital tools into traditional practices and leading to a surge in digital startups [19].

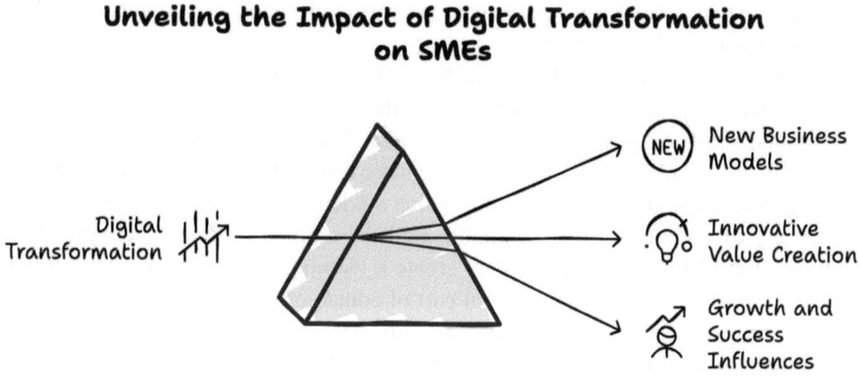

Fig. 1 The impact of digital transformation on SMEs, illustrating how technological advancements drive new business models, innovative value creation, and growth opportunities

This shift wasn't just limited to established businesses—social entrepreneurs also embraced digital technologies to improve their operations and reach more people. By using digital platforms, social enterprises were able to continue their work despite lockdowns and restrictions, making their efforts more efficient and impactful [19].

Overall, the pandemic acted as a catalyst for digital entrepreneurship, proving that businesses and organizations that adapt to digital transformation can thrive even in times of crisis (Fig. 1).

3 The Role of Digital Literacy and Innovation in a Technology-Driven World

In today's world, where technology evolves at an incredible pace, developing digital literacy and fostering innovation have become essential skills for both education and daily life. As we increasingly rely on digital tools for communication, learning, and problem-solving, individuals must be equipped with the knowledge and skills to navigate this digital landscape effectively.

Research highlights several key trends that underscore the importance of digital literacy in education. First, integrating digital literacy into learning environments helps students develop critical thinking, problem-solving, and technical skills necessary for academic and professional success. Second, technology plays a crucial role in enhancing research abilities by providing access to vast information sources, data analysis tools, and collaborative platforms. Lastly, digital tools have a significant impact on learning outcomes, offering interactive and personalized learning experiences that can improve student engagement and comprehension.

By understanding and embracing these trends, educators, and learners can harness the power of digital literacy and innovation to create more effective and inclusive educational experiences, preparing individuals for success in an increasingly technology-driven world.

A key development in modern education is the integration of digital literacy into thematic learning, particularly at the basic education level. This approach ensures that students not only use digital tools but also develop essential skills to engage with them effectively. Sari et al. emphasize three crucial factors for fostering digital literacy in schools: the frequency with which students use digital tools in learning, the variety of digital resources available, and the number of digital activities they participate in [20]. These aspects help create a learning environment where digital literacy becomes a natural and integral part of education.

This emphasis on digital literacy extends beyond basic education to higher education, where it plays a critical role in developing essential skills. Chen et al. highlight the importance of structured digital literacy frameworks in universities, showing how they promote critical thinking and problem-solving skills among students [21]. By incorporating digital tools into research and academic activities, students gain the ability to analyze and apply information more effectively.

Further supporting this perspective, Reyes and Morales present a systematic mapping of literature that demonstrates a growing link between research competencies and digital literacy. Their findings indicate that educational institutions are increasingly recognizing the need to equip students with digital literacy skills as a fundamental part of academic and professional development [22]. This shift reflects a broader understanding that digital literacy is not just an additional skill but a core component of modern education.

One of the most significant changes in modern education is the shift from traditional literacy methods, such as handwriting, to digital writing and learning tools.

This transformation reflects the growing importance of technology in classrooms, where digital tools are becoming essential for both teaching and learning.

Otsuka and Murai discuss how this global transition is reshaping education, with many schools prioritizing digital writing over handwriting as a primary method of communication and knowledge acquisition [23]. This shift is not just about convenience; it is part of a broader effort to integrate technology into education in a meaningful way.

A clear example of this transition is Japan's GIGA School Program, which aims to provide every student with a personal computer. By doing so, the program creates an environment where digital literacy can flourish, ensuring that students have access to the tools they need to succeed in a technology-driven world [24].

The impact of this shift goes beyond the classroom. As digital tools become more prevalent in education, students develop essential digital skills that prepare them for future careers in a workforce that increasingly relies on technology. This change highlights the growing need for digital literacy as a fundamental skill, just as important as reading and writing in the past.

Beyond the classroom, digital literacy is becoming increasingly important in various professional fields. As technology continues to shape the way we work, many industries now require employees to develop strong digital skills to stay competitive and effective in their roles.

One example of this shift can be seen in Thailand, where secretarial students are being trained to enhance their digital literacy. Hoaihongthong et al. highlight how these students are learning essential digital skills to support online learning initiatives and effectively communicate with students and colleagues through digital platforms [25]. This training not only prepares them for their future careers but also reflects the growing demand for digital competencies across multiple sectors.

This trend underscores a broader understanding that digital literacy is no longer just an academic skill—it is a professional necessity. As businesses and organizations continue to embrace digital tools, employees across different industries must adapt to new ways of working, collaborating, and problem-solving in a technology-driven world.

As technology continues to transform the way people manage money, digital literacy is becoming an essential skill in financial education. More than ever, individuals need to understand how to navigate digital banking, online transactions, and financial management tools to make informed decisions.

Recent research highlights the importance of digital financial literacy, particularly for students and small and medium enterprises (SMEs). Studies by Zaimovic et al. and Kusumawardhani et al. show that having strong digital financial skills enables individuals to engage effectively with online banking, digital payment systems, and financial planning tools [26, 27]. Without these skills, people may struggle to make secure transactions, manage their finances efficiently, or take advantage of digital financial services.

This growing focus on financial literacy in the digital age emphasizes the need for educational institutions to incorporate financial education into their curricula. By doing so, they can help students and professionals develop the knowledge necessary to navigate an increasingly complex financial landscape. As financial systems become more digitalized, equipping individuals with these skills will be crucial for both personal financial well-being and the success of businesses in the modern economy.

In our fast-paced digital age, cultivating robust digital literacy skills has become increasingly crucial. This entails more than merely instructing individuals on technology usage; it requires incorporating digital skills into educational curriculums, workplace training, and health literacy programs. By engaging with this integration, individuals can boost their confidence and skills in navigating the digital landscape, whether in educational, professional, or personal contexts.

As technology continues to advance, the need for digital competencies will only grow. From students learning how to research effectively online to professionals using digital tools in their jobs and individuals accessing health information through digital platforms, the impact of digital literacy spans multiple areas of life. Because of this, ongoing research and development in digital education and training will be essential to ensure that people have the skills needed to succeed in a world that increasingly relies on technology.

4 Empowering Students for the Future: Integrating Digital Tools and Entrepreneurial Training in Education

Amidst the seismic shifts of today's dynamic job market, educational institutions throughout the United States are fervently embracing digital innovations and weaving entrepreneurial training into their curricula. This forward-thinking approach not only empowers students to hone indispensable skills but also fuels their engagement in learning, meticulously preparing them to meet and exceed the exacting demands of the contemporary workforce.

By integrating technology into classrooms and providing hands-on entrepre neurial experiences, schools and universities equip students with problem-solving abilities, adaptability, and digital proficiency—skills that are highly valued in today's economy. Various case studies highlight how these initiatives have been successfully implemented, showcasing best practices and positive outcomes. These examples demonstrate how digital tools and entrepreneurial education can work together to create a dynamic learning environment that fosters creativity, innovation, and career readiness.

A compelling example of how digital literacy is transforming education comes from Quraishi et al., who conducted an in-depth case study on how higher education institutions integrate digital tools into their curricula. Their research highlights that when universities effectively incorporate digital literacy initiatives, students benefit in multiple ways—not only do they experience a more engaging learning environment, but their overall academic performance also improves.

Beyond simply using digital tools, the study emphasizes the importance of understanding how students and educators interact with these technologies and the challenges institutions face in implementing digital literacy programs. These insights provide valuable guidance for educators and policymakers seeking to design more effective digital education strategies [27].

This shift toward digital learning is not limited to higher education. Msambwa et al. explored the role of Information and Communication Technology (ICT) in secondary education and found that schools that successfully integrate digital tools see significant improvements in student engagement and learning outcomes. Their review supports the idea that well-planned digital literacy programs can positively impact students at various educational levels, helping them develop essential skills for the modern world [27].

Together, these studies illustrate a growing recognition of digital literacy as a key factor in academic success, reinforcing the need for continuous investment in digital education and technology-driven learning approaches.

Entrepreneurial education goes beyond simply teaching business concepts—it plays a crucial role in shaping students' confidence and attitudes toward starting their own ventures. Duong's research highlights that students' self-efficacy (belief in their ability to succeed) and positive attitudes are key factors influencing their entrepreneurial ambitions. The study emphasizes that educational institutions should do more than just deliver theoretical knowledge; they must actively encourage students to engage in entrepreneurial courses, hands-on training, and real-world experiences to build confidence and practical skills [28].

This idea is further supported by Valencia-Ar′ıas et al., whose research found that entrepreneurial training programs significantly enhance students' psychological traits, such as resilience, creativity, and risk-taking. As a result, these programs increase the likelihood that students will pursue entrepreneurship as a career path [29].

These findings highlight the need for educational institutions to create environments that foster entrepreneurial thinking, not just through lectures but through interactive experiences like business simulations, startup incubators, and mentorship programs. By providing opportunities for students to experiment, fail, and learn in a supportive setting, schools and universities can better prepare them to navigate the challenges of the business world with confidence.

The use of digital tools in entrepreneurial education is transforming the way students develop business skills. Many universities are adopting experiential learning approaches, which allow students to engage in real-world entrepreneurial scenarios through business simulations, virtual startup experiences, and interactive training platforms. These digital tools provide a hands-on learning environment, enabling students to experiment, make decisions, and understand the consequences—just as they would in a real business setting.

Research by An and Xu highlights the need to further improve entrepreneurship curricula by integrating more practical learning opportunities. They emphasize the importance of developing practice-based training programs that give students a deeper understanding of entrepreneurship beyond just theoretical knowledge [30].

By incorporating digital tools and realistic business challenges into the learning process, universities are not only equipping students with essential entrepreneurial skills but also preparing them for the unpredictable nature of the business world. This approach ensures that students gain both the confidence and the practical expertise needed to navigate the complexities of starting and managing a business in today's digital economy.

Moreover, the role of perceived university support in shaping students' entrepreneurial intentions is highlighted by Saeed et al., who propose a framework that includes educational, concept development, and business development support [31]. This framework suggests that a supportive university environment is crucial for fostering entrepreneurial self-efficacy, which in turn influences students' intentions to start their own businesses. This is particularly relevant in the context of US institutions that are increasingly recognizing the importance of creating supportive ecosystems for aspiring entrepreneurs.

Universities play a crucial role in shaping students' confidence and motivation to pursue entrepreneurship. A supportive university environment can make a significant difference in whether students feel prepared to start their own businesses. Saeed et al. propose a framework that highlights three key types of support that universities can provide:

- Support—Offering entrepreneurship courses, workshops, and mentorship programs that equip students with the knowledge and skills needed to succeed.
- Concept Development Support—Helping students refine and test their business ideas through incubators, pitch competitions, and collaboration with industry experts.
- Development Support—Providing resources such as funding opportunities, networking events, and access to startup accelerators that help turn ideas into viable businesses [31].

This framework suggests that when universities actively foster an entrepreneurial mindset, they help build students' self-efficacy—their belief in their ability to succeed as entrepreneurs. A strong sense of confidence, combined with practical experience and institutional support, significantly increases students' likelihood of launching their own ventures.

This is particularly relevant in the United States, where more institutions are recognizing the importance of creating entrepreneurial ecosystems. Many universities are now investing in innovation hubs, startup incubators, and industry partnerships to provide students with hands-on experiences that bridge the gap between academic learning and real-world business challenges. By fostering such environments, universities empower the next generation of entrepreneurs to take bold steps toward business creation and innovation.

The successful integration of digital tools and entrepreneurial training in US educational institutions involves a comprehensive approach that goes beyond traditional teaching methods. Schools and universities are combining digital literacy, hands-on learning experiences, and strong institutional support to create a dynamic learning environment that equips students with real-world entrepreneurial skills.

Case studies from various institutions highlight how these strategies boost student engagement, enhance skill development, and build confidence in aspiring entrepreneurs. By incorporating business simulations, startup incubators, mentorship programs, and interactive digital platforms, students gain valuable experience that prepares them for the challenges of launching and managing their own businesses.

Ultimately, this approach ensures that students are not only learning entrepreneurship in theory but also practicing it in real-world scenarios. As more institutions adopt these innovative methods, they are shaping a new generation of entrepreneurs who are equipped with the digital skills, problem-solving abilities, and business acumen needed to succeed in today's rapidly evolving economy (Fig. 2).

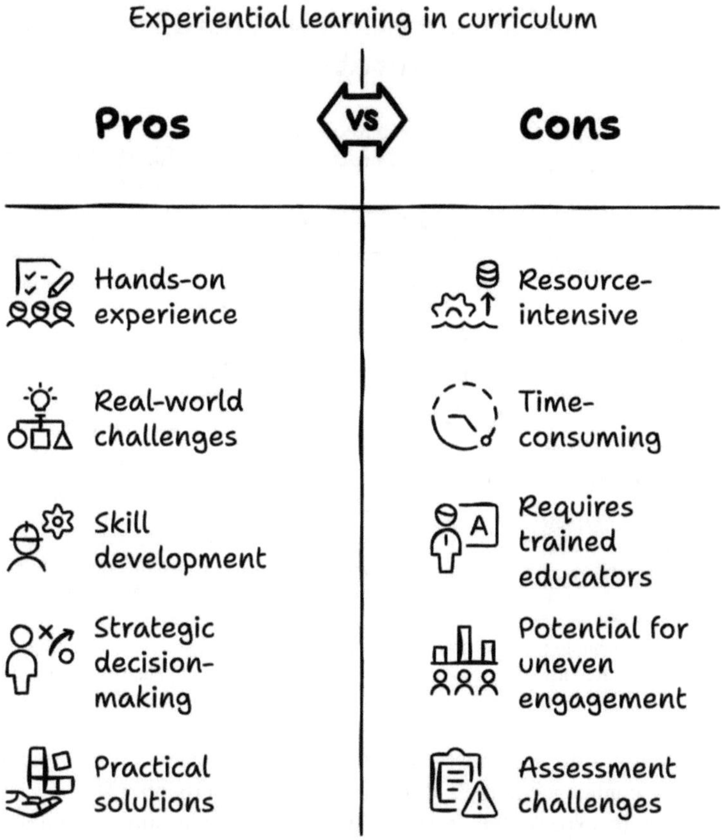

Fig. 2 Pros and cons of integrating experiential learning in the curriculum, highlighting benefits like hands-on experience and challenges such as resource intensity and assessment difficulties

5 Cultivating an Entrepreneurial Mindset in Higher Education: Strategies for Innovation and Resilience

One of the most effective ways to cultivate an entrepreneurial mindset among students is by incorporating experiential learning into the curriculum. Rather than relying solely on theoretical instruction, this approach actively engages students in hands-on projects that mirror real-world challenges. Through these experiences, students learn to identify problems, formulate hypotheses, make strategic decisions, and develop practical solutions—all essential skills for entrepreneurship. Research suggests that when students participate in interactive, real-world experiences, they significantly enhance their creativity, problem-solving abilities, and willingness to take risks—key attributes of successful entrepreneurs. By working on authentic business cases, launching small ventures, or collaborating on innovative projects,

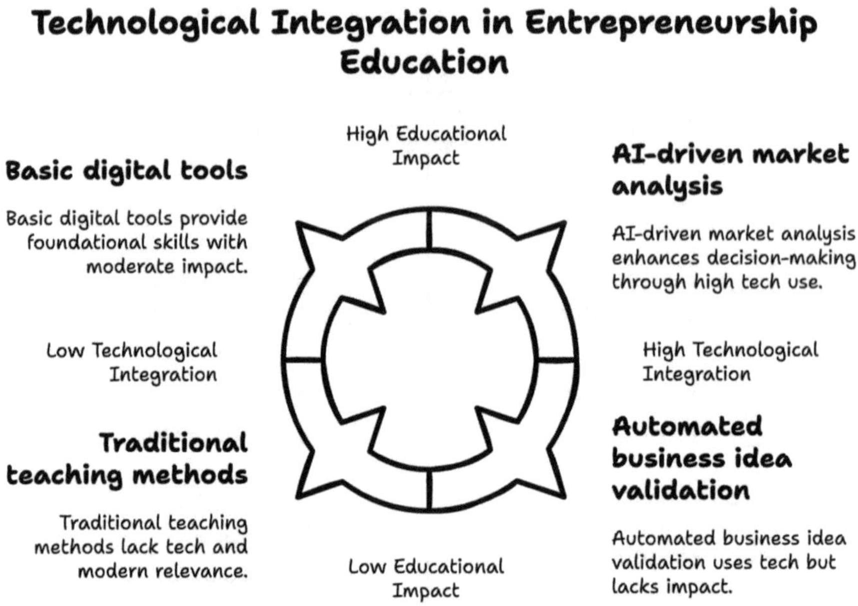

Fig. 3 Levels of technological integration in entrepreneurship education, comparing traditional methods, basic digital tools, AI-driven market analysis, and automated business idea validation

students develop a deeper understanding of the complexities of entrepreneurship while gaining confidence in their ability to navigate uncertainty [31].

In addition to experiential learning, many higher education institutions have established incubation and entrepreneurship programs to further support students (Fig. 3) in developing their entrepreneurial mindset. These programs offer mentorship, funding opportunities, and collaborative workspaces, creating an environment that fosters innovation, experimentation, and teamwork. Students involved in these programs benefit from both practical experience and a supportive network, which increases their motivation and readiness to pursue entrepreneurial endeavors [32]. By combining experiential learning with structured entrepreneurship programs, universities can empower students with the skills, mindset, and confidence needed to thrive in today's dynamic economic landscape.

A highly effective approach to developing an entrepreneurial mind set in students is the implementation of student-centered learning environments. This method shifts the traditional roles of both students and instructors, transforming the classroom into an interactive and engaging space where students take an active role in their own learning. Instead of simply receiving information through lectures, students participate in activities that encourage critical thinking, problem-solving, and creative exploration.

This pedagogical shift has been shown to significantly increase student motivation and engagement, as it creates a dynamic and stimulating atmosphere that fosters innovation. When students feel empowered to explore ideas, test solutions, and learn from real-world challenges, they become more confident in applying entrepreneurial principles. Research indicates that enhancing teaching methods and redesigning courses to be more interactive can inspire students to develop innovative thinking and problem-solving skills, which are essential for entrepreneurship. In addition to redesigning the learning environment, educators can further support entrepreneurial development by integrating diverse teaching methodologies. Approaches such as project-based learning and collaborative group work allow students to work on real or simulated business challenges, encouraging them to think creatively, collaborate effectively, and develop a deeper understanding of entrepreneurial concepts. These methods not only improve students' comprehension of course material but also prepare them for real-world entrepreneurial endeavors by fostering skills such as teamwork, adaptability, and strategic decision-making. By adopting student-centered learning approaches, educators can create an environment where students are actively engaged, motivated, and inspired to pursue innovative ideas, equipping them with the skills and mindset needed to succeed in entrepreneurship.

Beyond specific teaching strategies, it is crucial for educational institutions to cultivate a culture of innovation—one that encourages creativity, experimentation, and continuous learning. A key aspect of this culture is changing the perception of failure, shifting it from something to be avoided to a valuable learning experience. In entrepreneurship, failure is often an inevitable part of the journey, and students who learn to analyze their mistakes and adapt accordingly develop resilience, a critical trait for success in dynamic business environments [33].

To nurture this mindset, educators can integrate reflective practices into the curriculum, encouraging students to critically assess their experiences, challenges, and decision-making processes. Reflection helps students identify what worked, what didn't, and how they can improve in the future. By embedding these practices—such as structured self-assessments, peer feedback sessions, and case study discussions—educational institutions create a supportive learning environment where students feel empowered to take risks, experiment with new ideas, and build the confidence needed for entrepreneurship [34].

By fostering an educational culture that embraces innovation and resilience, institutions equip students with the mindset and skills necessary to navigate the uncertainties of the entrepreneurial world successfully.

6 Conclusion

This chapter has explored the transformative role of digital technologies, particularly artificial intelligence (AI), in reshaping both education and entrepreneurship. By analyzing key trends, challenges, and opportunities, we have highlighted how AI-driven tools and digital literacy initiatives are equipping students with critical thinking, problem-solving, and innovation skills essential for the modern economy.

A significant takeaway is the integration of digital tools into entrepreneurial education, which not only enhances learning experiences but also prepares students to navigate the complexities of a technology-driven workforce. Experiential learning approaches, including project-based learning and startup incubators, further reinforce entrepreneurial mindsets by fostering resilience, creativity, and adaptability.

Despite these advancements, the chapter also underscores critical challenges, including ethical considerations, access to technology, and the need for ongoing educator training. Addressing these issues is essential to ensure that AI and digital transformation remain inclusive and equitable across different educational and economic backgrounds.

Ultimately, the successful integration of AI, digital literacy, and entrepreneurship education will be a defining factor in preparing the next generation for the rapidly evolving digital landscape. By continuing to innovate and adapt, educational institutions can create dynamic learning environments that empower students to become future-ready leaders, entrepreneurs, and innovators.

References

1. Trang, N. T. Q., & Th, P. T. K. (2024). The role of AI in improving student learning outcomes: Evidence in Vietnam. *International Journal of Multidisciplinary Research and Analysis, 7*(6). https://doi.org/10.47191/ijmra/v7-i06-48
2. Zhao, T. (2023). *AI in educational technology*. Preprints. https://doi.org/10.20944/preprints202311.0106.v1
3. Qin, Q., & Ao, L. (2023). A new world of open space—Reflections and perspectives of Chinese scholars on the digital transformation of education and future education research. *Medicon Engineering Themes*. https://doi.org/10.55162/mcet.05.153
4. Zhou, L., Wei-jie, M., Wu, S., & Cheng, X. (2023). Development of digital education in the age of digital transformation: Citing China's practice in smart education as a case study. *Science Insights Education Frontiers, 14*(2), 2077–2092. https://doi.org/10.15354/sief.23.or095
5. Zhang, H., Lee, I., Ali, S., DiPaola, D., Cheng, Y., & Breazeal, C. (2022). Integrating ethics and career futures with technical learning to promote AI literacy for middle school students: An exploratory study. *International Journal of Artificial Intelligence in Education, 33*(2), 290–324. https://doi.org/10.1007/s40593-022-00293-3
6. Pan, S. (2024). Research on teaching strategies of immersive experiential teaching for collaborative learning in elementary and middle schools based on AI and VR. *Advances in Educational Technology and Psychology, 8*(5). https://doi.org/10.23977/aetp.2024.080503
7. Rakya, Z. H. (2023). Exploring the impact of artificial intelligence (AI) on learner-instructor interaction in online learning (literature review). *International Journal of Emerging Multidisciplinaries Computer Science & Artificial Intelligence, 2*(1), 1–4. https://doi.org/10.54938/ijemdcsai.2023.02.1.236
8. Glushkova, T., & Malinova, A. (2023). Advantages, problems and challenges in the application of AI technologies in school education. In *E-learning & artificial intelligence* (pp. 45–56). https://doi.org/10.34916/el.2023.15.04
9. Ng, D. T. K., Leung, J. K. L., Su, J., Ng, C. W., & Chu, S. K. W. (2023). Teachers' AI digital competencies and twenty-first century skills in the post-pandemic world. *Educational Technology Research and Development, 71*(1), 137–161. https://doi.org/10.1007/s11423-023-10203-6

10. Familoni, B. T., & Onyebuchi, N. C. (2024). Advancements and challenges in AI integration for technical literacy: A systematic review. *Engineering Science & Technology Journal, 5*(4), 1415–1430. https://doi.org/10.51594/estj.v5i4.1042
11. Judijanto, L., Mahmuddin, M., Fanggidae, H., & Erwin, E. (2024). Bibliometric analysis of the growth and challenges of digital entrepreneurship. *WSSHS, 2*(09), 1413–1428. https://doi.org/10.58812/wsshs.v2i09.1266
12. Xu, G., Hou, G., & Zhang, J. (2022). Digital sustainable entrepreneurship: A digital capability perspective through digital innovation orientation for social and environmental value creation. *Sustainability, 14*(18), 11222. https://doi.org/10.3390/su141811222
13. Elshaiekh, N. E. M., Al-Hijji, K., Shehata, A., & Alrashdi, S. M. A. (2023). An empirical analysis of factors motivating unemployed individuals to engage in digital entrepreneurship in Oman: Focus on technological infrastructure. *Sustainability, 15*(17), 12953. https://doi.org/10.3390/su151712953
14. Modgil, S., Dwivedi, Y., Rana, N., Gupta, S., & Kamble, S. (2022). Has COVID-19 accelerated opportunities for digital entrepreneurship? An Indian perspective. *Technological Forecasting and Social Change, 175*, 121415. https://doi.org/10.1016/j.techfore.2021.121415
15. Kalendzhyan, S. (2023). Digitalisation in the development of social entrepreneurship. *E3S Web of Conferences, 403*, 08030. https://doi.org/10.1051/e3sconf/202340308030
16. Sari, D. I. P., Prayitno, H. J., Rahmawati, L. E., & Prastiwi, Y. (2022). Culture of digital literacy in thematic learning at the basic education level. *Jurnal Ilmiah Sekolah Dasar, 6*(3), 467–475. https://doi.org/10.23887/jisd.v6i3.46334
17. Chen, C., Wang, N., Tu, Y., & Lin, H. J. (2021). Research trends from a decade (2011–2020) for informationliteracyinhighereducation:Contentandbibliometricmappinganalysis. *Proceedings of the Association for Information Science and Technology, 58*(1), 48–59. https://doi.org/10.1002/pra2.435
18. Reyes, C. E. G., & Morales, L. D. G. (2021). Research competencies mediated by technologies: A systematic mapping of the literature. *Education in the Knowledge Society, 22*, e23897. https://doi.org/10.14201/eks.23897
19. Otsuka, S., & Murai, T. (2023). The unique contribution of handwriting accuracy to literacy skills in Japanese adolescents. *Reading and Writing, 37*(5), 1183–1208. https://doi.org/10.1007/s11145-023-10433-3
20. Zaimovic, A., Torlakovic, A., Arnaut-Berilo, A., Zaimovic, T., Dedovic, L., & Meskovic, M. N. (2023). Mapping financial literacy: A systematic literature review of determinants and recent trends. *Sustainability, 15*(12), 9358. https://doi.org/10.3390/su15129358
21. Kusumawardhani, R., Ningrum, N. K., & Rinofah, R. (2023). Investigating digital financial literacy and its impact on SMEs' performance: Evidence from Indonesia. *International Journal of Professional Business Review, 8*(12), e04097. https://doi.org/10.26668/businessreview/2023.v8i12.4097
22. Quraishi, T., Ulusi, H., Muhid, A., Hakimi, M., & Olusi, M. R. (2024). Empowering students through digital literacy: A case study of successful integration in a higher education curriculum. *Journal of Digital Learning and Distance Education, 2*(8), 667–681. https://doi.org/10.56778/jdlde.v2i8.208
23. Msambwa, M. M., Kangwa, D., & Cai, L. (2023). A systematic literature review of ICT integration in secondary education: What works, what does not, and what next? *Discover Education, 2*(1). https://doi.org/10.1007/s44217-023-00070-x
24. Duong, C. D. (2023). A serial mediation model of the linkage between entrepreneurial education, self-efficacy, attitudes and intentions: Does gender matter? A multi-group analysis. *On the Horizon: The International Journal of Learning Futures, 31*(3/4), 174–195. https://doi.org/10.1108/oth-012023-0002
25. Valencia-Arias, A., Arango-Botero, D., & Sanchez-Torres, J. A. (2021). Promoting entrepreneurship´ based on university students' perceptions of entrepreneurial attitude, university environment, entrepreneurial culture and entrepreneurial training. *Higher Education, Skills and Work-based Learning, 12*(2), 328–345. https://doi.org/10.1108/heswbl-07-2020-0169

26. An, H., & Xu, Y. (2021). Cultivation of entrepreneurial talents through virtual entrepreneurship practice in higher education institutions. *Frontiers in Psychology, 12*, 690692. https://doi.org/10.3389/fpsyg.2021.690692
27. Atmani, S. E., Atmani, Z. E., & Madrane, M. (2023). Impact of pedagogical innovation on the development of entrepreneurial mindset among students in economics and management. *Journal for Educators Teachers and Trainers, 14*(6), 63–70. https://doi.org/10.47750/jett.2023.14.06.007
28. Carolis, D. M. D., & Litzky, B. E. (2019). Unleashing the potential of university entrepreneurship education. *New England Journal of Entrepreneurship, 22*(1), 58–66. https://doi.org/10.1108/neje-042019-0022
29. Nadelson, L. S., Palmer, A. D. N., Benton, T., Basnet, R. B., Bissonnette, M., Cantwell, L., & Lanci, S. (2018). Developing next generation of innovators: Teaching entrepreneurial mindset elements across disciplines. *International Journal of Higher Education, 7*(5), 114. https://doi.org/10.5430/ijhe.v7n5p114
30. Broy, M. (2002). Software engineering — From auxiliary to key technologies. In M. Broy & E. Dener (Eds.), *Software pioneers* (pp. 10–13). Springer.
31. Dod, J. (1999). Effective substances. In *The dictionary of substances and their effects*. Royal Society of Chemistry. Available via DIALOG. http://www.rsc.org/dose/title of subordinate document. Cited 15 Jan 1999
32. Geddes, K. O., Czapor, S. R., & Labahn, G. (1992). *Algorithms for computer algebra*. Kluwer.
33. Hamburger, C. (1995). Quasimonotonicity, regularity and duality for nonlinear systems of partial differential equations. *Annali di Matematica Pura ed Applicata, 169*, 321–354.
34. Slifka, M. K., & Whitton, J. L. (2000). Clinical implications of dysregulated cytokine production. *Journal of Molecular Medicine, 78*, 74–80. https://doi.org/10.1007/s001090000086

Part III
Intelligent Technologies and Their Cutting-Edge Applications

Advances, Challenges, and Future Directions in IoT Architectures for Smart Environments

Manuel J. C. S. Reis and Carlos Serôdio

1 Introduction

The Internet of Things (IoT) revolutionizes smart environments by enabling interconnected devices to communicate autonomously and perform tasks. IoT architecture in smart environments integrates sensors, networks, and platforms for efficient operations across sectors like manufacturing [1], transportation [2], agriculture [3], healthcare [4], industry [5], and smart cities [6]. However, challenges such as scalability, security, interoperability, and privacy persist due to the complex and fragmented IoT ecosystem. Addressing these issues is crucial for developing robust and efficient IoT systems. We have identified several open problems, including energy efficiency, data privacy, real-time data processing, edge computing, context awareness, and standardization, which continue to drive ongoing research.

This comprehensive review discusses the eight most pressing issues: scaling IoT architectures to accommodate device growth, ensuring security in the context of diverse and often unsecured devices, addressing privacy concerns due to sensitive data, and achieving interoperability across heterogeneous systems. Furthermore, challenges like optimizing energy consumption, real-time analytics, integrating edge computing, and creating adaptive systems are explored alongside the need

M. J. C. S. Reis (✉)
Engineering Department & IEETA, University of Trás-os-Montes e Alto Douro, Vila Real, Portugal
e-mail: mcabral@utad.pt

C. Serôdio
Department of Engineering, School of Sciences and Technology, University of the Trás-os-Montes e Alto Douro, Vila Real, Portugal

Center ALGORITMI, Universidade do Minho, Guimarães, Portugal

C. Rompante Cunha et al. (eds.), *Transformative Cloud Computing, IoT and Extended Reality*, Signals and Communication Technology,
https://doi.org/10.1007/978-3-032-03554-7_10

for global IoT standards. Solving these problems is critical to future-proofing IoT systems and ensuring seamless and efficient smart environments.

We will examine the core technical and operational challenges of integrating IoT devices in smart environments. These challenges include the need to support scalability and interoperability, the importance of energy efficiency, the critical role of data privacy and security, the demand for real-time data processing, the benefits and complexities of edge and fog computing, the requirement for context awareness and adaptability, the limitations due to lack of standardization, and the difficulties in network management and optimization.

As the number of IoT devices continues to grow, architectures must scale to accommodate them while maintaining cross-platform interoperability. Integrating different IoT protocols, such as MQTT and CoAP, across varied platforms and vendor ecosystems is difficult, given the heterogeneity of devices and communication standards [7, 8]. Energy efficiency is also crucial, especially for battery-operated devices in smart environments like cities, where extending the device's lifetime through efficient communication protocols, processing techniques, and low-power sensors is essential [9, 10].

Data privacy and security concerns are heightened in IoT systems due to the vast amounts of sensitive data they gather. Protecting this data, whether stored locally or in the cloud, is challenging because IoT systems are decentralized and often resource-constrained, complicating the deployment of robust security measures [11–13]. Furthermore, real-time data processing is crucial for traffic management and health monitoring applications. However, achieving low latency and handling high data volumes remains challenging, especially at the edge. Current systems often struggle with latency and the inability to process large data streams at the edge [14–17].

Integrating edge and fog computing helps reduce latency and bandwidth usage, yet implementing these layers in IoT architectures, particularly where resources are limited, poses additional difficulties. Ensuring optimal placement of computation and storage across the edge, fog, and cloud layers is an unsolved problem [18, 19]. Smart environments require IoT systems to be context-aware and adaptable to environmental changes, dynamically responding to shifts in context or user behavior to improve user experiences. However, developing predictive and adaptive architectures remains a research priority. Enabling systems to predict and adapt based on contextual information (e.g., user location, preferences) without user intervention is critical to enhancing user experience [20–22].

Standardization is another obstacle, as many IoT systems operate on proprietary technologies, limiting interoperability and scalability. The fragmentation in the IoT ecosystem across vendors, communication protocols, and platforms limits scalability and interoperability [23–25]. Managing and optimizing IoT networks with numerous interconnected devices is similarly complex, as maintaining optimal performance, low latency, and energy efficiency in environments like smart cities requires effective network management [26, 27].

To address these issues, machine learning techniques for scalability, blockchain-based solutions for security, and interoperability frameworks have been suggested.

Future research is expected to focus on AI-driven automation, resilient security frameworks, and open-source platforms to unlock the full potential of IoT in smart environments.

The aforementioned open problems in IoT architecture for smart environments represent ongoing challenges for researchers and engineers. As IoT continues to evolve and smart environments become more complex, addressing these issues will be crucial for realizing the full potential of IoT systems.

The following sections will discuss these problems, possible solutions, and future research directions. The final section of this study presents the main conclusions.

2 Scalability and Interoperability

In the Internet of Things (IoT) context, scalability denotes the system's capacity to accommodate an increasing number of connected devices, services, and data flows without a decline in operational efficiency. Conversely, interoperability pertains to the seamless collaboration between heterogeneous systems and devices, irrespective of differences in their architecture or communication standards.

With the growing number of connected devices in intelligent environments (e.g., smart cities or industrial IoT), ensuring systems can scale effectively is a significant challenge. Interoperability is also critical, as various devices often use different communication standards (e.g., ZigBee, Bluetooth, Wi-Fi) and must work together seamlessly [28–32].

Concerning challenges of Scalability, as the number of IoT devices continues to grow (with estimates of tens of billions of devices in the near future), managing large-scale networks becomes complex. Each additional device generates data, requires connectivity, and places a strain on processing resources. Traditional centralized architectures, where all data is sent to the cloud, struggle to cope with this increased load due to bandwidth, latency, and computational limitations.

Interoperability Challenges are typically related to the fact that different manufacturers with proprietary technologies and standards often design IoT devices. Integrating these into a unified system is difficult due to variations in communication protocols (e.g., MQTT, CoAP, Zigbee, Bluetooth), data formats, and security mechanisms. Achieving interoperability requires standardization, middleware solutions, and protocol translation mechanisms. Nonetheless, this diversity is present at each phase, from configuring devices to collecting and processing data, highlighting the need for a scalable, interoperable, and economical solution, like Service-Oriented Architectures (SOA) or other additional middleware [33–35]. IoT applications that employ SOA are gaining popularity, as they inherently support seamless interoperability and composability. With the emergence of SOA, microservices architectures have seen significant development, since these, in contrast to monolithic applications, are services deployed autonomously, offering a variety of benefits. However, migrating from a traditional concept to microservices is challenging [36].

Other solutions, based on syntactic and semantic interoperability, like the Open Messaging Interface (O-MI) and Open Data Format (O-DF), One Data Model (OneDM) Semantic, could be employed to obtain features like data balancing, interoperability, and data semantics in addition to scalability [37, 38].

Potential solutions to these problems may include responses on Decentralized Architectures, Middleware Solutions, and Standardized Protocols and APIs. Edge and fog computing (discussed below) help manage scalability by distributing processing closer to the data sources. Middleware platforms, like FIWARE or oneM2M, are proposed to provide interoperability across different IoT systems by abstracting underlying communication protocols and devices. Efforts to create interoperable APIs and standardized communication protocols (e.g., MQTT, CoAP, and DDS) are vital. Open-source IoT ecosystems like EdgeX Foundry can act as a glue between different devices and platforms.

In this context, the ability for different devices and data to work together across platforms facilitates interaction among IoT systems focused on specific and diverse areas. Once cross-platform interoperability is established, it guarantees cross-domain compatibility, allowing diverse platforms from different sectors to collaborate and create integrated IoT applications [39].

A different approach to supporting interoperability by IoT systems is proposed by [40], which proposes the use of a Model-Driven Architecture (MDA), standardized by the Object Management Group (OMG), in which a high level of abstraction to deal with heterogeneous devices, applying concepts such as extensibility and reusability, both to deal with Device-to-Device (D2D) and Device-to-Server (D2S) protocols, such as CoAP, HTTP, MQTT, AMQP, and XMPP.

Interoperability problems can also be solved at a data level, acting at the data level by defining ontologies, which facilitate tasks such as acquisition, representation, dissemination, and transfer of data between nodes, and ensure cross-platform and cross-device compatibility. Based on these principles, [41] proposes a distributed multilayered data platform with modular architecture on an OSGi layer, based on OM2M and RESTful to improve interoperability, making it dynamic and distributed, and improving issues such as system latency.

Integrating cloud and edge computing technologies greatly supports the expansion of IoT systems. Cloud platforms provide flexible storage and computational power, while edge computing enables localized data processing, minimizing delays and reducing bandwidth consumption. Combining these approaches ensures effective data handling and analysis, paving the way for scalable and high-performing IoT infrastructures.

Recent advances in artificial intelligence (AI), machine learning, and the deployment of 5G networks will play a distinct role in driving the growth of IoT. These innovations will enhance the intelligence and efficiency of IoT solutions, enabling improved connectivity and data processing capabilities. As these developments unfold, scalability will remain a key factor for the success of IoT initiatives.

However, frequently, interoperability and scalability work at odds. In other words, the greater the interoperability characteristics of an IoT system, the lower its scalability and overall performance. This idea is defended by [31, 38], who bases their study on scalability and support for heterogeneity. Your work translates into a trade-off study considering interfaces, data format (Open Connectivity Foundation (OCF), Open Mobile Alliance (OMA), Smart Applications Reference (SAREF), and Sensor Network Ontology (SSN)LWM2M data models over CoAP, JSON or UltraLight over HTTP/MQTT, OPC-UA, Sigfox LoRaWAN), communication protocols (MQTT brokers, OPC-UA servers, Sigfox Back-end, Modbus slaves, or CAN nodes), and supporting platforms (FIWARE/NGI, ThingsBoard, and Konker).

Future research directions will indeed include the development of scalable algorithms for managing large-scale IoT networks and global standardization efforts to ensure seamless interoperability.

Table 1 highlights key contributions to IoT research, focusing on scalability, interoperability, and technological advancements. It summarizes diverse approaches, including standards-based frameworks, blockchain integrations, distributed architectures, and microservices. The outcomes demonstrate significant advancements such as improved data sharing, enhanced device communication, and scalable platforms. However, the studies also reveal persistent challenges, including balancing security with scalability, addressing heterogeneous device integration, and overcoming computational overheads. These insights underscore the complexity of IoT system design and the need for further research to address open issues in achieving seamless, secure, and scalable IoT ecosystems.

3 Energy Efficiency

Energy efficiency in IoT refers to optimizing energy consumption in IoT devices and systems to prolong sensor battery life, reduce operational costs, and maintain system functionality.

Many IoT devices operate on constrained power sources, particularly in intelligent environments. The need to transmit, process, and store large amounts of data while maintaining low energy consumption presents a significant challenge. Techniques such as energy-efficient protocols, sleep scheduling, and edge computing are being explored to address this [42–46].

Consequently, energy efficiency, management, and control are now top priorities in the IoT field. The focus is on deploying solutions and algorithms that enhance battery efficiency and lower energy use in IoT sensors and devices, ultimately extending their operational life [47, 48].

An additional challenge involves enhancing the efficiency of routing mechanisms, focusing on minimizing energy consumption. Recent research efforts frequently incorporate Software-Defined Networking (SDN) with Artificial Intelligence across various computing layers—cloud, fog, and edge. These approaches aim to improve network scalability, optimize routing at lower network layers, and achieve effective load distribution through the application of heuristic algorithms such as Genetic Algorithms (GA), Particle Swarm Optimization (PSO), and Artificial Bee Colony (ABC).

Table 1 Summary of IoT research contributions: scalability, interoperability, and associated challenges

Paper	Scalability and interoperability	Technology solution	Outcomes	Challenges
24	Standards-based, scalable IoT data sharing	Cloud integration and monetization	Improved data sharing and scalability	Ensuring standards compliance
25	Cybersecurity and IoT certification	Proposed template for IoT certification	Framework for IoT certification processes	Complex IoT certification landscape
26	Challenges in IoT security certifications	Security certification strategies	Addressing security gaps	Lack of unified security approaches
27	IoT survey: Scalability and challenges	Architectures and frameworks	Enhanced understanding of IoT challenges	Addressing scalability issues
28	Conceptual IoT definition	IoT definition and components	Broad definition of IoT capabilities	Ambiguity in IoT definitions
29	Blockchain for IoT scalability and privacy	Blockchain and deep learning	Enhanced IoT threat intelligence	Balancing security and scalability
30	Enhanced MQTT protocol for IoT scalability	Distributed machine learning	Improved IoT security protocols	Computational overhead concerns
31	Framework for service discovery	Distributed IoT frameworks	Efficient service discovery	Managing distributed IoT components
32	Trade-offs in open IoT platforms	Platform optimization strategies	Scalable and interoperable platforms	Balancing interoperability and scalability
33	Guidelines for managing IoT scalability	NIST recommendations	Privacy and security risk management	Implementing comprehensive guidelines
34	Microservices for IoT device interoperability	Microservice-based architecture	Enhanced device communication	Complex microservice deployments
35	Technologies for critical infrastructure IoT	Middleware solutions	Improved monitoring solutions	Ensuring real-time monitoring
36	Platform for heterogeneous IoT devices	Scalable IoT platform	Support for diverse IoT devices	Handling device heterogeneity

37	Microservices using knowledge graphs	Knowledge graph extraction	Efficient system migration	Complexity of knowledge extraction
38	Standards for industrial IoT data exchange	Data exchange protocols	Reliable industrial IoT communication	Standard adoption in industries
39	Software architecture for IoT interoperability	Next-gen gateway solutions	Interoperability across devices	Integrating diverse device protocols
40	Taxonomies for IoT interoperability	Interoperability frameworks	Identification of open challenges	Addressing multiple interoperability gaps
41	Model-driven IoT device interoperability	Model-driven architecture	Enhanced device interactions	Implementing model-driven standards
42	Distributed IoT architecture	Distributed IoT systems	IoT advancements	Scalability in IoT systems

The first challenge faced by IoT is powering the devices. Many IoT devices, such as sensors and actuators, are deployed in environments where frequent battery replacement or recharging is impractical. For example, smart agriculture sensors might be placed in hard-to-reach fields, and in industrial IoT systems, downtime for maintenance can be costly. Contrasting with this is the fact that these devices are being exposed to more and more energy-intensive operations. Continuous data transmission, especially in real-time systems, consumes significant energy. The need for constant connectivity and high data throughput can rapidly drain battery life. While energy-saving mechanisms like sleep cycles and duty cycling reduce power consumption, they can introduce latency, affecting system responsiveness and performance. As such, energy vs. performance trade-offs is a real problem.

Potential solutions involve energy-efficient communication protocols and energy-harvesting methods. Protocols such as Low Power Wide Area Networks (LPWAN), Narrowband IoT (NB-IoT), SigFox, and 6LoWPAN are designed to reduce power consumption in IoT systems. Harvesting-energy technologies like solar power, wind, motion, and radio frequency signals potentially enable devices to generate their own power, diminishing the need for traditional battery reliance. Developing energy-efficient processors and sensors, such as ARM Cortex-M processors or Intel's Loihi neuromorphic chips, can help reduce energy usage without compromising performance. Additionally, adaptive power management systems that optimize when and how devices consume energy, such as Dynamic Voltage and Frequency Scaling (DVFS), can help improve battery life without affecting system responsiveness.

Future research directions include developing ultra-low-power processors, sensors, communication modules, and intelligent algorithms for dynamic power management and task scheduling that balance energy consumption and performance.

Table 2 synthesizes recent research on energy-efficient IoT systems, highlighting diverse technological solutions and their associated advantages and challenges. The studies cover strategies such as multi-hop routing, mathematical programming, cross-layer frameworks, and AI-driven approaches. The outcomes emphasize significant advancements in optimizing energy consumption, enhancing scalability, and improving resource utilization. Advantages include improved system adaptability, reliability, and energy savings across various IoT environments. However, the studies also reveal ongoing challenges, such as integrating new energy-efficient frameworks with existing systems, balancing security with energy optimization, and managing computational overhead in advanced routing algorithms. These insights illustrate the need for continued innovation to achieve sustainable and efficient IoT architectures.

4 Data Privacy and Security

Data privacy refers to ensuring that users' personal or sensitive data is collected, processed, and stored securely and protected against unauthorized access. Security

Table 2 Summary of research on energy efficiency in IoT systems: technologies, outcomes, and challenges

Paper	Energy efficiency	Technologies solutions	Outcomes	Advantages	Disadvantages	Challenges
9	Energy-efficient mobile distributed sensing platform	On-demand distributed sensing platform with location and activity awareness	Reduced energy consumption during sensing operations	Improved sensing efficiency and resource utilization	Complexity in platform deployment	Ensuring consistent energy efficiency at scale
10	Energy considerations in smart IoT systems	IoT frameworks tailored to smart campus applications	Optimized energy use across smart campus applications	Enhanced system scalability and adaptability	Challenges in integrating diverse IoT devices	Balancing energy optimization with performance demands
26	Energy efficiency in IoT architectures	Various energy-efficient IoT components	Improved IoT scalability and energy performance	Broader application of IoT energy principles	Scalability issues in older frameworks	Adapting architectures to modern IoT demands
27	Conceptual focus on energy optimization	IoT energy strategies and definitions	Clear focus on IoT energy-related goals	Systematic energy focus in IoT definitions	Limited technical details on energy optimization	Maintaining relevance with evolving technologies
43	Overview of energy-efficient industrial IoT	Routing and resource optimization algorithms	Efficient IoT systems for industrial applications	Optimized energy use in industrial IoT setups	Challenges in adapting to diverse industrial needs	Ensuring flexibility in industrial environments
44	Secure cross-layer IoT energy-efficient framework	Integrated cross-layer and secure framework	Enhanced security and energy savings	Cross-layer design improves both security and energy use	Implementation complexity in cross-layer models	Balancing security with energy efficiency
45	Mathematical programming for WSN design	Mathematical models for sensor network design	Improved network design for energy efficiency	Scalable and efficient WSN design	Potential computational overhead	Addressing heterogeneity in WSNs

(continued)

Table 2 (continued)

Paper	Energy efficiency	Technologies solutions	Outcomes	Advantages	Disadvantages	Challenges
46	Multi-hop routing for efficient resource allocation	Resource-efficient multi-hop routing protocols	Effective resource allocation for IoT networks	Better use of resources in constrained environments	Latency concerns in multi-hop networks	Optimizing routing under resource constraints
47	Energy-efficient routing algorithms in IoT	Routing techniques and algorithm surveys	Enhanced energy performance in routing	Comprehensive understanding of IoT routing	Complexity in deploying algorithms	Deploying efficient algorithms at scale
48	Optimizing energy in sensor networks	Mobile sinks for sensor optimization	Increased energy reliability in sensor networks	Enhanced reliability and sensor efficiency	Device dependency on mobile sinks	Handling scalability with sensor movement
49	Energy-efficient routing in IoT and fog computing	Fog-enabled routing in IoT applications	Improved energy use in IoT-fog environments	Energy savings with fog computing integration	Integration complexity with fog layers	Adapting routing for large fog-enabled networks
50	Optimized routing with SDN-AI in large-scale IoT	AI-driven SDN for large-scale industrial IoT	Efficient routing in large-scale IoT setups	Improved energy optimization at scale	High initial cost of implementation	Integrating AI with SDN for IoT challenges

in IoT systems involves securing data transmission and device authentication and protecting against threats such as data breaches or cyberattacks.

IoT systems collect massive amounts of sensitive data in real time, posing privacy, and security risks. Ensuring end-to-end encryption, authentication, and secure storage is essential, but resource constraints on IoT devices make implementing robust security challenging [49–57].

The first concern is related to Privacy. IoT devices often collect sensitive information, such as location data, health metrics, or personal habits (e.g., smart home cameras or wearables). In smart cities, IoT data could include traffic monitoring, which could reveal individuals' movements. Protecting this data from unauthorized access is critical to prevent privacy violations. Security issues remain a critical concern for devices with limited resources. Constraints on features such as processing capacity and memory result in many IoT devices struggling to support advanced security measures such as secure boot processes, comprehensive intrusion detection systems, or robust end-to-end encryption protocols. As a result, these devices are vulnerable to attacks such as distributed denial-of-service (DoS/DDoS), man-in-the-middle attacks, data breaches, information gathering, malware attacks, or even physical tampering. The infamous Mirai Botnet attack exploited unsecured IoT devices to launch one of the largest DDoS attacks in history.

Potential solutions include Lightweight Encryption Protocols, Blockchain for IoT Security, IoT Security Frameworks, and Privacy-Preserving Techniques. Encryption protocols, such as Elliptic Curve Cryptography (ECC) and Lightweight AES, provide robust security without overloading resource-constrained devices. Blockchain and distributed ledger technologies (DLTs) are being explored as decentralized mechanisms for securing IoT networks, ensuring data integrity, secure device identification, and trust without a central authority. Solutions like IoT Security Foundation (IoTSF) and Trusted IoT Alliance are working to establish best practices for security for IoT device manufacturers and service providers. Techniques such as federated learning, which trains AI models on local devices while keeping data decentralized, can be applied to IoT environments to protect user privacy while still enabling data analytics.

Future research directions will undoubtedly include the study of secure and efficient protocols for device authentication and data encryption that work in low-power environments and privacy-preserving machine learning techniques to analyze IoT data without compromising user privacy (e.g., federated learning).

Table 3 presents an overview of research contributions addressing data privacy and security in IoT systems, focusing on diverse technological solutions such as blockchain, federated learning, and privacy-preserving architectures. The studies highlight advancements like enhanced threat mitigation, improved data transport security, and increased resilience against vulnerabilities in IoT networks. Key outcomes include the development of trust frameworks, detailed security insights, and industry-specific protocols. Despite these achievements, the research reveals significant challenges, such as the scalability of security measures, blockchain

Table 3 Summary of research on data privacy and security in IoT systems: technologies, outcomes, and challenges

Paper	Data privacy and security	Technologies solution	Outcomes	Constraints	Challenges
11	Transport security in IoT	IP-based IoT transport security mechanisms	Enhanced IoT data transport protection	High implementation complexity	Ensuring universal applicability
12	Security, privacy, and trust in IoT	IoT trust and privacy frameworks	Improved trust and data security in IoT	Scalability of trust frameworks	Balancing privacy and performance
13	5G vulnerabilities and protections	5G-specific security protocols	Greater understanding of 5G security	Limited adoption of 5G standards	Securing dynamic 5G environments
51	Comprehensive IoT security overview	AI and IoT integration for security	Detailed security and privacy insights	Resource-intensive security measures	Maintaining security in diverse IoT apps
52	Blockchain-based IoT security	Blockchain for IoT threat mitigation	Increased IoT network resilience	Blockchain scalability issues	Adapting blockchain to IoT constraints
53	IoT privacy and countermeasures	Privacy mechanisms for IoT	Mitigated privacy risks in IoT	Complexity in IoT integration	Protecting sensitive data
54	Privacy-preserving trends in IoT	Privacy-preserving IoT architectures	Identification of trends and solutions	Cost and implementation challenges	Overcoming privacy-preservation limitations
55	Advanced security measures in IoT	Best practices for IoT security	Advanced security solutions for IoT	Adoption barriers in IoT ecosystems	Scalability of advanced IoT security
56	Security in industry 4.0	Industry 4.0-specific security protocols	Enhanced industry 4.0 security measures	Integration challenges with industry 4.0	Meeting industry 4.0 security needs
57	Cybersecurity dataset for IoT and IIoT	Federated learning for IoT security	Realistic data for IoT threat modeling	Limited real-world dataset availability	Real-world applicability of datasets
58	Industrial IoT security requirements	Fog computing for IoT security	Better industrial IoT security measures	Fog deployment costs and complexity	Managing fog and edge security
59	Blockchain for IoT security	Blockchain applications in IoT	Improved security in decentralized IoT	Blockchain energy consumption concerns	Addressing blockchain vulnerabilities

energy consumption, and the integration of fog and edge security solutions. These findings emphasize the need for ongoing research to create scalable, efficient, and robust privacy-preserving IoT systems.

5 Real-Time Data Processing and Analytics

Real-time data processing refers to the capability of processing data streams and providing analysis, feedback, and decisions almost instantly as IoT devices generate the data.

IoT data's sheer volume and velocity in intelligent environments require frameworks and architectures to handle real-time processing without causing significant delays. Latency issues in real-time analytics can lead to critical failures in applications like traffic control or emergency response systems [52, 58–60].

One of the most critical problems in IoT applications is Latency-Sensitive. Many IoT applications (e.g., autonomous vehicles, industrial automation, and smart grids) require real-time responses. Data transmission, processing, or decision-making delays can lead to catastrophic outcomes such as traffic accidents or machine failures. Another critical problem is the Data Explosion. IoT systems generate massive amounts of data, often requiring real-time or near-real-time processing. The data must be filtered, processed, and acted upon in milliseconds. Still, centralized cloud architectures can introduce delays due to bandwidth limitations and the distance from data sources, which is problematic for critical applications like autonomous vehicles, smart grids, and real-time health monitoring. Centralized processing in cloud systems can introduce latency and bandwidth constraints, as all data must be sent to a remote data center for processing.

The potential solutions may involve Edge and Fog Computing, In-Memory Computing, and Stream Processing Frameworks. Performing data processing nearer to its point of origin through edge and fog computing significantly minimizes latency and facilitates prompt decision-making. For instance, in smart city environments, edge nodes can analyze footage from traffic cameras to dynamically control traffic signals, eliminating the need to transmit all data to centralized cloud servers. Additionally, In-Memory Computing enhances the speed of real-time analytics by utilizing main memory for both storage and computation, thereby avoiding the slower access times associated with disk-based systems. Systems like Apache Kafka or Apache Flink allow real-time data streaming, processing, and analysis, helping IoT architectures handle high-velocity data without bottlenecks.

Future research directions will include developing distributed, scalable algorithms for real-time IoT data analytics and machine learning models that adapt to dynamic IoT data streams.

Table 4 summarizes recent research on real-time data processing and analytics in IoT systems, focusing on technologies such as privacy-preserving techniques, integrated platforms, and cloud-based frameworks. The outcomes showcase advancements in improving privacy during data analytics, enhancing data-driven decision-making, and optimizing IoT operations through big data tools. However, significant constraints remain, including the high computational cost of privacy-preserving methods, scalability issues in integrated platforms, and latency concerns in cloud-dependent frameworks. These challenges underscore the need for innovative solutions that balance performance, scalability, and privacy in real-time IoT analytics.

6 Edge Computing and Fog Computing Integration

Edge computing brings data processing closer to IoT devices to reduce latency and bandwidth usage. Fog computing extends cloud computing by providing services such as storage, computing, and networking between the edge and the cloud.

Integrating edge and fog computing into IoT architecture helps reduce latency and optimize resource consumption. Still, the design of these systems must balance the computational load between the cloud, fog, and edge nodes. This is incredibly challenging in resource-constrained environments [61–64].

Deciding what tasks should be processed at the edge, fog, or cloud layer is nontrivial. Balancing factors like computational load, latency, and energy consumption requires dynamic and intelligent algorithms. Processing data at the edge (i.e., near the devices generating data) helps to minimize latency, improve responsiveness, and reduce bandwidth consumption by sending only critical information to the cloud. For example, in healthcare, edge devices can monitor patient data locally and only send critical alerts to healthcare providers. Fog computing extends edge computing by adding a processing layer between the edge and the cloud. Fog nodes might process intermediate results and handle tasks that are too computationally expensive for edge devices but too time-sensitive for the cloud.

One of the most critical challenges IoT devices face is related to Distributed Processing Complexity. Efficient management of resources across edge, fog, and cloud layers is complex. Edge devices often have limited computational power and storage, while fog nodes serve as intermediaries and can be geographically distributed. Managing tasks across cloud, fog, and edge layers can be complex, especially when optimizing the placement of tasks and managing resource allocation dynamically. Additionally, security must be assured at multiple layers. Data must be protected as it traverses from edge to fog to cloud, increasing the complexity of implementing consistent security policies.

The potential solutions include Task Offloading, Edge AI, and Hierarchical Architectures. In Task Offloading, adaptive algorithms that intelligently decide where to offload tasks (based on network conditions, processing requirements, and energy availability) are essential for optimizing edge-fog-cloud architectures. For

Table 4 Summary of research on real-time data processing and analytics in IoT systems: technologies, outcomes, and challenges

Paper	RT data processing and analytics	Technologies solutions	Outcomes	Constraints	Challenges
54	Trends in privacy-preserving IoT data processing	Privacy-preserving techniques for IoT analytics	Improved privacy in IoT data analytics	High computational cost for privacy preservation	Balancing privacy and analytics performance
60	IoT-enabled evolution and analytics	Integrated IoT platforms for data analytics	Enhanced data-driven decision-making	Scalability limitations in integrated platforms	Integration across diverse IoT systems
61	Data processing with IoT and cloud	Cloud-based IoT frameworks	Optimized operations	Cloud dependency and latency issues	Ensuring real-time processing at scale
62	Big data management and real-time analytics	Big data frameworks for real-time analytics	Advanced tools for managing and analyzing data	Complexity in big data system implementations	Handling the volume and variety of big data

example, machine learning models can be dynamically offloaded to the cloud during high-demand periods but run on edge devices during low-demand periods. Embedding AI capabilities in edge devices allows for localized, real-time decision-making. For example, AI-based object detection in smart cameras can operate entirely on the edge without sending video feeds to the cloud. Using hierarchical architectures for managing resources ensures that computational tasks are distributed optimally across different system layers (edge, fog, and cloud). For example, fog nodes can process medium-complexity tasks, while the cloud handles intensive analytics.

Another important feature of IoT networks related with interoperability and scalability is the capacity of elasticity, which means the capability to adjust and adapt resources flexibly based on demand. This functionality can be achieved through the inclusion of cloud technology, which depending on the strategy used (edge, fog, cloud) allows IoT ecosystems to efficiently manage resource allocation to accommodate constantly changing workloads, maintaining high performance and minimizing costs.

Future research directions will include optimizing fog-to-cloud and edge-to-cloud communication models for efficiency and security and developing fault-tolerant architectures for real-time task offloading and distributed processing.

Table 5 provides an overview of research on the integration of edge and fog computing in IoT systems, focusing on technologies such as energy-aware routing algorithms, secure cross-layer frameworks, and middleware for seamless IoT integration. The studies highlight outcomes like improved energy efficiency, optimized routing, and enhanced real-time processing capabilities. Despite these advancements, several constraints remain, including the complexity of implementing cross-layer models, high computational demands in heterogeneous networks, and challenges in scaling edge solutions across diverse IoT environments. Addressing issues like latency in multi-hop routing and ensuring compatibility with legacy systems will be critical for maximizing the potential of edge and fog computing in IoT ecosystems.

7 Context Awareness and Adaptability

Context awareness refers to the ability of IoT systems to sense and respond to the environment or user-specific parameters (e.g., location, preferences). Adaptability refers to the system's capacity to change behavior dynamically based on context.

In smart environments, IoT systems must understand and adapt to changing environmental factors and user needs in real time. This requires advanced algorithms for machine learning, context sensing, and dynamic decision-making [65–67].

Other essential features are automatic orchestration and collaboration. Since specific components include proprietary data, software, computing responsibilities, and hardware resources, it is necessary to integrate all these elements and facilitate coordinated communication between them [68].

Table 5 Summary of research on edge and fog computing integration in IoT: technologies, outcomes, and challenges

Paper	Edge and fog computing integration	Technologies solutions	Outcomes	Constraints	Challenges
43	Energy-efficient edge and fog solutions for IIoT	Resource optimization algorithms in edge/fog IIoT	Improved energy efficiency in industrial IoT	Limited scalability in some IIoT settings	Balancing energy efficiency and performance
44	Cross-layer framework for secure fog and edge IoT	Secure cross-layered fog/edge framework	Enhanced security and energy efficiency	Complexity in implementing cross-layer models	Ensuring scalability and security
45	Mathematical models for edge-enabled WSNs	Programming approaches for heterogeneous networks	Scalable and efficient WSN design	High computational demands in edge systems	Handling heterogeneity in IoT networks
46	Survey on multi-hop routing for fog/edge IoT	Protocols for resource-efficient multi-hop routing	Optimized routing and resource allocation	Latency concerns in multi-hop routing setups	Managing latency in fog-enabled networks
47	Energy-efficient IoT routing in fog environments	Algorithms for energy-aware routing in fog setups	Better energy utilization in IoT environments	Complexity in adapting to fog environments	Deploying algorithms in large IoT networks
63	Comprehensive architecture survey of fog computing	Architectural models for scalable fog computing	Comprehensive understanding of fog architectures	Integration challenges in legacy systems	Addressing interoperability in fog systems
64	Middleware and stack design for fog/edge IoT	Interface technologies for fog/edge computing	Streamlined IoT integration and middleware design	Cost-intensive middleware implementation	Ensuring middleware compatibility
65	Edge computing for internet of everything	Edge-driven IoT integration frameworks	Broader IoT ecosystem integration	Resource allocation in edge-heavy setups	Scaling edge solutions across IoT devices
66	Edge computing architectures for IoT applications	IoT-specific edge architecture models	Enhanced real-time processing capabilities	Design complexity for diverse IoT needs	Designing flexible architectures for IoT

Context Awareness is a major aspect. IoT systems in smart environments must sense and interpret contextual data—such as user location, device status, and environmental conditions—and use it to make decisions dynamically. For example, a smart thermostat might adjust heating based on occupancy or external weather conditions. Another critical facet is adaptability. IoT systems should adapt to changing environments or user preferences without requiring manual input. For example, smart lighting systems adapt to human presence or ambient light.

The significant challenges include Accurate Context Sensing, Dynamic Adaptation, and Resource Constraints. Context-aware IoT systems must constantly gather and interpret data about their environment (e.g., user location, temperature, device status), which requires sophisticated sensors and processing algorithms. A key challenge is capturing context accurately and in real time without overwhelming the system. IoT systems must adapt to changing environments without human intervention. For example, a smart home system should adjust lighting or heating based on occupancy and external weather conditions. However, making intelligent decisions based on this context requires advanced AI and machine learning models. However, continuous context sensing and real-time adaptation consume significant resources (e.g., energy and processing power). Managing these constraints while providing intelligent, real-time responses is difficult, especially for resource-limited devices.

The potential solutions will undoubtedly include developments in Context-Aware Middleware, Adaptive Machine Learning Models, and Efficient Context Sensing. Middleware solutions that provide context abstraction and management (e.g., CA-RISMA and SOCAM) help IoT devices and applications understand and act on contextual data without requiring the devices themselves to handle complex processing. AI models that can adapt dynamically to changing contexts, such as reinforcement learning, can help IoT systems make better decisions based on their current environment. These models can also learn from user preferences over time. Additionally, using efficient context-sensing techniques (e.g., event-driven sensing) reduces the amount of data that needs to be continuously processed. For example, a smart system can switch between low-power and high-power modes based on detected events.

Future research directions will devise new techniques for real-time context sensing and dynamic decision-making in resource-constrained IoT environments and integrate AI and context awareness to enable predictive services in smart environments.

Table 6 summarizes recent advancements in context-aware and adaptable IoT systems, focusing on technologies such as dynamic service migration, cognitive QoS management, and automated context-handling frameworks. The studies demonstrate outcomes like improved service continuity, sustainable communication, and enhanced decision-making in IoT applications. Advantages include better service reliability, energy efficiency, and reduced manual intervention. However, constraints such as latency in service migration, integration complexity, and scalability challenges remain significant barriers. Addressing these issues requires innovative approaches to balance computational costs, ensure adaptability at scale, and maintain performance under varying conditions in diverse IoT environments.

Table 6 Summary of research on context awareness and adaptability in IoT: technologies, outcomes, and challenges

Paper	Context awareness and adaptability	Technologies solutions	Outcomes	Advantages	Constraints	Challenges
21	Dynamic service migration in wearable IoT	Edge-based dynamic service migration techniques	Improved service continuity and adaptability	Better service reliability	Latency in service migration processes	Ensuring real-time migration without disruption
22	Cognitive QoS management for networking	Cognitive algorithms for QoS-aware transmission	Enhanced video transmission quality	Optimized QoS for real-time video streaming	High computational requirements	Maintaining QoS under varying conditions
67	Surveys on context-aware intelligent IoT services	Context-aware service frameworks for IoT	Broad review of IoT context-aware services	Comprehensive understanding of context-aware IoT	Diverse implementation challenges	Addressing heterogeneity in IoT environments
68	Context-aware messaging for sustainable IoT	Messaging protocols for adaptive IoT communication	Sustainable communication in IoT networks	Energy efficiency and improved communication	Integration complexity in IoT networks	Ensuring sustainability of messaging frameworks
69	Context-aware computing in IoT environments	Computing models for context-aware IoT	Improved decision-making in IoT applications	Enhanced user-centric IoT experiences	Scalability in large-scale IoT setups	Balancing computational costs and performance
70	Automation framework for context-aware IoT systems	Automated context-handling frameworks	Increased automation and adaptability in IoT	Reduction in manual intervention for IoT	Limited automation standards in IoT	Implementing context-aware frameworks at scale

8 Standardization

Standardization in IoT refers to establishing standard protocols, formats, and interfaces that allow diverse IoT devices and systems to communicate and work together without proprietary restrictions.

The lack of widely adopted standards in IoT creates fragmentation, making it challenging to integrate different systems and technologies in smart environments. International efforts, such as developing standard communication protocols and certification frameworks, are ongoing, but challenges remain [69–72].

A lack of universally adopted standards in IoT results in fragmented ecosystems where different devices and platforms cannot easily communicate. This creates "vendor lock-in," making it hard to scale IoT deployments. Without standard communication, data exchange, and security protocols, it is hard to build large-scale, interoperable IoT systems across different industries and domains (e.g., smart homes, smart cities, industrial IoT).

As such, one of the most critical challenges is the existence of multiple standards. Organizations such as the IEEE, IETF, and ISO are working on IoT standards, but competing standards from different alliances (e.g., Thread, ZigBee, Z-Wave) create confusion and slow down adoption.

Potential solutions will certainly rely on developing common standards and frameworks such as oneM2M, Open Connectivity Foundation (OPC-UA), or OMA-LWM2M that provide consistent communication and data-sharing protocols. Platforms like FIWARE, EdgeX Foundry, and OpenIoT promote open, standardized IoT ecosystems. By adopting these platforms, developers can build inherently compatible applications with multiple devices and services. Government-backed initiatives, such as the European IoT Large-Scale Pilots (IoT-LSP) or NIST Cybersecurity Framework, aim to promote common standards across industries and borders.

Future research directions shall develop efforts toward global standardization, with collaborations between government bodies, industry, and research communities, as well as protocols that seamlessly integrate various IoT ecosystems while allowing future expansion.

Table 7 provides an overview of research addressing standardization challenges in IoT, covering areas such as cloud integration mechanisms, IoT certification frameworks, and interoperability standards for industrial IoT and smart cities. The outcomes demonstrate improved integration, scalable data-sharing practices, and enhanced IoT security through standard protocols. Key advantages include structured certification processes, better interoperability, and clarity in IoT system definitions. However, the research also highlights constraints such as the complexity of integrating cloud and IoT systems, incompatibilities in existing protocols, and scalability challenges in urban IoT. Achieving seamless standardization requires innovative solutions that harmonize global frameworks, address gaps in implementation, and ensure adaptability to the evolving IoT landscape.

Table 7 Summary of research on standardization in IoT: technologies, outcomes, and challenges

Paper	Standardization	Technologies solutions	Outcomes	Constraints	Challenges
7	Integration of cloud and IoT standards	Cloud and IoT integration mechanisms	Improved integration of cloud and IoT systems	Complexity in integrating cloud and IoT	Achieving seamless standard integration
8	Security protocol standardization for IoT	Existing IoT security protocols	Enhanced IoT security through standard protocols	Incompatibilities in existing protocols	Ensuring protocol compatibility in IoT
23	Standards-based scalable IoT data sharing	Standardized data-sharing ecosystems	Efficient and scalable data-sharing practices	Scalability limitations in data-sharing	Balancing scalability and standardization
24	IoT certification frameworks	Template-based IoT certification schemes	Structured IoT certification landscapes	Limited coverage of certification schemes	Harmonizing EU standards with global needs
25	Challenges in IoT security certifications	Approaches for security certifications	Identification of security gaps and solutions	Fragmentation in certification practices	Developing unified certification approaches
26	Standard frameworks for IoT definitions	Definitional frameworks for IoT systems	Improved understanding of IoT structures	Ambiguities in IoT standard definitions	Maintaining relevance with evolving IoT tech
27	Broad IoT standardization definitions	Conceptual IoT models for standards	Greater clarity in IoT standard definitions	Lack of universal IoT standards	Addressing interoperability across devices
71	Interoperability standards for IIoT	Taxonomies and frameworks for IIoT	Enhanced interoperability in industrial IoT	Adapting standards to diverse IIoT needs	Creating inclusive IIoT standard frameworks
72	IETF standardization efforts in IoT	Protocols developed by IETF for IoT	Increased adoption of IoT standards	Gaps in implementation of IETF standards	Overcoming barriers in IoT standard adoption
73	IoT standard practices	IoT practices tailored to smart cities	Better IoT practices for setups	Resource constraints in smart city IoT	Aligning standards with IoT demands
74	IoT applications in smart city standards	Smart city IoT and service frameworks	Comprehensive view of IoT city services	Scalability challenges in urban IoT	Implementing standards in large urban areas

9 Network Management and Optimization

Network management in IoT involves controlling, coordinating, and optimizing the communication infrastructure that connects IoT devices. Optimization refers to techniques that reduce latency, minimize energy usage, and ensure reliability.

IoT networks in smart environments, such as smart cities, often consist of a massive number of devices transmitting data over wireless networks. Managing and optimizing these networks to ensure performance and reliability is a significant challenge, particularly as the number of devices grows [73–76].

Smart environments often involve massive, dynamic networks of IoT devices communicating over various wireless technologies. Managing these networks—optimizing for latency, throughput, and energy—requires intelligent coordination. IoT networks must guarantee specific QoS parameters (e.g., reliability, delay, and bandwidth) to ensure the smooth operation of time-sensitive applications (e.g., real-time monitoring in healthcare or smart grids).

IoT environments involve different network technologies (e.g., cellular, Wi-Fi, Bluetooth) operating together. Managing these heterogeneous systems while ensuring consistent performance remains a challenge. Additionally, in smart cities, for example, traffic from emergency services should be prioritized over regular traffic, but implementing such QoS guarantees is challenging. IoT networks are often dynamic, with devices constantly entering or leaving the network (e.g., mobile devices in smart cities). Maintaining a robust, resilient, and self-healing network in such dynamic environments is challenging.

The potential solutions include Software-Defined Networking (SDN), Network Function Virtualization (NFV), and Machine Learning-Based Optimization. SDN can dynamically manage IoT traffic, optimize routing, and prioritize critical data. NFV allows virtualized network functions to be created, reducing reliance on physical network infrastructure and enabling better resource allocation and scaling. It can help deploy services on demand based on real-time traffic patterns. AI-based algorithms can optimize network resources by learning patterns in network usage and traffic demands.

Software-Defined Networking (SDN) and Network Functions Virtualization (NFV) are increasingly influential within network security. They are essential components for the on-demand delivery of security solutions, in line with the Security-as-a-Service (SECaaS) model [77].

Future research directions will undoubtedly persist in developing intelligent, self-optimizing networks that autonomously manage IoT devices, dynamically allocate resources, and in researching new communication protocols designed for large-scale, heterogeneous IoT environments.

Table 8 summarizes the network management and optimization research in IoT, highlighting key technologies such as IoT certification frameworks, model-driven architectures, deep learning for QoS enhancement, and SDN/NFV mechanisms for secure networks. The outcomes emphasize improved network reliability, enhanced understanding of IoT network dynamics, and optimized QoS for real-time

Table 8 Summary of research on network management and optimization in IoT: technologies, outcomes, and challenges

Paper	Network management and optimization	Technologies solutions	Outcomes	Constraints	Challenges
25	Security-focused network optimization in IoT	IoT certification and security frameworks	Improved security and network reliability	Fragmentation in security certification approaches	Developing unified security certifications
26	Foundational concepts for IoT network optimization	IoT network protocols and standardizations	Enhanced understanding of IoT network dynamics	Lack of uniformity in IoT standards	Maintaining relevance with evolving IoT tech
40	Interoperability-driven network architecture for IIoT	Model-driven architectures for seamless IoT networks	Greater interoperability in industrial IoT systems	Complexity in implementing model-driven systems	Ensuring interoperability across varied devices
41	Distributed network architectures	Scalable and interoperable IoT architectures	Efficient management of urban IoT networks	Scalability issues in large-scale IoT deployments	Balancing scalability and efficiency in networks
1	Automotive IoT network management strategies	Integration of smart materials and IoT networks	Streamlined IoT networks in automotive applications	High costs of automotive IoT implementation	Integrating IoT seamlessly into vehicles
75	Quality of service (QoS) improvements in IoT	QoS protocols for real-time IoT applications	Enhanced QoS in IoT applications	Latency concerns in real-time QoS management	Addressing latency and reliability in QoS protocols
76	Deep learning applications for IoT QoS enhancement	Deep learning models for IoT QoS optimization	Optimized QoS using advanced machine learning	Computational overhead in applying deep learning	Efficient deployment of deep learning models
77	Comprehensive network optimization in IoT	Optimization algorithms for IoT networks	More efficient IoT network designs	Challenges in handling diverse IoT data streams	Adapting optimization techniques to real-world use
78	Reviews on IoT network optimization techniques	Network resource management techniques	Practical insights into IoT resource management	Resource allocation inefficiencies in IoT networks	Ensuring consistent network performance in IoT
79	SDN and NFV mechanisms for secure IoT networks	SDN and NFV frameworks for IoT security	Strengthened IoT network security via SDN/NFV	Implementation challenges in SDN and NFV	Overcoming adoption barriers for SDN/NFV

applications. These advancements offer greater interoperability, efficient resource management, and improved security in IoT systems. However, challenges persist, such as fragmentation in security certification approaches, scalability issues in large IoT deployments, and the complexity of integrating deep learning models in real-time systems. Overcoming these constraints is essential for ensuring the consistent performance and seamless operation of IoT networks at scale.

10 Conclusions

The IoT is reshaping smart environments by allowing connected devices to communicate and perform various tasks autonomously. IoT architecture in these environments integrates sensors, networks, and platforms to optimize operations in industries such as healthcare, smart cities, and manufacturing. However, due to its fragmented nature, IoT faces major hurdles, including scalability, security, privacy, and interoperability. Tackling these issues is essential for creating efficient and secure IoT systems and driving ongoing research to develop innovations and standards.

This review highlights challenges such as managing expanding networks, securing heterogeneous devices, ensuring interoperability, and safeguarding sensitive data. Potential solutions include using machine learning for scalability, blockchain for security, and global frameworks for standardization. Future research focuses on advancing AI-driven automation, building stronger security frameworks, and improving open-source platforms to foster interoperability and privacy.

Addressing these challenges is crucial as IoT systems continue to grow in complexity and become embedded in critical real-world applications. The evolving nature of IoT systems requires continued research, new technologies, and standardized solutions to realize their full potential in smart environments. These ongoing efforts reflect the intricate task of building scalable, secure, and adaptable IoT architectures.

References

1. Ammar, M., Haleem, A., Javaid, M., Bahl, S., Garg, S. B., Shamoon, A., & Garg, J. (2022). Significant applications of smart materials and internet of things (IoT) in the automotive industry. *Materials Today: Proceedings, 68*, 1542–1549. https://doi.org/10.1016/j.matpr.2022.07.180
2. Ushakov, D., Dudukalov, E., Kozlova, E., & Shatila, K. (2022). The internet of things impact on smart public transportation. *Transportation Research Procedia, 63*, 2392–2400. https://doi.org/10.1016/j.trpro.2022.06.275
3. Sinha, B. B., & Dhanalakshmi, R. (2022). Recent advancements and challenges of internet of things in smart agriculture: A survey. *Future Generation Computer Systems, 126*, 169–184. https://doi.org/10.1016/j.future.2021.08.006

4. Kashani, M. H., Madanipour, M., Nikravan, M., Asghari, P., & Mahdipour, E. (2021). A systematic review of IoT in healthcare: Applications, techniques, and trends. *Journal of Network and Computer Applications, 192*, 103164. https://doi.org/10.1016/j.jnca.2021.103164
5. Kalsoom, T., Ahmed, S., Rafi-ul-Shan, P. M., Azmat, M., Akhtar, P., Pervez, Z., Imran, M. A., & Ur-Rehman, M. (2021). Impact of IoT on manufacturing industry 4.0: A new triangular systematic review. *Sustainability, 13*, 12506. https://doi.org/10.3390/su132212506
6. Ghazal, T. M., Hasan, M. K., Alshurideh, M. T., Alzoubi, H. M., Ahmad, M., Akbar, S. S., Al Kurdi, B., & Akour, I. A. (2021). IoT for smart cities: Machine learning approaches in smart healthcare—A review. *Future Internet, 13*, 218. https://doi.org/10.3390/fi13080218
7. Botta, A., de Donato, W., Persico, V., & Pescapé, A. (2016). Integration of cloud computing and internet of things: A survey. *Future Generation Computer Systems, 56*, 684–700. https://doi.org/10.1016/j.future.2015.09.021
8. Granjal, J., Monteiro, E., & Sa Silva, J. (2015). Security for the internet of things: A survey of existing protocols and open research issues. *IEEE Communications Surveys & Tutorials, 17*, 1294–1312. https://doi.org/10.1109/COMST.2015.2388550
9. Perera, C., Talagala, D. S., Liu, C. H., & Estrella, J. C. (2015). Energy-efficient location and activity-aware on-demand Mobile distributed sensing platform for sensing as a service in IoT clouds. *IEEE Transactions on Computational Social Systems, 2*, 171–181. https://doi.org/10.1109/TCSS.2016.2515844
10. Domínguez-Bolaño, T., Barral, V., Escudero, C. J., & García-Naya, J. A. (2024). An IoT system for a smart campus: Challenges and solutions illustrated over several real-world use cases. *Internet of Things, 25*, 101099. https://doi.org/10.1016/j.iot.2024.101099
11. Garcia-Macias, J. A. (2023). Transport in the IP-based internet of things: Status report. *Procedia Computer Science, 224*, 18–25. https://doi.org/10.1016/j.procs.2023.09.006
12. Sicari, S., Rizzardi, A., Grieco, L. A., & Coen-Porisini, A. (2015). Security, privacy and Trust in Internet of things: The road ahead. *Computer Networks, 76*, 146–164. https://doi.org/10.1016/j.comnet.2014.11.008
13. Sousa, A., & Manuel, J. C. S. (2024). Reis 5G security features, vulnerabilities, threats, and data protection in IoT and Mobile devices: A systematic review. *Evolutionary Studies in Imaginative Culture*, 414–427. https://doi.org/10.70082/esiculture.vi.1054
14. Oladimeji, D., Gupta, K., Kose, N. A., Gundogan, K., Ge, L., & Liang, F. (2023). Smart transportation: An overview of technologies and applications. *Sensors, 23*, 3880. https://doi.org/10.3390/s23083880
15. Hussain, A., Ali, T., Althobiani, F., Draz, U., Irfan, M., Yasin, S., Shafiq, S., Safdar, Z., Glowacz, A., Nowakowski, G., et al. (2021). Security framework for IoT based real-time health applications. *Electronics, 10*, 719. https://doi.org/10.3390/electronics10060719
16. Yeri, V., & Shubhangi, D. C. (2020). IoT based real time health monitoring. In *Proceedings of the 2020 second international conference on inventive research in computing applications (ICIRCA)* (pp. 980–984). IEEE.
17. Gubbi, J., Buyya, R., Marusic, S., & Palaniswami, M. (2013). Internet of things (IoT): A vision, architectural elements, and future directions. *Future Generation Computer Systems, 29*, 1645–1660. https://doi.org/10.1016/j.future.2013.01.010
18. Chiang, M., & Zhang, T. (2016). Fog and IoT: An overview of research opportunities. *IEEE Internet of Things Journal, 3*, 854–864. https://doi.org/10.1109/JIOT.2016.2584538
19. Yi, S., Li, C., & Li, Q. (2015, June 21). A survey of fog computing: Concepts, applications and issues. In *Proceedings of the proceedings of the 2015 workshop on Mobile big data; ACM: Hangzhou China* (pp. 37–42).
20. Baker, S. B., Xiang, W., & Atkinson, I. (2017). Internet of things for smart healthcare: Technologies, challenges, and opportunities. *IEEE Access, 5*, 26521–26544. https://doi.org/10.1109/ACCESS.2017.2775180
21. Raza, I., Hussain, S. A., & Latif, S. (2024, July 29). EdgeMob: A context-aware dynamic service migration for wearable IoT in healthcare. In *Proceedings of the 2024 international conference on computing, internet of things and microwave systems (ICCIMS)* (pp. 1–6). IEEE.

22. Cheng, B., Wang, M., Lin, X., & Chen, J. (2021). Context-aware cognitive QoS Management for Networking Video Transmission. *IEEE/ACM Transactions on Networking, 29*, 1422–1434. https://doi.org/10.1109/TNET.2021.3066262
23. Figueredo, K., Seed, D., & Wang, C. (2022). A scalable, standards-based approach for IoT data sharing and ecosystem monetization. *IEEE Internet of Things Journal, 9*, 5645–5652. https://doi.org/10.1109/JIOT.2020.3023035
24. Khurshid, A., Alsaaidi, R., Aslam, M., & Raza, S. (2022). EU cybersecurity act and IoT certification: Landscape, perspective and a proposed template scheme. *IEEE Access, 10*, 129932–129948. https://doi.org/10.1109/ACCESS.2022.3225973
25. Cirne, A., Sousa, P. R., Resende, J. S., & Antunes, L. (2022). IoT security certifications: Challenges and potential approaches. *Computers & Security, 116*, 102669. https://doi.org/10.1016/j.cose.2022.102669
26. Atzori, L., Iera, A., & Morabito, G. (2010). The internet of things: A survey. *Computer Networks, 54*, 2787–2805. https://doi.org/10.1016/j.comnet.2010.05.010
27. Minerva, R., Biru, A., & D. (2015). *Rotondi towards a definition of the internet of things (IoT)* (Vol. 1, pp. 1–86).
28. Kumar, P., Kumar, R., Gupta, G. P., Tripathi, R., & Srivastava, G. (2022). P2TIF: A Blockchain and deep learning framework for privacy-preserved threat intelligence in industrial IoT. *IEEE Transactions on Industrial Informatics, 18*, 6358–6367. https://doi.org/10.1109/TII.2022.3142030
29. Alotaibi, N. S., Sayed Ahmed, H. I., Kamel, S. O. M., & ElKabbany, G. F. (2024). Secure enhancement for MQTT protocol using distributed machine learning framework. *Sensors, 24*, 1638. https://doi.org/10.3390/s24051638
30. Sivrikaya, F., Ben-Sassi, N., Dang, X.-T., Gorur, O. C., & Kuster, C. (2019). Internet of Smart City objects: A distributed framework for service discovery and composition. *IEEE Access, 7*, 14434–14454. https://doi.org/10.1109/ACCESS.2019.2893340
31. Ottolini, D., Zyrianoff, I., & Kamienski, C. (2022, January 8). Interoperability and scalability trade-offs in open IoT platforms. In *Proceedings of the 2022 IEEE 19th Annual Consumer Communications & Networking Conference (CCNC)* (pp. 1–6). IEEE.
32. Boeckl, K., Fagan, M., Fisher, W., Lefkovitz, N., Megas, K. N., Nadeau, E., O'Rourke, D. G., Piccarreta, B., & Scarfone, K. (2019). *Considerations for managing internet of things (IoT) cybersecurity and privacy risks* (p. NIST IR 8228). National Institute of Standards and Technology.
33. Rath, C. K., Mandal, A. K., & Sarkar, A. (2023). Microservice based scalable IoT architecture for device interoperability. *Computer Standards & Interfaces, 84*, 103697. https://doi.org/10.1016/j.csi.2022.103697
34. Alonso, L., Barbarán, J., Chen, J., Díaz, M., Llopis, L., & Rubio, B. (2018). Middleware and communication Technologies for Structural Health Monitoring of critical infrastructures: A survey. *Computer Standards & Interfaces, 56*, 83–100. https://doi.org/10.1016/j.csi.2017.09.007
35. Javed, A., Malhi, A., Kinnunen, T., & Framling, K. (2020). Scalable IoT platform for heterogeneous devices in smart environments. *IEEE Access, 8*, 211973–211985. https://doi.org/10.1109/ACCESS.2020.3039368
36. Li, Z., Shang, C., Wu, J., & Li, Y. (2022). Microservice extraction based on knowledge graph from monolithic applications. *Information and Software Technology, 150*, 106992. https://doi.org/10.1016/j.infsof.2022.106992
37. Madhikermi, M., Yousefnezhad, N., & Framling, K. (2018, November). Data exchange standard for industrial internet of things. In *Proceedings of the 2018 3rd international conference on system reliability and safety (ICSRS)* (pp. 53–61). IEEE.
38. Resende, C., Moreira, W., & Almeida, L. (2022, September 6). NextGenGW: A software-based architecture targeting IoT interoperability. In *Proceedings of the 2022 IEEE 27th international conference on emerging technologies and factory automation (ETFA)* (pp. 1–4). IEEE.
39. Noura, M., Atiquzzaman, M., & Gaedke, M. (2019). Interoperability in internet of things: Taxonomies and open challenges. *Mobile Networks and Applications, 24*, 796–809. https://doi.org/10.1007/s11036-018-1089-9

40. Amjad, A., Azam, F., & Anwar, M. W. (2022, October 21). Device interoperability for industrial IoT using model-driven architecture. In *Proceedings of the 2022 24th international multitopic conference (INMIC)* (pp. 1–6). IEEE.
41. Pranavasri, V., Francis, L., Mogadali, U., Pal, G., Vaddhiparthy, S. S. S., Vattem, A., Vaidhyanathan, K., & Gangadharan, D. (2023, October 12). Scalable and interoperable distributed architecture for IoT in smart cities. In *Proceedings of the 2023 IEEE 9th world forum on internet of things (WF-IoT)* (pp. 01–06). IEEE.
42. Mao, W., Zhao, Z., Chang, Z., Min, G., & Gao, W. (2021). Energy-efficient industrial internet of things: Overview and open issues. *IEEE Transactions on Industrial Informatics, 17*, 7225–7237. https://doi.org/10.1109/TII.2021.3067026
43. Mustafa, R., Sarkar, N. I., Mohaaghegh, M., & Pervez, S. (2024, January 17). A secure and energy-efficient cross-layer framework for internet of things. In *Proceedings of the 2024 international conference on information networking (ICOIN)* (pp. 661–666). IEEE.
44. Taparci, E., Olcay, K., Akmandor, M. O., Kabakulak, B., Sarioglu, B., & Gokdel, Y. D. (2024). A mathematical programming approach for IoT-enabled, energy-efficient heterogeneous wireless sensor network design and implementation. *Sensors, 24*, 1457. https://doi.org/10.3390/s24051457
45. Awais, M., Mateen, A., Khan, Z. A., & Javaid, N. (2018, November). A survey on multi-hop routing protocols for efficient resource allocation in IoTs. In *Proceedings of the 2018 fifth HCT information technology trends (ITT)* (pp. 239–244). IEEE.
46. Gamal, M. (2023). Energy efficiency routing algorithms in IoT: A survey. In A. E. Hassanien, V. Snášel, M. Tang, T.-W. Sung, & K.-C. Chang (Eds.), *Proceedings of the 8th international conference on advanced intelligent systems and informatics 2022* (Lecture notes on data engineering and communications technologies) (Vol. 152, pp. 668–679). Springer International Publishing. ISBN 978-3-031-20600-9.
47. Ali, S., Nadeem, M., Ahmed, S., & Tahir, M. (2024). IoT-based framework for optimizing energy efficiency and reliability in acoustic sensor networks using Mobile sinks. *Scientific Reports, 14*, 24122. https://doi.org/10.1038/s41598-024-74664-2
48. Kaur, L., & Kaur, R. (2021). A survey on energy efficient routing techniques in WSNs focusing IoT applications and enhancing fog computing paradigm. *Global Transitions Proceedings, 2*, 520–529. https://doi.org/10.1016/j.gltp.2021.08.001
49. Aouedi, O., Vu, T.-H., Sacco, A., Nguyen, D. C., Piamrat, K., Marchetto, G., & Pham, Q.-V. (2024). A survey on intelligent internet of things: Applications, security, privacy, and future directions. *IEEE Communications Surveys & Tutorials, PP*(99), 1–1. https://doi.org/10.1109/COMST.2024.3430368
50. Khan, M. A., & Salah, K. (2018). IoT security: Review, Blockchain solutions, and open challenges. *Future Generation Computer Systems, 82*, 395–411. https://doi.org/10.1016/j.future.2017.11.022
51. Elhoseny, M., Thilakarathne, N. N., Alghamdi, M. I., Mahendran, R. K., Gardezi, A. A., Weerasinghe, H., & Welhenge, A. (2021). Security and privacy issues in medical internet of things: Overview, countermeasures, challenges and future directions. *Sustainability, 13*, 11645. https://doi.org/10.3390/su132111645
52. Safaei Yaraziz, M., Jalili, A., Gheisari, M., & Liu, Y. (2023). Recent trends towards privacy-preservation in internet of things, its challenges and future directions. *IET Circuits, Devices and Systems, 17*, 53–61. https://doi.org/10.1049/cds2.12138
53. USENIX Association (Ed.). (2017). *Proceedings of the 26th USENIX Security Symposium: August 16–18, 2017, Vancouver, BC, Canada*. USENIX Association. ISBN 978–1–931971-40-9.
54. Jamai, I., Ben Azzouz, L., & Saidane, L. A. (2020, June). Security issues in industry 4.0. In *Proceedings of the 2020 international wireless communications and Mobile computing (IWCMC)* (pp. 481–488). IEEE.
55. Ferrag, M. A., Friha, O., Hamouda, D., Maglaras, L., & Janicke, H. (2022). Edge-IIoTset: A new comprehensive realistic cyber security dataset of IoT and IIoT applications for centralized and federated learning. *IEEE Access, 10*, 40281–40306. https://doi.org/10.1109/ACCESS.2022.3165809

56. Tange, K., De Donno, M., Fafoutis, X., & Dragoni, N. (2020). A systematic survey of industrial internet of things security: Requirements and fog computing opportunities. *IEEE Communications Surveys & Tutorials, 22*, 2489–2520. https://doi.org/10.1109/COMST.2020.3011208
57. Wang, Q., Zhu, X., Ni, Y., Gu, L., & Zhu, H. (2020). Blockchain for the IoT and industrial IoT: A review. *Internet of Things, 10*, 100081. https://doi.org/10.1016/j.iot.2019.100081
58. Bauer, M., Sanchez, L., & Song, J. (2021). IoT-enabled smart cities: Evolution and outlook. *Sensors, 21*, 4511. https://doi.org/10.3390/s21134511
59. Su, P., Chen, Y., & Lu, M. (2022). Smart City information processing under internet of things and cloud computing. *The Journal of Supercomputing, 78*, 3676–3695. https://doi.org/10.1007/s11227-021-03972-5
60. Li, K.-C., Jiang, H., & Zomaya, A. Y. (Eds.). (2017). *Big data management and processing* (1st ed.). Chapman and Hall/CRC. ISBN 978-1-315-15400-8.
61. Habibi, P., Farhoudi, M., Kazemian, S., Khorsandi, S., & Leon-Garcia, A. (2020). Fog computing: A comprehensive architectural survey. *IEEE Access, 8*, 69105–69133. https://doi.org/10.1109/ACCESS.2020.2983253
62. Ali, O., Ishak, M. K., Bhatti, M. K. L., Khan, I., & Kim, K.-I. (2022). A comprehensive review of internet of things: Technology stack, Middlewares, and fog/edge computing Interface. *Sensors, 22*, 995. https://doi.org/10.3390/s22030995
63. Kong, X., Wu, Y., Wang, H., & Xia, F. (2022). Edge computing for internet of everything: A survey. *IEEE Internet of Things Journal, 9*, 23472–23485. https://doi.org/10.1109/JIOT.2022.3200431
64. Hamdan, S., Ayyash, M., & Almajali, S. (2020). Edge-computing architectures for internet of things applications: A survey. *Sensors, 20*, 6441. https://doi.org/10.3390/s20226441
65. Gil, D., Ferrández, A., Mora-Mora, H., & Peral, J. (2016). Internet of things: A review of surveys based on context aware intelligent services. *Sensors, 16*, 1069. https://doi.org/10.3390/s16071069
66. Almagrabi, A. O., & Al-Otaibi, Y. D. (2020). A survey of context-aware messaging-addressing for sustainable internet of things (IoT). *Sustainability, 12*, 4105. https://doi.org/10.3390/su12104105
67. Perera, C., Zaslavsky, A., Christen, P., & Georgakopoulos, D. (2014). Context aware computing for the internet of things: A survey. *IEEE Communications Surveys & Tutorials, 16*, 414–454. https://doi.org/10.1109/SURV.2013.042313.00197
68. Chegini, H., & Mahanti, A. (2019, December 2). A framework of automation on context-aware internet of things (IoT) systems. In *Proceedings of the proceedings of the 12th IEEE/ACM international conference on utility and cloud computing companion* (pp. 157–162). ACM.
69. Hazra, A., Adhikari, M., Amgoth, T., & Srirama, S. N. (2023). A comprehensive survey on interoperability for IIoT: Taxonomy, standards, and future directions. *ACM Computing Surveys, 55*, 1–35. https://doi.org/10.1145/3485130
70. Ishaq, I., Carels, D., Teklemariam, G., Hoebeke, J., Abeele, F., Poorter, E., Moerman, I., & Demeester, P. (2013). IETF standardization in the field of the internet of things (IoT): A survey. *Journal of Sensor and Actuator Networks, 2*, 235–287. https://doi.org/10.3390/jsan2020235
71. Syed, A. S., Sierra-Sosa, D., Kumar, A., & Elmaghraby, A. (2021). IoT in smart cities: A survey of technologies, practices and challenges. *Smart Cities, 4*, 429–475. https://doi.org/10.3390/smartcities4020024
72. Rafiq, I., Mahmood, A., Razzaq, S., Jafri, S. H. M., & Aziz, I. (2023). IoT applications and challenges in smart cities and services. *The Journal of Engineering, 2023*, e12262. https://doi.org/10.1049/tje2.12262
73. Singh, M., & Baranwal, G. (2018, February). Quality of service (QoS) in internet of things. In *Proceedings of the 2018 3rd international conference on internet of things: Smart innovation and usages (IoT-SIU)* (pp. 1–6). Bhimtal.
74. Kimbugwe, N., Pei, T., & Kyebambe, M. (2021). Application of deep learning for quality of service enhancement in internet of things: A review. *Energies, 14*, 6384. https://doi.org/10.3390/en14196384

75. Srinidhi, N. N., Kumar, S. M. D., & Venugopal, K. R. (2019). Network optimizations in the internet of things: A review. *Engineering Science and Technology, an International Journal, 22*, 1–21. https://doi.org/10.1016/j.jestch.2018.09.003
76. Srivastava, A., & Kumar, A. (2022). A review of network optimization on the internet of things. In H. S. Saini, R. Sayal, A. Govardhan, & R. Buyya (Eds.), *Innovations in computer science and engineering* (Lecture notes in networks and systems) (Vol. 385, pp. 49–57). Springer Singapore. ISBN 9789811689864.
77. Farris, I., Taleb, T., Khettab, Y., & Song, J. (2019). A survey on emerging SDN and NFV security mechanisms for IoT systems. *IEEE Communications Surveys & Tutorials, 21*, 812–837. https://doi.org/10.1109/COMST.2018.2862350

Mobility Management in Intelligent Transportation Systems

Virender Kumar and Mohammad Derawi

1 Introduction

To improve security, safety, and transportation efficiency, a novel intelligent system was required. Increasing vehicle traffic congestion and vehicle road accidents have led to emerge an intelligent transportation system (ITS) [1] and some other applications that enhance traffic management efficiency, highway security, and can provide infotainment on-demand services to users. An ITS is a technology, platform, or application that enhances the quality of transportation and relies a lot on collected data and on its analysis. Various applications that are related to transportation systems are referred as ITS, which provide security, safety, good transportation efficiency, safer roads for users, and efficient mobility management among vehicles. To make possibility of these applications, automobiles are well outfitted with communication devices and sensors so that they are able to exchange and gather information in vehicular environment to maintain vehicle traffic efficiency and improve road safety. Vehicles well equipped with sensors and communication devices and adopting specific wireless technology to facilitate communication between moving vehicles form a wireless communication network known as vehicular ad hoc networks (VANETs). To support various ITS applications ranging from vehicle safety on road to infotainment on-demand services for users, VANETs have emerged as promising wireless communication network in recent years [2].

V. Kumar (✉)
Department of Electronics and Communication Engineering, Chandigarh University, Mohali, Punjab, India
e-mail: virendergangotra@gmail.com

M. Derawi
Department of Electronic Systems, NTNU Gjøvik, Gjøvik, Norway
e-mail: mohammad.derawi@ntnu.no

C. Rompante Cunha et al. (eds.), *Transformative Cloud Computing, IoT and Extended Reality*, Signals and Communication Technology,
https://doi.org/10.1007/978-3-032-03554-7_11

The attention toward VANETs by both industrial and research communities has reflected the interest of government and reputed organizations all over world. To improve transportation efficiency and road safety, car manufacturers in Europe have instituted car-to-car communication consortium (C2C-CC). Besides, several consortiums have been instituted to standardize proposed algorithms and protocols. For vehicular networks, a spectrum of 75 MHz in band of 5.9 GHz has been approved by U.S. Federal Communication Commission (FCC). In 2008, a spectrum of 30 MHz in same band if ITS is also allocated by European Telecommunications Standards Institute. In 2014, various steps have been taken by the U.S. Department of Telecommunications (DoT) National Highway Traffic Safety Administration (NHTSA) to enable vehicle-to-vehicle communication for reasons of safety. Emerging applications of VANETs are perfectly matched with the aim of ITS since it offers these kinds of platforms for minimum losses and safer roads. Driven by its great possibilities, many stake holders such as automotive industries, academia, and government have taken great initiative in the development of VANETs, with several projects [3].

1.1 Intelligent Transportation Systems and Applications

In ITS, data is collected through hardware such as: GPS, cameras and sensors, and the data type, which is to be collected as speed and time, surveillance, vehicle weight, delays, traffic count, etc. As shown in Fig. 1, there is a real-time data

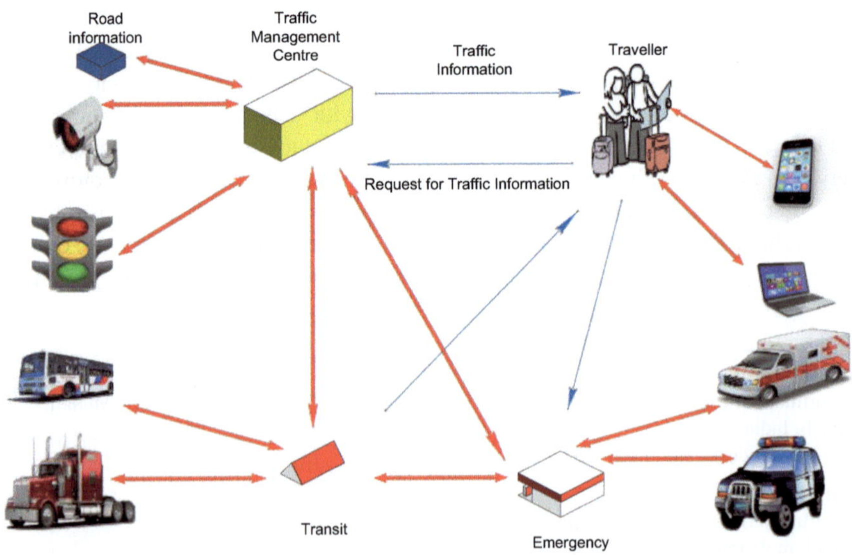

Fig. 1 ITS working [1]

transmission from road to traffic management center (TMC). TMC is a center where all data analysis occurs, and it coordinates all transmission to or from users. In TMC, data analyses, such as cleaning of data, rectification of error, adaptive logical analysis, and data synthesis, are done. This analysis of data is done to achieve intelligent information related to vehicular environment. In the next transmission of data (TMC to road), the collected real-time data is transmitted rapidly from TMC to users or travelers. In ITS, intelligent represents the collection of information and transmission of this information to travelers after analysis of data in TMC. Examples of intelligent information are real-time information, delay, travel speed, travel time, accidents on road, work zone condition, diversion or change in routes, etc. [4].

Due to rising population of urban area, there is a lot of impact on city roads in the form of congestion, crowd, or collision of vehicles. To cope up with these hurdles, ITS provides main application of smart traffic management, and some of other applications are given as: electronic toll collection, automated road speed enforcement, emergency vehicle alerting systems, real-time parking management, RFID in freight transportation, dynamic traffic light sequences, collision prevention systems, variable speed limits, etc.

2 Vehicular Ad Hoc Networks

In recent years, due to developments in wireless communication, a new infrastructure has emerged out referred as VANETs. Various applications of ITS are supported by VANETs. In vehicular networks, vehicles move in high speed, and they make self-organized networks such as similar to mobile ad hoc networks (MANETs). In VANETs the main communication nodes are vehicles, so, to provide safe and efficient transportation, several mobility management strategies are presented [5]. Unlike MANETs, fixed infrastructure units are deployed in VANETs along with road referred to as road-side units (RSUs). VANETs can be identified apart from other ad hoc networks because of its unique elevated mobility characteristic and hybrid network architecture. Figure 2 illustrates a VANET network architecture in which vehicles are communicating nodes and outfitted with onboard units (OBUs) and/or a single application unit (AU) or several. The OBUs are sensor device featuring interfaces for wireless networking and enable automobiles for communication. On the other hand, AUs are devices, which run applications and help OBUs to make communication with other vehicles. In VANETs, vehicles communicate to each other and with the deployed stationary road-side infrastructure RSUs. Intervehicle communication is referred as vehicle-to-vehicle (V2V) communication and vehicle-to-infrastructure (V2I) communication [6]. There is other type of communication referred as hybrid vehicle (HV) communication in which communication occurs using both V2I and V2V.

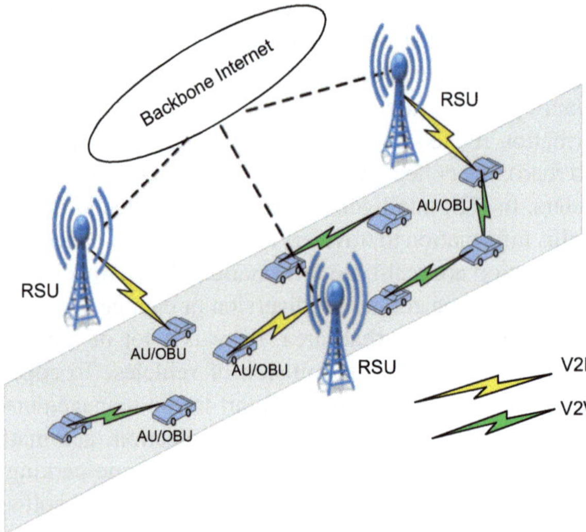

Fig. 2 An illustrative VANETs network architecture [6]

2.1 VANETs Characteristics

Due to some unique characteristics of VANETs, investigations and achievement conducted in MANETs are unable to apply directly to VANETs. Vehicular networks can be differentiated with MANETs in terms of vehicles' mobility pattern, dynamic network architecture, energy constraints of RSUs or OBUs, and real-life applications. Besides, the radio range allotted for VANETs is separated into different diverts as in dedicated short-range communication (DSRC), with the end goal that the control channel is utilized for transmitting control information, and safety messages and infotainment applications can be run by utilizing service channels from providers to users [7].

Network Density and Topology In VANETs, vehicles are characterized with high mobility leading to frequent occurrence of connection loss from RSUs. Besides, vehicles' density can vary from sparse mode network to very dense distribution. Therefore, variation in network density and topology has to be tackled by VANETs protocols.

Scalability In VANETs, the number of vehicles may increase from few to huge scale, particularly, in dense areas where frequent occurrence of multilane roads and intersections. Therefore, there is high probability of interference and wireless collisions during data dissemination among vehicles.

Mobility Model Due to dynamic topological behavior of VANETs, the mobility models required for VANETs are different from other ad hoc networks. Due to the restriction of vehicle mobility by predefined direction, road, and the number of lanes, VANETs topology is not entirely random. Advance prediction of trajectories and specific constraints imposed in vehicular environment such as traffic lights, obstacles, and speed limitation influences strictly on mobility model and network density.

Data Security and Privacy in the Network VANETs encounter a main challenge of privacy and data security due to use of wireless communication, which is susceptible to attacks, because attacks do not require physical access to infrastructure in vehicular networks. Hence, a designing of robust VANETs is needed.

Energy and Processing Capacities Unlike other ad hoc networks, in VANETs, moving vehicles are equipped with OBUs having enough and great computing powers in VANETs. These OBUs support wireless technologies such as global system for mobile communication GSM, Bluetooth, Wireless Fidelity (WIFI), and Worldwide Interoperability for Microwave Access (WiMAX), etc.

2.2 *VANETs Applications*

Vehicular communication offers immense capability for increasing user awareness and road safety. In VANETs, vehicles exchange messages or information through a radio channel to support mobile distributed applications. These applications are divided into three categories on the basis of their objectives. We concentrate only on few key applications in each category [8].

Applications for Road Safety To minimize road accidents and to enhance vehicles' or its deriver's safety by providing alarming messages are aims of this category of applications. There are certain applications, which are aimed to reduce the likelihood of traffic accidents and the death of vehicle occupants and are known as applications for road safety. In these applications, various mobility management schemes are used to avoid a number of accidents. These applications provide assistance and information for drivers to stay away from collision with other vehicles by using mobility management schemes. In road safety applications, alarming messages may be regarding of weather condition, collision warning, road damage, road sign alarms, and alarms in case of approaching vehicle, etc.

Critical Location Warning Getting information in advance about road characteristics and events while driving is interesting for vehicles or users for a specific area at certain time with enough anticipation. The location on the road generally denotes a work zone, a sharp curve in road, slippery road, etc. RSUs located on this area distribute alarming messages to vehicles before entering specific area. Critical location warning has been shown in Fig. 3.

V2V Cooperative Communication VANETs offer sharing of information by broadcasting from one vehicle to other neighboring vehicles within its transmission range when it encounters a critical situation via auditory display or visual and

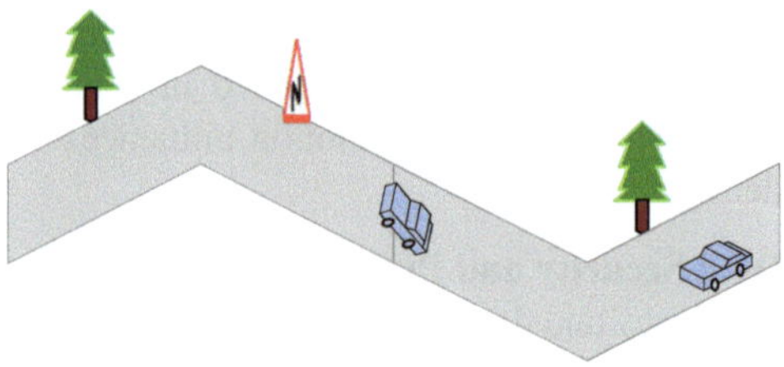

Fig. 3 Critical location warning [9]

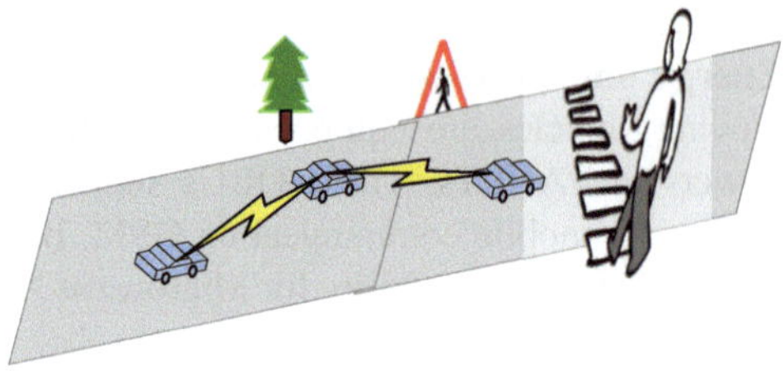

Fig. 4 Vehicle-to-vehicle awareness [9]

transmits alarming message to vehicles moving behind it in V2V communication mode. As shown in Fig. 4, when critical situation is found on road, driver is warned by algorithm, and alert messages are transmitted to other vehicles through relayed vehicle to other vehicles until the geographic zone is reached by the information.

Traffic Signal Extension As illustrated in Fig. 4, as vehicle approaches to traffic sign, information regarding location of traffic light and timing of signals is provided to vehicle. Hence, driver is notified by algorithm for optimal speed of vehicle or simply an alert display.

Applications for Traffic Efficiency and Management In these kinds of applications, traffic information is shared by RSUs among moving vehicles on road aiming of optimized traffic flow and improving the user experience. In these applications, the main focus is given on improving the flow of traffic and traffic assistance and updating the local information such as road condition, congestion, and accidental scenario. There are two areas in which management is done, i.e., speed management and cooperative navigation.

Optimized Traffic Flow Control As shown in Fig. 5, RSU monitors street, collects traffic-related data, and predicts congestion level on roadways within their transmission range. Further, RSU disseminates all gathered information to vehicle entering into its range and as a result optimized traffic flow occurs in VANET.

V2V Merging Assistance Application Merging of vehicles on main road without disrupting flow of vehicles in V2V communication has been shown in Fig. 6. Occurrence of prior communication among vehicles at sharp road curve makes traffic management efficient in VANETs scenario.

Applications for Information The main aim of this group of applications is to offer drivers with advertisements, entertainment, and requested data during

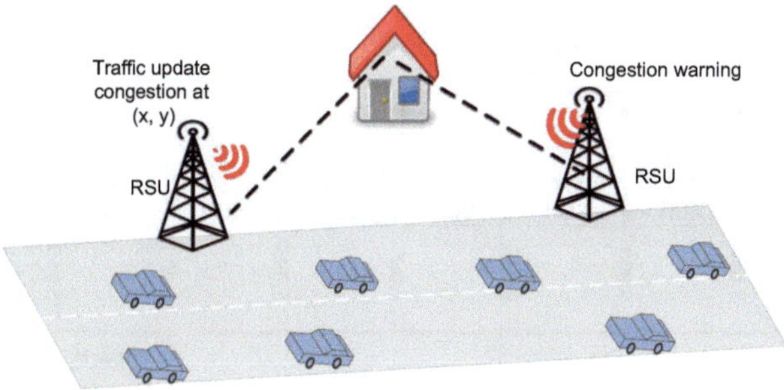

Fig. 5 Road congestion alert [10]

Fig. 6 V2V merging application [10]

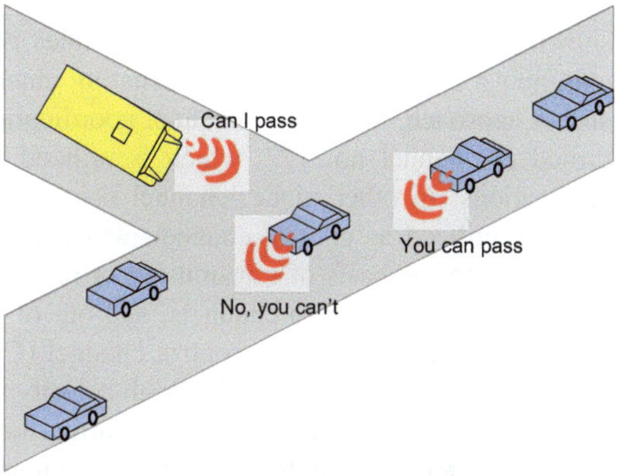

their journey. Internet access, video streaming and file sharing, online gaming, custom information services, and driver interest services are some examples of infotainment applications [11]. Since these applications are luxury services, it can tolerate some delay. (i) Cooperative local services: These applications concentrate on providing infotainment through locally based services such media downloads, point-of-interest notifications, and local electronic commerce. (ii) Global Internet services: Information gleaned from these sources is the main focus. ITS station life cycle, which concentrates on software and data updates, fleet management, parking zone management, and communities services, which include insurance and financial services, are typical examples.

3 Dedicated Short-Range Communication

A 75 MHz spectrum has been set aside by FCC in USA for DSRC to handle communication in vehicular environment [12]. On the basis of 802.11, the 802.11p is the protocol, which is utilized for DSRC. This wireless communication technology is outfitted with frequency band of 5.9 GHz in a total bandwidth of 75 MHz, as

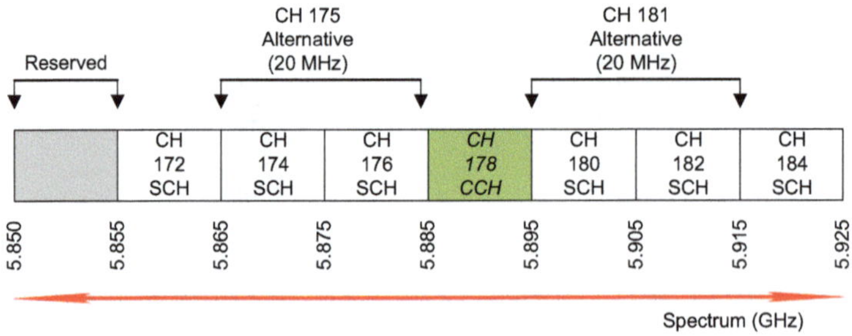

Fig. 7 DSRC spectrum [14]

shown in Fig. 7. The DSRC standard defines two approaches for channel access: continuous channel access and alternating channel access. In continuous channel access approach, there is no channel coordination required, and this mode always turned to channel no. 172, which is utilized for broadcasting and other safety applications. In other mode, channel coordination is required in which channel no. 178 utilized as control channel, which controls the management of channel switching on the basis of transmitting or receiving messages. In DSRC spectrum, there are a total of seven channels, and out of these seven channels, one channel (channel no. 178) named as control channel (CCH). In vehicular communication, safety messages are considered as high-priority messages, and the channel reserved for these safety messages is CCH. Furthermore, other six channels are called as service channels (SCHs), which are assigned to serving data communication purpose in vehicular networks [13].

In European Union (EU), ITS communication system utilized 5.9 GHz band for its operation. This spectrum is categarized into A to D. For traffic and safety applications, ITS-G5A with the frequency band (5.875–5.905 MHz) of three channels of 10 MHz is used. For non-safety applications, ITS-G5B with 20 MHz (two channels) and ITS-G5D (5.905–5.925 MHz) are set away for use in the future with traffic applications [16]. For ITS applications, two-band solution is reserved in Japan. The government of Japan has initiated 5.8 GHz in ITS applications utilizing a communication of 760 MHz, which is not compatible with protocol (802.11p) used in the Europe and USA.

4 Wave Layered Architecture

A standard architecture for occurrence of V2V and V2I communication in VANET is established by IEEE referred as wireless access in vehicular environment (WAVE). This communication architecture WAVE belongs to standard IEEE 802.11p. The demonstration of layered architecture of WAVE with 802.11p standard has been shown in Fig. 8. The WAVE standards comprised of IEEE 1609 suit, IEEE

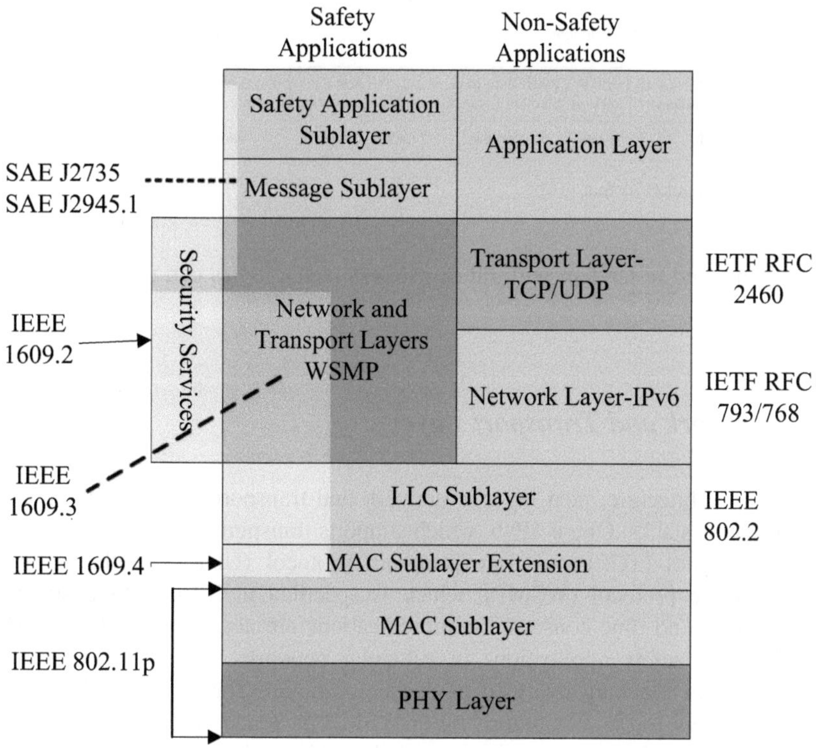

Fig. 8 WAVE layered architecture [16]

802.11p, Society of Associative Engineer J2735 and J2945. IEEE 1609 contains a set of standards of IEEE 1609.2 and IEEE 1609.3. For security services, IEEE 1609.2 standard is planned, and for network and transport layers' wave short message protocol (WSMP) services, a standard IEEE 1609.3 is assigned. IEEE 1609.4 standard is assigned for multichannel operations. Each layer is related with some specific standard. In the following part of this section, an overview of standard in layered presentation similar to open system interconnection (OSI) model is given [15–17].

4.1 Message Sublayer

This layer is at the top of the architecture, and it is defined in standards SAE J2735 and SAE J2945. Standard SAE J2735 specifies various kinds of messages to support various vehicle-based applications. The messages specified by SAE J2735 are used for both V2I and V2V communications in VANET scenario. Applications that may

WSMP-N-Header					WSMP-T-Header			WSMP Data
Subtype	WSMP-N-Header Option Indicator	WSMP Version	WAVE Information Element Extension	TPID	Address Information	WAVE Information Element Extension	WSM Length	WSMP Data
4 bits	1 bit	3 bits	Variable	1 Octet	Variable	Variable	Variable	

Fig. 9 WSMP packet format [16]

be implemented in tandem with other wireless technologies can also make use of this standard.

4.2 Network and Transport Layer

In WAVE architecture, next layer is network and transport layer, which supports two protocols stacks. One is IPv6, which supports transport protocols transmission control protocol (TCP) and user datagram protocol (UDP). Another is wave short message protocol (WSMP), which is specified in IEEE 1609.3 standard. High-priority and time-constraint communications are accommodated by WSMP protocol, i.e., safety applications in vehicular networks. WSMP is created for the purpose of reducing overload of packet IPv6. The format of WSMP packet has been shown in Fig. 9. Comprising of WSMP-N-Header and WSMP-Y-Header gives WSMP header as shown in figure below. Networking protocol features are selected by 4 bits subtype in WSMP-N-Header. Other parts of WSMP-N-Header are WSMP-N-Header option indicator, WSMP version, WAVE information element extension, and TPID. WSMP-T-Header consists of address information, WAVE information element extension, and WSM length. A WAVE management entity (WME) is also defined by IEEE 1609.3 standard to manage networking services of WAVE. Management information base (MIB) that has required status information and configuration is maintained by WME. On the basis of service requests, access of service channels by WAVE devices is also done by WME.

4.3 Logical Link Control Layer

Standard IEEE 802.2 specifies logical link control (LLC) layer whose purpose is to ensure and manage the reliability of data transmission. Three types of services are supported by LLC such as: (1) connectionless service without acknowledgement, (2) connection logical service with acknowledgement, and (3) connectionless service with acknowledgement. WSMP utilizes type 1 LLC service as specified in IEEE 802.2.

4.4 Media Access Control Layer

The MAC layer is defined by two standards: IEEE 802.11p and IEEE 1609.4, which covers MAC multichannel operations. The MAC multichannel sublayer extends the functionality of the MAC layer by enabling systems with one or multiple radios, such as those operating in the U.S. 5.9 GHz band, to efficiently switch between channels. A 75 MHz portion of the spectrum is designated for WAVE applications, with 5 MHz reserved for the guard band, 10 MHz allocated for the CCH, and six SCHs each with a 10 MHz bandwidth. There are two standard methods for channel access. The first one is the continuous access mode, which operates without the need for channel coordination. In this mode, channel 172 is utilized for broadcasting basic safety messages (BSM) as well as time-sensitive and high-priority data, such as for crash avoidance safety applications. The second mode, alternating access, necessitates channel coordination. In this mode, channel 178 serves as a control channel responsible for orchestrating channel switches to facilitate message transmission on other service channels. In this mode, at least one service channel and control channel are used by a typical WAVE device. The channel switching mechanism employs time division to separate CCH traffic, such as service advertisements, from potential interference on SCHs.

4.5 Physical Layer

The IEEE 802.11p standard defines the physical and MAC functions derived from the DSRC-ASTM E2213-03 standard originally published by the American Society for Testing and Materials (ASTM). IEEE 802.11p is a draft that improves upon the IEEE 802.11 standard, initially released in 2010. Including IEEE 802.11 within the IEEE family aims to incentivize chip vendors supporting 802.11p, thereby potentially lowering costs. The physical layer is segmented into sublayers. The initial sublayer is the physical medium-dependent (PMD) sublayer, which interfaces directly with the wireless medium and employs the orthogonal frequency-division multiplexing (OFDM) technique found in 802.11a. The second sublayer is the physical layer convergence procedure (PLCP). It establishes the relationship between the MAC frame and the fundamental PHY layer data unit by incorporating PHY layer overhead into the MAC frame, resulting in the creation of the PHY protocol data unit (PPDU) during transmission. On the receiver side, the PLCP reverses this process to extract a MAC frame from the PPDU. The structure of PPDU fields has remained consistent in the IEEE 802.11 standard for an extended period and has not changed in the IEEE 802.11p amendment.

5 Relay Vehicle Selection Protocols in VANETs

In V2I communication within range of RSU, there is a direct communication between vehicle and RSU. RSU can send information to vehicle, and at the same time, vehicle can request RSU for some data referred as requested data. But, due to high mobility of vehicles and limited locations of RSUs due to cost constraint, there is frequent occurrence of connection loss of vehicles from RSUs, and vehicles enter in uncovered area as shown in Fig. 10. Uncovered area may be defined as distance between two consecutive neighboring RSUs in which vehicles are not connected with RSUs. Therefore, RSUs cannot communicate or provide requested data directly to vehicles which were within its covered area or in its range, just few moments before, and now these are in uncovered area. Now, vehicle has to wait to get demanded data till to enter in next RSU's range.

In V2V communication, each vehicle has its transmission range within which it can make connection with other vehicles, but, beyond that transmission range, it cannot transmit messages to other vehicles. In this type of vehicular scenario, the concept of relay vehicle (RV) selection is a challenging task, and various schemes or algorithms are proposed by researchers for RV selection in VANETs. As shown in Fig. 10, vehicle "B" had requested for data to RSU 1, and due to its high mobility, vehicle "B" remains for a very short time within RSU1's range and enters in uncovered area and becomes destination vehicle (DV). Therefore, RSU1 could not provide requested data to vehicle "B" and selects vehicle "C" as RV to forward

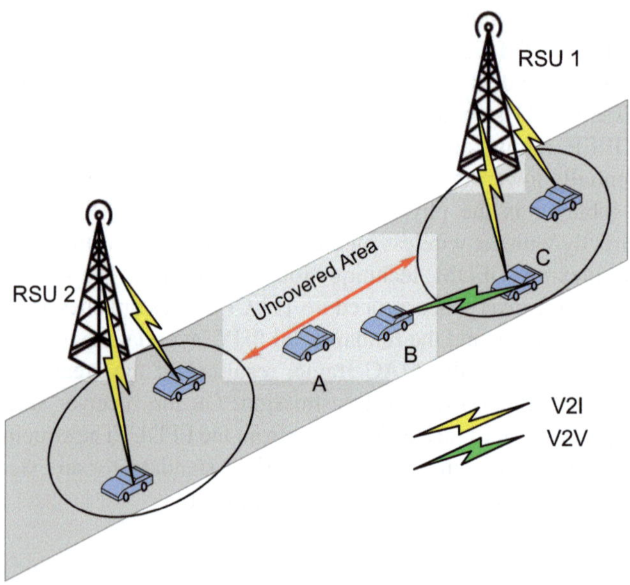

Fig. 10 Vehicles in uncovered area [6]

data to vehicle "B." If the requesting vehicle were "A," then its preceding vehicle "B" may also be selected as RV by RSU. An efficient relay vehicle selection and efficient broadcasting method is required for rapid, reliable, and scalable content or data dissemination, and to select optimal RV in vehicular network, several protocols have been presented. Some of the basic protocols are given below.

5.1 Linear-Based Protocols

In these protocols, the decision of forwarding messages to destination vehicle (DV) is made by itself DV on the basis of how much far is preceding forwarder vehicle, its velocity, and message direction. Hence, by examining neighboring vehicles' statistical data, suitable vehicle is selected as RV. DDT (distance defer transmission) is based on this type of approach [18]. In DDT, farthest vehicle retransmits message to DV, after when defer time is expired from other neighboring vehicles. The algorithm presented in [19] also assumes linear-based protocols in which DV defers the transmission time proportional to time required to receive full message and enters into wait state.

5.2 Area-Based Protocols

The concept of these protocols was initiated for MANETs in [20]. In these protocols, area covered in previous transmission and area that would be covered in new transmission is calculated by algorithm. Depending on threshold value, relay node is selected for message transmission. Recently, some area-based protocols are presented in [21, 22] in which each vehicle is enabled independently to take decision of transmitting or suppressing messages assuming covered new transmission area and heterogeneous transmission range.

5.3 Network Topology-Based Protocols

In these protocols, on the basis of prior information collected from beaconing process, the decision of selecting suitable relay node is taken, and data forwarding is performed either at sender end or at destination side. This approach observes low latency because of no requirement of waiting or time out at each hop. But this approach must be correlated with beaconing process, since it might be suffering from lower reception rates due to not receiving messages at relay node. The beaconing process keeps updating data by refreshing whenever required.

5.4 Clustering-Based Protocols

In cluster-based protocols, relay node is selected after formation of cluster of vehicles and choosing cluster heads (CHs) of various clusters. The main advantage of clustering is that it maintains the stability of vehicular network because a CH has knowledge of each and every vehicle's location, speed, and time of leaving cluster. Throughput and latency time improved in clustering-based protocols due to stability of network.

5.5 Cooperative Communication-Based Protocols

In these protocols, neighboring vehicles play key role for message forwarding. There is one-hop or more-hop message forwarding depending upon vehicle's speed and location. In cooperative communication-based protocol, neighboring vehicles cooperate in forwarding data to destination vehicle from source vehicle or RSU. The best example of these kinds of protocols is relay vehicle-based data transmission in vehicular networks [23, 24].

6 Mobility Management Based on Relay Vehicle Selection in VANETs

Several ideas have been discussed in the literature of mobility management in vehicular communication in order to maintain connectivity, to obtain scalability and reliability in VANETs. Various literatures can be found that reflects that relay vehicle plays an important role to manage mobility in vehicular communication. VANETs differ from MANETs in terms of their unique characteristics such as high mobility of vehicles and their dynamic network topology. Therefore, mobility management schemes proposed for MANETs are not applicable to VANETs. In recent years, various mobility management schemes are proposed to enhance the network stability and performance without using any relay vehicle in VANETs. In the undertaken research work, the emphasis has been laid upon selecting the relay vehicle to enhance network performance and successfully content delivery. In this chapter, mobility management in a vehicle-to-vehicle communication, vehicle-to-infrastructure communication, hybrid vehicle communication, and various mobile IP protocols used in wireless communication has been reviewed.

6.1 Relay Selection-Based Mobility Management for V2V Communication in VANETs

Although in vehicular scenarios, mobility of vehicles is managed by repairing, recovering, maintaining, and discovering new routes, but RV selection also plays a key role in managing mobility of vehicles while transmitting data from RV to DV in V2V communication. An emphasis on various contents that are part of mobility management in the V2V communication scenario has been mentioned below. Figure 11 illustrates V2V and V2I communications in VANETs.

Source Vehicle-Driven Relay Selection Schemes In source vehicle-driven relay vehicle selection scheme, current transmitting vehicle (CTV) selects RV in its neighborhood [26]. CTV selects RV on the basis of the available information regarding neighboring vehicles' locations in its transmission range. Chou et al. [27] present a scheme that transmits emergency messages in a multi-hop VANET environment. An approximate vehicular emergency dissemination scheme is proposed in which CTV neighboring vehicles maintain their routing table containing information about their speed, inter-vehicle distance, a ranking of RVs, and stability. Besides, each potential vehicle (each vehicle that receives information from SV) is assigned wait time (WT) based on the ranking of RVs given by CTV. After the expiry of WT, potential vehicles become RVs and retransmit information to the next vehicle. Simulation results of AVED outperform other schemes such as flooding, weighted P-Persistence, and Slotted 1-Persistence. Oliveira et al. [28] present a broadcasting scheme adaptive data dissemination protocol (ADDP) for emergency messages in multiple VANET scenarios. In this approach, CTV calculates the gain (G) of all neighboring vehicles on the basis of local density and distance of vehicles from itself. The vehicle bearing the highest gain is selected as RV. Simulation results of

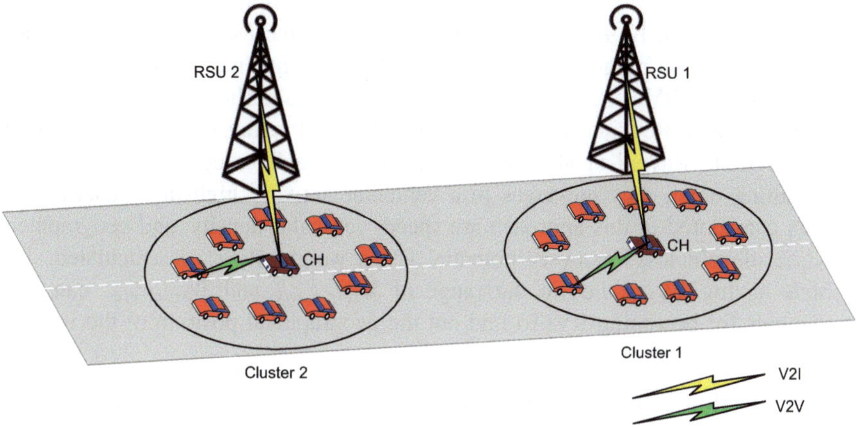

Fig. 11 V2V and V2I communication scenarios [25]

ADDP give better efficiency than DOT, PG, and TBN schemes. Rayeni et al. [29] present a density-based broadcasting scheme in which CTV calculates information about its neighborhood vehicle density and inter-vehicle space using a GPS system. Inbuilt OBUs of each vehicle play an important role in gathering this information. The basic concept of this methodology is to select the farthest vehicle from CTV, as RV. Wu et al. [30] present an approach in which antenna height, density, and velocity of vehicles are considered for RV selection process by CTV. It is considered in this approach that every vehicle knows about its velocity, antenna height, and density, and this information is shared with neighborhood vehicles.

Destination Vehicle-Driven Relay Selection Schemes In this type of RV selection scheme, independent selection criteria are used for each receiving vehicle for becoming RV. Rehman et al. [31] present the schemes in which every receiving vehicle shows the capability of becoming RV based on independent assessment function. Each vehicle has this IAF function, and upon receiving information, the vehicle having the highest IAF function is selected as relay vehicle. D. Tian et al. [32] present a protocol based on the distributed position for the broadcasting of emergency messages in a vehicular environment for both urban and highway scenarios. The RV selection process in this model considers the position, speed, and direction of potential RV from CTV with the help of GPS. A flag (FL) is given a value of 1 or -1 depending upon the direction of potential RV from CTV for calculation of waiting time (WT). After the expiry of WT, potential RV retransmits messages toward DVs. Any parameter for retransmission is not considered. Sung et al. [33] present a mechanism for RV selection in the vehicular environment based on road-layout information. In this mechanism, it is assumed that each vehicle knows road-layout information and their position from CTV. Depending upon the difference of position between the potential RV and CTV, an FT (forwarding timer) of potential RV is calculated. After the expiry of FT, potential RV can retransmit the messages on being selected as RV. Besides, another parameter intersection-forwarding timer (IFT) is calculated for those vehicles approaching or at the edge of the intersection. IF T_{max} is determined based on the location of potential RV in the transmission range of the vehicle. In the same manner, after the expiration of IFT, potential RV is selected as RV. Simulation results show better efficiency in comparison with ReC and ABSM in terms of no. of retransmission and network coverage time. Rayeni et al. [34] present an approach based on cross-layer for transmitting emergency messages in a vehicular network highway scenario. The factors considered in this approach are speed, vehicular density, and geographical position of vehicles. For every potential RV, a weight W_{relay} is calculated. The vehicle having W_{relay} lies in the range of $W_{min.relay}$ and $W_{max.relay}$ can give credentials for becoming RV. To find out the geographical position of the vehicle, GPS is used.

Relay Selection in Linear-Based Protocol In these types of protocols, moving vehicles take decision of becoming RV based on its location and direction of moving vehicle with respect to SV. So, every vehicle has different probability of becoming RV due its continuous changing position in vehicular environment. Sun et al.

[35] present distance-defer transmission (DDT) linear-based approach in which the process of retransmitting messages is deferred by receiving vehicle for a time that is inversely proportional to distance from SV. Therefore, the probability of becoming RV of farthest receiving vehicle is high, and it retransmits first in its neighborhood vehicles. Its simulated results show that DDT outperforms network topology-based approaches in sparse networks. Slavik et al. [36] present distance-to-mean (DTM) scheme in which wait time is considered. This wait time is directly proportional to the distance between SV and DV for observing those neighborhood vehicles that can retransmit a specific message. In DTM, spatial mean is calculated using positional information of all neighboring vehicles, and then distance is computed from receiving vehicle to spatial mean. The vehicle having larger distance is selected as RV because large area will be covered for retransmitting the messages. A threshold value is set in DTM, and any vehicle that measures the distance greater than threshold level will rebroadcast the messages. Naumov et al. [37] present the approach for RV selection based on the use of power of the sensed signal level. All vehicles are classified in one category out of three categories: IN group, OUT group, and preferred group (PG). Vehicles having signal power greater than OUT group and less than IN group are placed in PG group. Vehicles of IN group have signal power level stronger than PG group vehicles. After classification, vehicles wait for hold-off time to rebroadcast the messages. Thereafter, priority is assigned to groups with higher priority PG group. Next priority is given to OUT group. Based on hold-off time and priority of rebroadcasting, vehicle is selected as RV.

Relay Selection in Network Topology-Based Protocols In these protocols, on the basis of prior information collected from beaconing process, the decision of selecting suitable relay vehicle is taken, and data forwarding is performed either at sender end or at destination side. This approach observes low latency because of no requirement of waiting or time out at each hop. But this technique must be correlated with beaconing process, since it might be suffering from lower reception rates due to not receiving messages at relay node. The beaconing process keeps updating data by refreshing whenever required. Palazzi et al. [38] present an architecture that prioritizes the farthest vehicle from SV as forwarding vehicle. In this approach the defer time is determined based on contention window (CW) and varies between CWmin and CWmax depending upon the distance between SV and its estimated transmission range. The beaconing process is used to compute estimated transmission range. Huang et al. [39] present a packet-forwarding scheme based on vehicle density that uses the concept of defer time computed by CW in IEEE 802.11p MAC protocol. In this approach, instead of choosing farthest vehicle from transmitter, the optimal hop distance is used from transmitter to receiver. By opting optimal hop distance, the end-to-end broadcast delay is minimized and is determined by considering distance, transmission range, and density. Vehicle density is calculated by beaconing messages by counting the number of vehicles in the particular transmission range of vehicle.

Relay Selection in MAC-Based Cooperative Communication Protocols In recent years, cooperative communication has been leveraged to facilitate various improvements in throughput, reliability, and reduced transmission delay in vehicular networks. Bharti et al. in [40] propose a cooperative ad hoc MAC (CAH-MAC) scheme for MAC control in VANETs on the basis of distributed TDMA. In this scheme, RV is selected only on few conditions, (1) if transmission of packets fails, (2) if the RV is able to receive the packets, (3) if DV is in transmission range of RV, (4) if there is availability of time slot. The relay node that announces for retransmitting the packets becomes RV in case of multiple potential RVs, and retransmission occurs only in unused time slots. Thus, retransmission of packets in unused time leads to increase in throughput of the vehicular network. The drawback of CAH-MAC scheme is that RV consumes unused time slot even if there is no collision of packets leading to unavailability of unused time slots for other nodes or vehicles. This problem is presented in [41] by Bharti et al., and they investigated that if time slots are reserved for relay node, then due to cooperation collision network, performance degrades. To sort out this issue of reservation unused time slots collisions, same authors propose enhanced version of CAH-MAC scheme that is eCAH-MAC [42]. In eCAH-MAC, the RV suspends retransmission of packets if any one-hop neighborhood vehicle (SV or DV) is attempting to transmit packets. RV performs cooperative transmission only when no neighborhood vehicle in cooperative communication and the utilization of unreserved time slots occurs efficiently. Simulation results of above said cooperative communication schemes demonstrate that other existing schemes on MAC-based protocols are outperformed in terms of PDR (packet drop ratio) and PTD (packet transmission delay).

6.2 Relay Selection-Based Mobility Management for V2I Communication in VANETs

Vehicles interact with fixed RSUs beside roads in V2I communication scenario in vehicular networks as shown in Fig. 11. In this vehicular environment, each RSU has its own coverage range. On arrival of vehicles in RSU's range, beaconing messages are transmitted from RSU to vehicle and/or from vehicle to RSU before any transmission occurs. There are frequent disconnections between vehicles, and RSUs occur due to high speed of vehicles and limited deployment of RSUs beside the road. Hence, different schemes for the control of mobility are offered in the literature, and few of these are detailed in this section.

Adaptive Multiple Relay Selection in Vehicular Delay Tolerant Networks Ahamed et al. [43] present a selection scheme based on opportunistic relay node selection that uses various vehicles diversity in fading channel to identify relay vehicles in random manners. RSU adopts an optimal rate with which packets are transmitted to every vehicle moving within its coverage area. Out of many relay candidates, the potential vehicle that first successfully decodes received packets is selected as RV, but the condition is that it must be closest to DV. A relaying scheme is presented by Gt et al. [44] who utilize an analytical model for choosing multi-hop relays. This approach assumes the capacity of path link and vehicle's location in vehicular network. The optimal relay nodes are selected on the basis of higher path link capacity and closest location from destination vehicle. Wu et al. [45] introduce a multi-hop VDTN adaptive carry forward technique. In this scheme, one vehicle is selected to prestore the requested data by RSU out of many passing by vehicles. DV can regulate its speed with respect to RV speed by adopting interior point technique to extend the connection time between RV and DV in an uncoverage area. Besides, the vehicle whose outage time is minimum is selected as RV. However, there are some drawbacks that if the amount of missing data is larger than one, RV is insufficient to prestore this missing information. Controlling the speed of vehicles in vehicular networks is unfair when vehicles are moving with random speed. Ahmed et al. [46] present an adaptive multiple RV selection algorithm that allows RSUs selecting RVs based on most relevant criterion such that large amount missing or requested data is received by DV successfully. When vehicle enters in coverage range of RSU, it requests to RSU for data, and if RSU becomes aware that target vehicle is not able to receive all the data between available communication time, it selects RVs in such a manner that all the requested data can be transmitted to DV within communication time. The novel contribution of this research work is in allocation of missing requested data to multiple relay vehicles in an uncovered area. Missing requested data is allocated to those multiple RVs which are closest to DV in an uncovered area. Simulation results of this work show better performance in terms of data retrieval from selected RVs with respect to conventional schemes.

Clustering-Based Relay Selection in Vehicular Networks Chai et al. [47] present clustering-based data transmission scheme in vehicular network that considers selected CH as relay node in RSU's transmission range. In this research work, CH selection and clustering switching algorithms are proposed that jointly consider vehicle degree, availability of total resource bandwidth of CHs candidates, and relative difference of velocity between total CHs and cluster vehicles or vehicle mobility. Vehicle degree is the total number of vehicles to which a CH can communicate directly. The vehicle bearing higher vehicle degree, resource bandwidth, and mobility is selected as CH or relay. There is one drawback in this work that is CH or relay retransmits messages only within its transmission range. Once a DV leaves CH's coverage area, CH is not able to provide messages to DV. Sou et al. [48] present an RSU advertising analytical model to enhance VANETs connectivity that improves the safety message delivery in vehicular environment. When an SV meets with an accident, a safety message is broadcasted to concerned RSU deployed along with the direction of SV. Thereafter, RSU advertises these safety messages periodically to coming vehicles in the same direction until it gets a signal that to stop advertising safety message. If RSU is not found in SV's transmission range, then the closest vehicle coming from the opposite direction is selected as RV and takes on the responsibility to broadcast safety messages to rehealing node. This rehealing node could be the RSU or vehicle coming toward the SV. Thus, through reducing number of hops for safety message transmission, unnecessary traffic of messages is discarded. Ji et al. [49] present a cluster-based reliable and efficient data transmission algorithm based on link-reliability clustering in vehicular environment. In order to eliminate redundant unstable neighboring vehicles, a link lifetime neighboring sampling algorithm is proposed, and stable neighboring vehicle is selected as relay node for further clustering process. The proposed clustering approach is based on LREL and named as link-reliability clustering algorithm (LRCA). Simulation results of this approach show better results in terms of packet delivery ratio, clustering stability, control overhead, and end-to-end delay.

Relay-Based Network Access Selection in VANETs To support various service requirements for users, multiple communication technologies such as WLAN and cellular system are accessed by vehicle in vehicular networks. When a vehicle comes under an overlapped coverage range of multiple technologies, it has to select one suitable technology. Thus, to access network selection is also an important area for researchers, and relay vehicles play a significant role in accessing suitable network. Navarro et al. [50] present an algorithm for network selection based on MDP (Markov decision process) to achieve optimal decision in heterogeneous environment. MDP is basically used for selection network and vertical handoff in integrated networks. To characterize the status of network bandwidth of integrated network, a Markov model is proposed. The main factors, which are considered to present Markov model, are call completion, call arrival, and different network selection techniques on network resource occupation. Based on the Markov model, Huang et al. [51] present a network selection algorithm based on cost function without using any multi-hop communication, and the networks that are integrated are UMTS and WLAN. The number of hops for relay node access is two and the Markov model is three dimensional. The main traffic models of vehicles that are modeled are service arrival rate, vehicular user's dwell time, and channel holding time. Simulation results show that relay vehicle-based network access selection scheme provides better blocking probability compared to conventional schemes.

Relay-Based Cooperative Link Scheduling Protocols The selection of specific links in a specific manner so that moving vehicles can simultaneously use the cooperative connections while broadcasting, without interrupting the reception of incoming packets, is referred as link scheduling. Relay node plays a significant role in cooperative link scheduling in vehicular networking. Zheng et al. [52] present scheduling scheme for double-hop vehicular environment based on 2D-MCKP. This scheme chooses coordinator vehicles for each DV, and radio resources are assigned V2I and V2V links to define the problem of maximum utility function of summation. The enhancement of average utility with computational complexity is also performed by this scheme. However, the data packets arrival process and requirements of multiple services are not considered in this scheme. Pan et al. [53] present an investigation on throughput problems in cognitive vehicular environment with specific constraints taken, and cooperative communication aware link scheduling is proposed. In this scheme, a network model is designed as a graph in which regular connections are expanded with the help of dummy relay nodes to ensure that RV-based cooperative communication is compatible with direct link communication. Each vertex of a graph is assumed as resource point to make scheduled and shown as a pair of extended connection channels. Thereafter, a three-dimensional collaborative graph is developed to show interruption in collaborative expanded link. A maximization problem for throughput is formulated based on three-dimensional cooperative graphs. This optimal maximization problem for throughput is solved by linear programming which provides feasible using heuristic algorithm. Zheng et al. [54] present a scheduling scheme based on bipartite graph for vehicular networks. This scheme comprises three phases: (1) weighted solutions bipartite graph formation, (2) maximum weighted matching solution, and (3) optimization of the number of relay vehicles. In first phase, each vertex of bipartite graph is represented as vehicles and is split up into two groups such as single group of one-hop vehicles and second group of two-hop vehicles. In second phase, Kuhn Munkres algorithm is used to solve weighted matching problem of bipartite graph. In third phase, a search algorithm is performed to compute optimum separation. This link scheduling based on bipartite graph algorithm has lower complexity as compared to exhaustive search algorithm. This scheme does not employ vehicle's departure and arrival process, which is significant in vehicular networking. Zhang et al. [55] present performance analysis of multi-hop cognitive vehicular environment focused on energy consumption. In this analysis, energy efficiency of cognitive VANETs is formulated as problem, and according to an optimization issue, a cooperative relay scheduling scheme is proposed. The RV selection is based on the distance between potential relay and SV. Simulation results show the improvement of network performance in terms of energy consumption.

Traffic Management System in VANETs RSUs and vehicles are well equipped with specific intelligent system in traffic management systems. Milanes et al. [56] present a solution based on intelligent V2I communication-based traffic management. Avoidance of accidents and a regulating traffic flow are achieved by this solution by keep on giving information about conditions of traffic. A control room is responsible for calculating the location of vehicle, speed of vehicle, direction of vehicle, and roads layout. Traffic conditions are evaluated based on fuzzy-control algorithm. This algorithm is designed in such a way so that driving state, reference speed, and distance of vehicles can be easily evaluated leading to determine traffic situation. A system known as vehicle infrastructure integration (VII) is proposed in Ma et al. [57]. It determines the future journey time based on data such as the present travel time and the flow and density of VII-enabled vehicles. It does this by using two artificial intelligence (AI) paradigms: artificial neural networks (ANNs) and support vector regression (SVR).

6.3 *Relay Selection-Based Mobility Management in HV Communication in VANETs*

To extend the range in V2I communication scenarios, HV communication systems are proposed. In these HV communication scenarios, vehicles communicate with

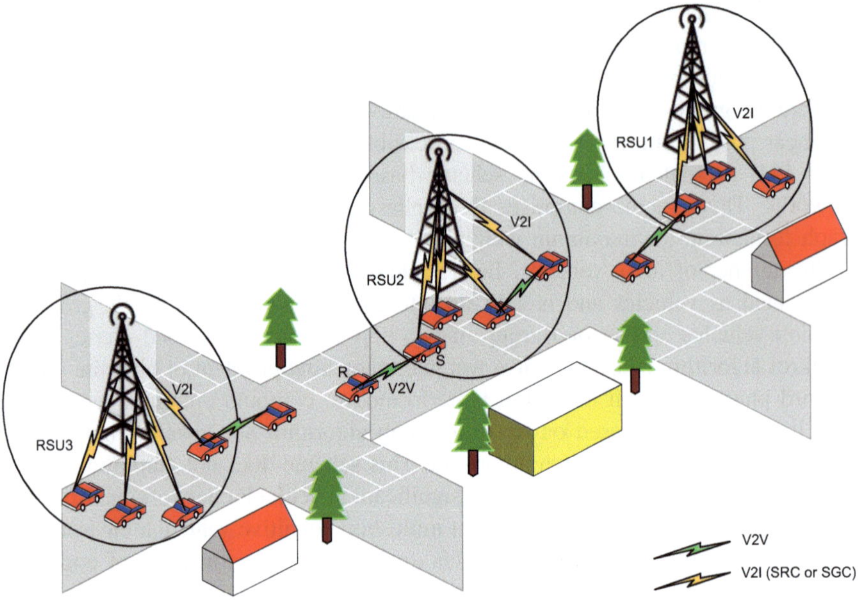

Fig. 12 HV communication scenario [60]

RSUs, while when they are not in the direct range of other vehicles. The HV communication system allows the similar applications or uses as system for V2I communication having a greater range of transmission. This part of chapter provides the few of mobility management schemes in HV communication in vehicular network. Figure 12 shows the scenario of vehicular communication in HV communication. At last, mobility management in heterogeneous environment is discussed.

Liu et al. [58] present a scheduling algorithm to study the cooperative data broadcasting system characteristics. In this scheme, RSU broadcast disseminates the information to passing by vehicles with V2I connectivity mode, and vehicles can additionally disseminate this information to its neighborhood vehicles in V2V communication mode. In three phases of this scheme, firstly, all vehicles advertise their presence and neighborhood vehicle information to RSU through heartbeat message broadcasting. Secondly, all vehicles transmit updated information regarding itself and neighborhood vehicle to RSU. Finally, vehicles may change their operational mode based on scheduling decision algorithm made by RSU. Besides, to improve the network coding, impact cache strategy is proposed. Although this scheme improves the data dissemination performance, end-to-end latency is increased, which is its drawback. Mehar et al. [59] present dissemination protocol to optimize bandwidth in heterogeneous vehicular environment. In this protocol, the farthest vehicle moving in each direction is selected as RV for fast data dissemination. Besides, this dissemination protocol is capable to cope up with road design and automobile networking scenarios. This algorithm maximizes message transmission, and to reduce the effect of disconnection in vehicular network, an additional store and forward mechanism has been taken. Simulation results show low end-to-end delay, high delivery ratio, and low bandwidth.

7 Conclusion

VANETs distinct from other kinds of wireless networks regarding communication patterns, mobility, as well as different networks architectures. It gets really difficult for safety messages disseminations based on relay vehicle in vehicular networks. In light of the review analysis, it has been noted that more work has been carried out in V2V communication networks in contrast to V2I communication networks. One likely explanation for this finding is that compared to V2I, the broadcasted communication prototype is more suitable for V2V scenarios. Safety messages dissemination through relay vehicle is more relied in V2V communication rather than V2I communication, so orientation of investigators will be toward safety applications only. This chapter provides an extensive cutting-edge review of relay vehicle-based mobility management in V2V, V2I, and HV communication scenarios in VANETs. The survey achieved a vast assortment of literature, which is most pertinent to the research topics mentioned above. This review study's primary objective was to categorize schemes based on traits and thoroughly examine several study options and the direction of future research. The extremely dynamic topology and frequently disconnected network, which are seen as significant issues, are addressed by the majority of the surveyed schemes.

References

1. Manaswini, R., Saikrishna, B., & Gupta, N. (2021). Architecture and framework enabling internet of vehicles towards intelligent transportation system. In A. Khanna & D. Gupta, A. E. Hassanien, S. Anand, & A. Jaiswal (Eds.), *Proceedings of international conference on innovative computing and communications (ICICC 2020)* (vol. 2, pp. 973–984). Springer: Singapore. https://doi.org/10.1007/978-981-15-5148-2-81
2. Delicato, F. C., Fuentes, L., Gámez, N., & Pires, P. F. (2009). A middleware family for VANETs. In P. M. Ruiz, & J. J. Garcia-Luna-Aceves (Eds.), *Ad-Hoc, mobile and wireless networks*. Lecture Notes in Computer Science (vol. 5793, pp. 409–422). Springer: Berlin. https://doi.org/10.1007/978-3-642-04383-3-31
3. Seth, I., Guleria, K., Panda, S. N. (2024). A comprehensive review on vehicular ad-hoc networks routing protocols for urban and highway scenarios, research gaps and future enhancements. *Peer-to-Peer Networking and Applications, 17*, 2090–2122. https://doi.org/10.1007/s12083-024-01683-1
4. Haghighat, A. K., Ravichandra-Mouli, V., & Chakraborty, P., et al. (2020). Applications of deep learning in intelligent transportation systems. *Journal of Big Data Analytics in Transportation, 2*, 115–145. https://doi.org/10.1007/s42421-020-00020-1

5. Kumar, V., & Manchanda, R. (2024). A Relay Selection based Mobility Management in VANETs: A Review, 3rd Indo-European Conference on Sustainable Materials, Environment, and Construction, In AIP Conference Proceedings (vol. 3100(1)).
6. Kumar, V., & Dahiya, P. K. (2020). Clustering-based optimal relay vehicle selection scheme for vehicular ad-hoc networks (VANETs). *International Journal of Mobile Computing and Multimedia Communications (IJMCMC), 11*(4), 67–83.
7. Kumar, V., & Dahiya, P. K. (2016). Mobility management in vehicular ad-hoc networks: a review. *IOSR Journal of Electronics and Communication Engineering, 11*(1), 85–96.
8. Zeadally, S., Hunt, R., Chen, Y.S., Irwin, A., & Hassan, A. (2012). Vehicular ad hoc networks (VANETs): status, results, and challenges. *Telecommunication Systems, 50*, 217–241. https://doi.org/10.1007/s11235-010-9400-5
9. Jakubiak, J., & Koucheryavy, Y. (2008). State of the art and research challenges for VANETs. In *Proceeding of the 5th IEEE consumer communications and networking conference (CCNC).*
10. Hartenstein, H., & Laberteaux, K. (2008). A tutorial survey on vehicular ad hoc networks. *IEEE Communications Magazine, 46*(6), 164–171.
11. Soldo, F., Casetti, C., Chiasserini, C. F., & Chaparro, P. A. (2011). Video streaming distribution in VANETs. *IEEE Transactions on Parallel and Distributed Systems, 22*(7), 1032–1039.
12. Kenney, J. B. (2011). Dedicated short-range communications (DSRC) standards in the United States. *Proceedings of the IEEE, 99*(7), 1162–1182.
13. Liu, Y., Dion, F., & Biswas, S. (2005). Dedicated short-range wireless communications for intelligent transportation system applications: state of the art. *Transportation Research Record, 1910*(1), 29–37. https://doi.org/10.1177/0361198105191000104
14. Kumar, R., Pal, R., Prakash, A., & Tripathi, R. (2019). A collective scheduling algorithm for vehicular ad hoc network. In A. Khare, U. Tiwary, I. Sethi, & N. Singh (Eds.), *Recent trends in communication, computing, and electronics*. Lecture Notes in Electrical Engineering (vol. 524, pp. 187–196). Singapore: Springer. https://doi.org/10.1007/978-981-13-2685-1-18
15. Uzcátegui, R. A., Sucre, A. J. D., & Acosta-Marum, G. (2009). WAVE: a tutorial. *IEEE Communications Magazine, 47*(5), 126–133.
16. IEEE Standard 1609.3-2016 (2016). IEEE standard for wireless access in vehicular environments (WAVE)—networking services. In *IEEE Std 1609.3-2016 (Revision of IEEE Std 1609.3-2010)* (pp. 1–160).
17. IEEE Standard 1609.4-2016 (2016). IEEE standard for wireless access in vehicular environments (WAVE)—multichannel operation. In *IEEE Std 1609.4-2016 (Revision of IEEE Std 1609.4-2010)* (pp. 1–94).
18. Kosch, T., Adler, C. J., Eichler, S., & Schroth, C. (2006). The scalability problem of vehicular ad hoc networks and how to solve it. *IEEE Wireless Communications, 13*, 22–28.
19. Mahajan, A., Potnis, N., Gopalan, K., & Wang, A. (2006). Urban mobility models for VANETs. In *Proceedings of the IEEE workshop on next generation wireless networks.*
20. Qiu, L., & Wang, J. (2015). Research of the model of traffic information acquisition and distribution based on VANET. In *Proceedings of the 2nd international conference on civil, material and environmental sciences* (pp. 1–4).
21. Alotaibi, M. M., & Mouftah, H. T. (2015). High speed multi-hop data dissemination for heterogeneous transmission ranges in VANETs. In *Proceedings of the IEEE international conference on ubiquitous wireless broadband (ICUWB), Montreal, QC, Canada*, pp. 1–7.
22. Alotaibi, M. M., & Mouftah, H. T. (2015). Probabilistic area-based dissemination for heterogeneous transmission ranges in vehicular ad-hoc networks. In *Proceedings of the IEEE international conference on computer and information technology; Ubiquitous Computing and Communications; Dependable, Autonomic and Secure Computing; Pervasive Intelligence and Computing (CIT/I-UCC/DASC/PICOM)* (pp. 1101–1108). IEEE.
23. Zhang, J., Zhang, Q., & Jia, W. (2009). VC-MAC: A cooperative MAC protocol in vehicular networks. *IEEE Transactions on Vehicular Technology, 58*(3), 1561–1571.
24. Chen, Y.-S., & Hung, K.-W. (2012). VC2-MAC: A two-cycle cooperative MAC protocol in vehicular networks. *Computer Communications, 35*(7), 861–874.

25. Gupta, N., Prakash, A., & Tripathi, R. (2017). Clustering-based enhanced safety message dissemination medium access control protocol for vehicular ad hoc network. *International Journal of Ad Hoc and Ubiquitous Computing, 24*(1–2), 76–89.
26. Li, G.-S., Wang, W.-L., & Yao, X.-W. (2012). An adaptive and opportunistic broadcast protocol for vehicular ad hoc networks. *International Journal of Automation and Computing, 9*(4), 378–387.
27. Chou, Y.-H., Chu, T.-H., Kuo, S.-Y., & Chen, C.-Y. (2018). An adaptive emergency broadcast strategy for vehicular ad hoc networks. *IEEE Sensors Journal, 18*(12), 4814–4821.
28. Oliveira, R., Montez, C., Boukerche, A., & Wangham, M. S. (2017). Reliable data dissemination protocol for VANET traffic safety applications. *Ad Hoc Networks, 63*, 30–44.
29. Rayeni, M. S., Hadi, A., & Sahu, P. K. (2015). Dynamic spatial partition density based emergency message dissemination in VANETs. *Vehicular Communications, 2*(4), 208–222.
30. Wu, C., Ohzahata, S., Ji, Y., & Kato, T. (2014). Multi-hop broadcasting in VANETs integrating intra-flow and inter-flow network coding. In *Proceedings of the 80th IEEE vehicular technology conference (VTC Fall)* (pp. 1–6).
31. Rehman, O. M. H., Bourdoucen, H., & Ould-Khaoua, M. (2015). Performance impact of relay cardinality on sender-driven messages dissemination in VANETs. In *Proceedings of the IEEE 8th GCC conference & exhibition (GCCCE)* (pp. 1–6).
32. Tian, D., et al. (2018). A distributed position-based protocol for emergency messages broadcasting in vehicular ad hoc networks. *IEEE Internet of Things Journal, 5*(2), 1218–1227.
33. Sung, Y., & Lee, M. (2018). A road layout based broadcast mechanism for urban vehicular ad hoc networks. *Wireless Communications and Mobile Computing, 2018*(1), 1565363.
34. Rayeni, M. S., Hadi, A., & Sahu, P. K. (2015). Dynamic spatial partition density-based emergency message dissemination in VANETs. *Vehicular Communications, 2*(4), 208–222.
35. Sun, M.-T., Feng, W. C., Lai, T.-H., Yamada, K., Okada, H., & Fujimura, K. (2000). GPS-based message broadcast for adaptive inter-vehicle communications. In *IEEE vehicular technology conference (VTC'2000)* (vol. 6, pp. 2685–2692).
36. Slavik, M., Mahgoub, I., & Alwakeel, M. M. (2014). Analysis and evaluation of distance-to-mean broadcast method for VANET. *Journal of King Saud University, Computer and Information Sciences, 26*(1), 153–160.
37. Naumov, V., Baumann, R., & Gross, T. (2006). An evaluation of inter-vehicle ad hoc networks based on realistic vehicular traces. In *Proceedings of the 7th ACM international symposium on Mobile ad hoc networking and computing (MobiHoc'06)* (pp. 108–119). ACM.
38. Palazzi, C. E., Roccetti, M., & Ferretti, S. (2010). An inter-vehicular communication architecture for safety and entertainment. *IEEE Transactions on Intelligent Transportation Systems, 11*(1), 90–99.
39. Huang, J., Huang, Y., & Wang, J. (2014). Vehicle density based forwarding protocol for safety message broadcast in VANET. *The Scientific World Journal*. https://doi.org/10.1155/2014/584164
40. Bharati, S., & Zhuang, W. (2013). CAH-MAC: Cooperative ADHOC MAC for vehicular networks. *IEEE Journal on Selected Areas in Communications, 31*(9), 470–479.
41. Bharati, S., Thanayankizil, L. V., Bai, F., & Zhuang, W. (2013). Effects of time slot reservation in cooperative ADHOC MAC for vehicular networks. In *Proceedings of the IEEE international conference on communications (ICC)* (pp. 6371–6375).
42. Bharati, S., Zhuang, W., Thanayankizil, L. V., & Bai, F. (2017). Link-layer cooperation based on distributed TDMA MAC for vehicular networks. *IEEE Transactions on Vehicular Technology, 66*(7), 6415–6427.
43. Ahmed, S. H., Kang, H., & Kim, D. (2015). Vehicular delay tolerant network (VDTN): Routing perspectives. In *Proceedings of the IEEE consumer communications and networking conference (CCNC)* (pp. 898–903).
44. Ge, Y., Wen, S., Ang, Y.-H., & Liang, Y.-C. (2010). Optimal relay selection in IEEE 802.16j multihop relay vehicular networks. *IEEE Transactions on Vehicular Technology, 59*(5), 2198–2206.

45. Wu, D., Zhu, G., & Zhao, D. (2013). Adaptive carry-store forward scheme in two-hop vehicular delay tolerant networks. *IEEE Communications Letters, 17*(4), 721–724.
46. Ahmed, S. H., Mu, D., & Kim, D. (2018). Improving bivious relay selection in vehicular delay tolerant networks. *IEEE Transactions on Intelligent Transportation Systems, 19*(3), 987–995.
47. Chai, R., Yang, B., Li, L., Sun, X., & Chen, Q. (2013). Clustering-based data transmission algorithms for VANET. In *International conference on wireless communications and signal processing (WCSP)*. https://doi.org/10.1109/WCSP.2013.6677258
48. Sou, S. I., & Tonguz, O. K. (2011). Enhancing VANET connectivity through roadside units on highways. *IEEE Transactions on Vehicular Technology, 60*(8), 3586–3602. https://doi.org/10.1109/TVT.2011.2165739
49. Ji, X., Yu, H., Fan, G., Sun, H., & Chen, L. (2018). Efficient and reliable cluster-based data transmission for vehicular ad hoc networks. *Mobile Information Systems*, 1–15. https://doi.org/10.1155/2018/9826782
50. Stevens-Navarro, E., Lin, Y., & Wong, V. W. S. (2008). An MDP-based vertical handoff decision algorithm for heterogeneous wireless networks. *IEEE Transactions on Vehicular Technology, 57*, 1243–1254.
51. Huang, Q., Yang, Y., Chai, R., & Chen, Q. (2012). Relay vehicle based access network selection scheme for Vehicular Ad Hoc Network. In *International conference on wireless communications and signal processing (WCSP)*. https://doi.org/10.1109/WCSP.2012.6542985
52. Zheng, Q., Zheng, K., Chatzimisios, P., & Liu, F. (2015). Joint optimization of link scheduling and resource allocation in cooperative vehicular networks. *EURASIP Journal on Wireless Communications and Networking, 2015*(1), 170.
53. Pan, M., Li, P., Fang, Y. (2012). Cooperative communication aware link scheduling for cognitive vehicular networks. *IEEE Journal on Selected Areas in Communications, 30*(4), 760–768.
54. Zheng, K., Liu, F., Zheng, Q., Xiang, W., Wang, W. (2013). A graph-based cooperative scheduling scheme for vehicular networks. *IEEE Transactions on Vehicular Technology, 62*(4), 1450–1458.
55. Zhang, L., Guo, C.-L., Liu, X., & Feng, C.-Y. (2014). An energy efficient cooperation relay scheduling scheme in multi-hop cognitive vehicular networks. In *Proceedings of the IEEE international conference on wireless communication and sensor network (WCSN)* (pp. 468–473).
56. Milanés, V., Villagrá, J., Godoy, J., Simó, J., Pérez, J., & Onieva, E. (2012). An intelligent V2I-based traffic management system. *IEEE Transactions on Intelligent Transportation Systems, 13*(1), 49–58.
57. Ma, Y., Chowdhury, M., Sadek, A., & Jeihani, M. (2012). Integrated traffic and communication performance evaluation of an intelligent vehicle infrastructure integration (VII) system for online travel-time prediction. *IEEE Transactions on Intelligent Transportation Systems, 13*(3), 1363–1372.
58. Liu, K., Ng, J. K.-Y., Wang, J., Lee, V. C. S., Wu, W., & Son, S. H. (2016). Network-coding-assisted data dissemination via cooperative vehicle-to-vehicle/-infrastructure communications. *IEEE Transactions on Intelligent Transportation Systems, 17*(6), 1509–1520.
59. Mehar, S., Senouci, S.-M., & Remy, G. (2012). Dissemination protocol for heterogeneous cooperative vehicular networks. In *Proceedings of the IEEE wireless days (WD)* (pp. 1–6).
60. Pal, R., Gupta, N., Prakash, A., & Tripathi, R. (2018). Adaptive mobility and range based clustering dependent MAC protocol for vehicular ad hoc networks. *Wireless Personal Communications, 98*, 1155–1170.

Anonymity and Privacy-Enhancing Mechanisms for AI-Based e-Health Systems

Kyriaki Tsantikidou, Nicolas Sklavos, and Iraklis Symeonidis

1 Introduction

Healthcare infrastructures are rapidly evolving, with real-time interactions between patients and health services becoming more and more feasible, even across EU borders. This progress is mainly due to the adoption of advanced technologies, such as Artificial Intelligence (AI), Machine Learning (ML), Deep Learning (DL), and Big Data, that enable large volume of real-data collection and transmission, high-speed processing, and highly accurate analysis. Specifically, AI-based models and classifiers play a pivotal role in optimizing traditional healthcare systems. They offer smart services that enhance the quality of life by interacting with patients in real time from the convenience of their homes, while simultaneously ensuring better utilization of hospital resources. Patients receive continuous care without the need for frequent hospital visits. The accuracy of medical prognosis and data analytics also increases.

Cloud computing enables seamless data access and facilitates on-demand resource sharing from any location worldwide. The continuous expansion of cloud structures has led to the efficient training and deployment of machine learning models on cloud platforms [1]. Once deployed, users can utilize these models for predictions without the need to manage or maintain the underlying model architecture. Microsoft Azure ML [2] and Google Prediction API [3] are two

K. Tsantikidou · N. Sklavos (✉)
SCYTALE Group, Computer Engineering and Informatics Department, University of Patras, Patra, Greece
e-mail: k.tsantikidou@upatras.gr; nsklavos@upatras.gr

I. Symeonidis
NSS, KTH Royal Institute of Technology, Stockholm, Sweden
e-mail: iraklis.symeonidis@ri.se

C. Rompante Cunha et al. (eds.), *Transformative Cloud Computing, IoT and Extended Reality*, Signals and Communication Technology,
https://doi.org/10.1007/978-3-032-03554-7_12

great examples of this concept called Machine Learning as a Service (MLaaS). In MLaaS, both training and classification processes can be outsourced to cloud servers. A common MLaaS architecture involves a cloud-based pre-trained ML mode, in which clients submit their queries for inference. Overall, this approach is widely applied in healthcare systems, including disease diagnosis, and other domains, including fraud detection, and social media platforms.

Data from various sources, such as information regarding patients' biometrics and surroundings, is continuously collected, processed and potentially shared with relevant third parties for medical purposes [4]. Thus, the handling of such sensitive information results in significant privacy and security challenges, as it is being regulated by the General Data Protection Regulation (GDPR). The GDPR mandates the thorough protection of sensitive data, especially patient-related data, throughout its lifecycle [5]. This means that data must be encrypted at the instant it is collected and remain encrypted during all subsequent transmission and processing by third parties. Third parties cannot be trusted with the private information of patients even though they offer smart and efficient health services [6, 7]. Only patients and their respective doctors should have full access. Failure to meet these standards can result in severe consequences, including compromised patient trust, legal and ethical complications and potential misuse of data.

Extensive research regarding robust, privacy-preserving frameworks, mechanisms and techniques must be performed to address the above-mentioned challenges. The desirable developed systems must be capable of maintaining data privacy and anonymity without hindering the effectiveness and accuracy of AI-based health services [8]. Moreover, their integration and implementation into the resource-constrained devices, which are commonly employed for smart health systems, must be considered. Appropriate solutions that achieve a more balanced trade-off between security, accuracy, resource consumption, and performance efficiency must be researched and evaluated.

This chapter aims to provide a comprehensive guide for healthcare service developers in order to better understand and adopt anonymity and privacy-enhancing mechanisms for AI-based e-Health systems. First, the complete framework of healthcare structures, which adopt AI-based data analysis models, will be described. Furthermore, the privacy vulnerabilities and most common threat models will be demonstrated and analyzed. The threat models will focus on the vulnerabilities that are introduced with AI-based systems and their impact on data privacy, anonymity, and the requirement of de-identification.

Second, the most utilized anonymity and privacy-enhancing mechanisms, such as Homomorphic Encryption (HE), Secure Multi-Party Computation (SMPC), Zero-Knowledge Proof, Federated Learning, and Data Anonymization Techniques, will be researched and listed. Each mechanism's functionality and methodologies for privacy enhancement will be highlighted. Moreover, their integration with AI-based tools will be discussed. Published implemented architectures showcasing each mechanism's integration with AI-based health data analytics will be presented, effectively highlighting the advantages and disadvantages of each proposed solution.

The chapter will focus on the mechanisms' applicability to smart healthcare applications, their level of security, and mitigation capability.

Finally, the chapter will discuss future directions based on the previously highlighted drawbacks of current published related research. As the constant evolution of current frameworks and the introduction of new techniques and mechanisms necessitate the creation of detailed guides, this chapter will facilitate policy-oriented and technical stakeholders in implementing efficient privacy-dedicated e-Health systems. The overall contribution of this chapter will be the categorization of threats, attacks, privacy-enhancing mechanisms, and future directions based on different hardware, security, and performance parameters.

2 Complete Framework of AI-Based e-Health Systems

In this section, the framework of AI-based healthcare systems will be analyzed. Its characteristics and components will be discussed. The security requirements as well as the integrated vulnerabilities will also be highlighted. Finally, the general threat model employed for these structures will be presented.

2.1 Structure of AI-Based e-Health Systems

The basic structure of smart healthcare systems, which mostly relies on the Internet of Things (IoT) technology, consists of three main units, as depicted in Fig. 1. The first unit includes IoT devices, healthcare sensors, and hospital equipment, in case the system is applied to a hospital infrastructure. All these components have resource constraints and are responsible for collecting data either of the patients or of their environment in real time. In terms of common IoT communication protocols, such as Message Queue Telemetry Transport (MQTT), these components are called Publishers, as they collect data and publish them to other authorized devices and users when requested [9].

The second unit consists of edge/fog computing and gateway devices, which have more resources and high computing capabilities. According to the MQTT protocol, these components are the Brokers of the communication, meaning they are responsible for routing requests or data and overall connecting the first unit with Cloud services, which is the third unit. Communication between the first and second units is performed via local area networks, such as IPv6 over Low Power Wireless Personal Area Networks (6LoWPAN), ZigBee® 14, and Wireless HART, that have small computation and energy requirements. Contrary to this, communication between the second and third units is performed via wide area networks, such as mobile phone networks, which are more general-purpose networks with higher computing capabilities. This differentiation exists due to the inability of the resource-constrained Publishers to correspond to wide area networks computing demands and big data transmissions.

The third unit, namely cloud services, has storing and processing abilities and communicates remotely with authorized users. As Artificial Intelligent (AI) mechanisms are developed, more and more healthcare systems utilize them to efficiently perform their desired services [1, 10]. These AI-based processing mechanisms are mostly performed by third-party vendors. As described by [11], IoT-based healthcare systems connect all scattered healthcare resources and patients and organize them in a decentralized manner. The users, whose data is being collected, are the data owners and third-party vendors are various organizations, such as universities and labs, that provide advanced services, such as data analysis, machine learning modeling, disease classification, and prediction. This creates a privacy problem for healthcare systems, as third-party services cannot be trusted with sensitive healthcare information or even session secrets for decrypting data [1, 8]. Therefore, even though data sharing can assist the development of the medical industry, most medical institutions, who abide by different administrative management, are reluctant to take the risk of sharing private information and violating privacy ethics or possibly losing economic benefits [12].

2.2 *Privacy Vulnerabilities*

E-Health systems, based on IoT technology, present various vulnerabilities throughout their architecture. Each basic unit of an e-health system, as depicted in Fig. 1, is affected by a variety of attacks targeting the system's hardware, firmware,

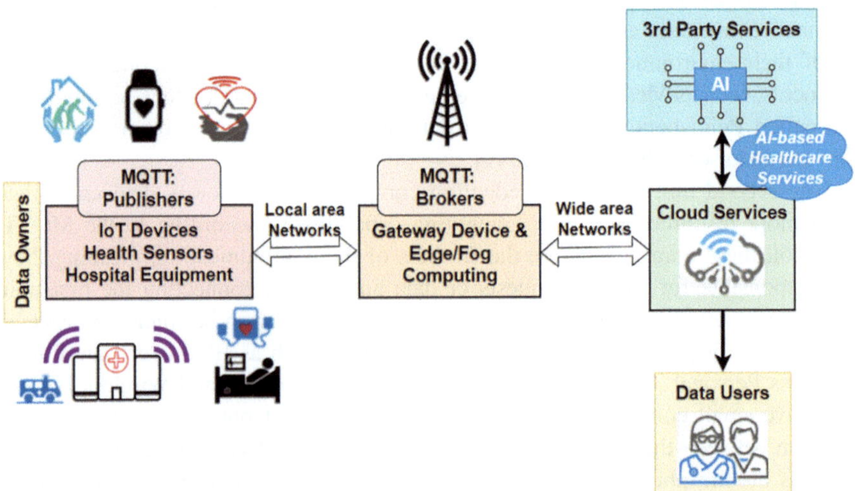

Fig. 1 An MQTT-based structure of AI-based e-Health systems

software, network protocol, or application processes [13]. Hardware vulnerabilities encompass security risks associated with a hardware device throughout its entire lifecycle, namely from its initial development within the supply chain to its eventual decommissioning. In contrast, vulnerabilities at the software and firmware levels stem from weaknesses in the application code, data injection mechanisms, and update failures. Network-level vulnerabilities originate from deficiencies in communication protocols and inadequate handling of traffic load. Process-level vulnerabilities refer to logical weaknesses within the system's design and programming methodologies. Depending on the components employed, each unit is susceptible to one or more of these vulnerabilities.

Various security and mitigation solutions have been proposed throughout the years, efficiently securing a simple e-health system. However, as AI solutions are added to the systems, and consequentially outside parties and high-processing vendors are engaged, privacy enforcement becomes more difficult to control [11]. Third parties require for their services unencrypted health information as raw materials, leading to the creation of authorization backdoors [11]. Moreover, third-party services require different access privileges, according to the utilized device, asset or virtual machine, in order to perform their computations. As IoT systems are highly scalable and include heterogeneous components, tracking and managing all these authorizations and access privileges is extremely difficult.

According to [6], there are three categories of cloud servers that offer smart, high-processing healthcare services. A trusted server is a server that ensures complete confidentiality, no information leakage or unauthorized access to stored private data. Nevertheless, threats to these types of servers may still arise from internal adversaries. Semi-trusted servers operate under the assumption of honesty but exhibit a degree of curiosity, potentially gaining access to sensitive data by colluding with malicious users. Lastly, untrusted servers lack inherent security assurances and require robust privacy-preserving mechanisms to safeguard data. Without these mechanisms, they remain susceptible to attacks originating from both internal and external adversaries.

In 2021, a health organization, called Broward Health and based in Florida, was attacked by a bad actor who had gained access to their private information by successfully intruding into a third-party vendor's system. The extracted health information included social security numbers, financial account information, phone numbers, birth names, medical records, and many more. Hacking and social engineering are two popular attacks employed for bypassing implemented authentication mechanisms. A successful example is the Carbanak malware, which was discovered in Kaspersky Lab in 2014. The adversaries gained access to the internal computers of financial institutions through spear phishing, afterwards infecting them with the malware. Finally, the uncontrolled access and management of electronic health records (EHRs) by multiple parties has also caused data breaches, as was demonstrated in 2020 by the largest Health Insurance Portability and Accountability Act (HIPAA) violation penalty being imposed on a health insurer, called Premera Blue Cross [14].

2.3 General Threat Model

By examining the e-Health structure at a first glance, there are three primary attack methods targeting different units [13, 15, 16]. The first type of attack focuses on the hardware components of the units, where adversaries can gain physical access to the IoT devices with the purpose of installing malware, tamper with their functionality, or even physically destroy them. Additionally, attackers can exploit authorization mechanisms to gain remote access, compromising both the availability and security of transmitted data. Due to resource limitations, these devices often lack high complexity cryptographic mechanisms, leading to easy encryption key extraction, data decryption, and unauthorized operations. The second category of attacks target gateway devices, which have similarities to a control center in Industry 4.0 systems. Both physical and cyber vulnerabilities are exploited to disturb operations or manipulate data and software. Even though stronger security mechanisms can be employed in this unit, prevention and mitigation of attacks can still be challenging as ongoing technological advancements and system scalability introduce new vulnerabilities. The third attack focuses on cloud services, primarily exploiting SQL or authorization vulnerabilities rather than physical breaches, which are difficult for external adversaries. However, as third-party vendors gain more authorization, they can be considered insiders, thus their threats must also be included into the general threat model.

By further analyzing the e-health structure and considering AI-based analytic tools provided by cloud servers and third-party vendors, the privacy threats can be divided into two categories [6, 17, 18]. The first category includes Data-Level privacy threats, which target private information, such as raw data, identity, and statistical properties of publicly shared information. After the transmission of sensitive data for processing, its usage and storage become difficult to control, thus increasing the risk of privacy leakages. There are two approaches for stealing sensitive information when a data owner publishes data. The first approach, called identification attack, aims to deanonymize the data by linking it to specific individuals [19]. Advanced machine learning techniques can be employed to recognize obfuscated images, geological patterns, and social network information. The second approach, called inference attack, extracts hidden sensitive information by performing reasoning or learning on data which may seem unrelated. Biometric data, such as fingerprints or gestures, can be reconstructed, which can then be utilized to compromise authentication systems [20].

The second category includes Model-Level privacy threats. These types of attacks target the machine learning model, seeking to extract critical information, such as architectural details, parameters, and training data. One common method is the model extraction attack, where adversaries repeatedly query the targeted model with the purpose of obtaining its output information and then utilizing the input-output pairs to replicate its behavior and steal its attributes. In the past, attackers successfully recreated neural networks, decision trees, and other AI models with these attacks against cloud-based machine learning services [21]. Another attack method is the model inversion attack, which infers sensitive information or properties of the training dataset by exploiting the model's confidence values or the difference in behavior when interacting with their training data and completely new data [22].

3 Anonymity and Privacy-Enhancing Mechanisms

Privacy enhancement in data-driven systems, specifically in those employing machine learning and cloud computing services, requires robust security mechanisms to protect sensitive information from unauthorized access and exploitation. Privacy-preserving and anonymization techniques are crucial in ensuring data confidentiality while maintaining the functionality and efficiency of computation processes. In recent years, global organizations, such as the National Institute of Standards and Technology (NIST), have increased their efforts in creating high security level standards regarding privacy-enhancing cryptography, by developing reference material and promoting advanced cryptographic tools for achieving their set goals [23]. Therefore, in an effort to assist privacy engineers at fulfilling privacy requirements in AI-based e-health systems, in this section, the most common anonymity and privacy-enhancing mechanisms are analyzed. Their applicability, methodologies, advantages, and disadvantages are highlighted.

3.1 Homomorphic Encryption (HE)

Homomorphic Encryption (HE) is a technique that enables the processing of encrypted data without requiring decryption [24]. It allows mathematical operations to be performed directly on encrypted information. Thus, the sensitive data always remains in an encrypted state. This characteristic makes HE highly valuable for processing sensitive data while safeguarding privacy, as it ensures secure analysis and processing of confidential health information. The primary goal of HE is to protect the confidentiality of information while enabling mathematical computations to be carried out on it. By employing HE, the original data's integrity and confidentiality are preserved, allowing computations to be performed on encrypted inputs, with decryption applied only to the final output. The operation flow of an AI-based health system that employs HE schemes is depicted in Fig. 2.

Homomorphic encryption (HE) encompasses various types, each designed for specific functionalities [25]. These include partially homomorphic encryption (PHE), somewhat homomorphic encryption (SHE), leveled homomorphic encryption (LHE), and fully homomorphic encryption (FHE). Partially homomorphic encryption (PHE) enables either addition or multiplication, a single type of gate, on encrypted data, but not both simultaneously. Somewhat homomorphic encryption (SHE) provides a middle ground by permitting a limited number of addition

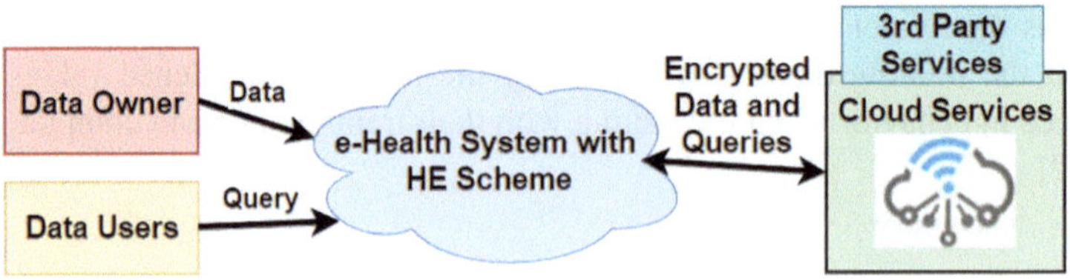

Fig. 2 Operation flow of an e-Health system with homomorphic encryption schemes

and multiplication operations on encrypted data. However, the ciphertext's size increases in accordance with every computation performed. Leveled homomorphic encryption (LHE) is similar with SHE with the exception that the ciphertext's size does not change as it is independent of specific homomorphic parameters. Fully homomorphic encryption (FHE) is the most advanced form, allowing both addition and multiplication operations of unbounded depth. The employment of an HE algorithm depends on the specific needs of the application. While PHE and SHE provide higher efficiency for particular operations, FHE offers broader functionality at the expense of greater computational overhead.

The algorithms for FHE mostly rely on the Ring Learning With Errors (RLWE) hardness assumption problem, which presents quantum-level of hardness. The two main and most secure algorithms are the BGV (Brakerski/Fan-Vercauteren scheme) [26] and BFV (Brakerski-Gentry-Vaikuntanathan) [27], which only support integers. The CKKS (Cheon-Kim-Kim-Song scheme) [28] enables computation on real numbers, but it is not by definition a HE. CKKS can perform approximate calculations, thus the result of computation on encrypted data and plain data are not the same. This can have various consequences for healthcare models, especially for those employed for emergencies or disease prognosis.

There are various libraries that include either one or more of the mentioned HE algorithms [29–32]. Moreover, there are various published works that propose HE-based systems for securing AI-based models for health data analytics [1, 11, 33–37]. These proposed systems utilize various open-sourced databases for their computations and validation testing. Lastly, contradicting those works that mostly adopt a more software-oriented approach, there are other published HE-based architectures that are implemented in hardware-dedicated devices, such as FPGAs [38–42]. Resource utilization, execution time, and security parameters are the three basic parameters whose trade-offs are being examined.

While homomorphic encryption schemes offer a promising solution for secure computations in e-health systems, they also present significant drawbacks that hinder their practical implementation. One primary challenge is the high computation overhead due to the substantial processing power and memory resources the HE schemes require. Thus, they are inefficient for real-time e-health applications. The encryption and decryption process of HE is computationally expensive, leading to increased latency in healthcare services, such as disease diagnosis, remote patient monitoring, and medical data analytics. Moreover, HE schemes often produce large-sized ciphertexts, significantly increasing transmission costs and storage that are problematic for resource-constrained e-health components. Another drawback is the high complexity of key management. HE schemes secure the communication flow and access control by employing sophisticated key distribution and maintenance mechanisms. Finally, as the mathematic complexity that HE schemes cover is low, they struggle to support more advanced machine learning models, thus limiting their applicability and scalability.

3.2 *Secure Multi-Party Computation (SMPC)*

Secure Multi-Party Computation (SMPC) is a cryptography technique that enables multiple parties who are interconnected to jointly execute various processes over their private inputs while ensuring that no individual party can gain access to the others' data. Specifically, each party holds a part of a secret in a distributed manner. After the interactive computation, in which the correctness of the outputs and the privacy of inputs is ensured, each party obtains the desired output with no secret disclosure. Similar to the cloud servers' categories, corrupt parties are also divided into semi-honest and malicious [43]. The first category implies that even though the parties try to learn the secret inputs, they otherwise faithfully follow the protocol. The second category refers to the parties that may present deviations in their behavior during the execution of the protocol. One of the basic advantages of SMPC schemes is their ability to perform secure computations in a decentralized manner and distinguishing the utilization of third-party vendors. The operation flow of an AI-based health system that employs SMPC schemes is depicted in Fig. 3.

There are various SMPC schemes that try to achieve the following security requirements. First, privacy is preserved as all parties know nothing except their own output and the following information that is derived from this output. Second, all produced outputs should be correct with no exception. Third, the input selected

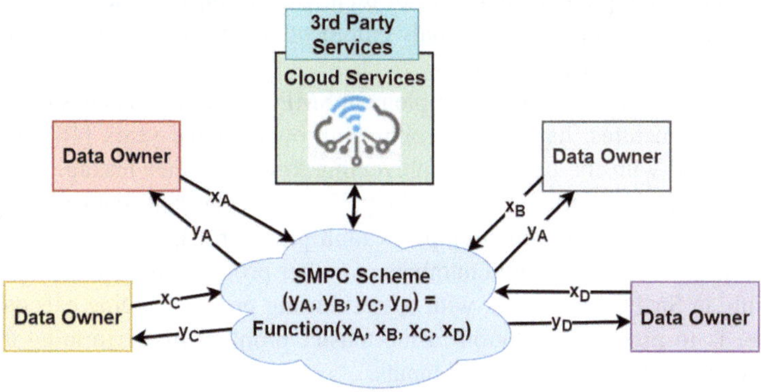

Fig. 3 Operation flow of an e-Health system with a secure multi-party computation scheme

by dishonest or compromised communication participants should be independent of the inputs of the authorized participants. Each party should have the freedom to choose their own input without external interference. Fourth, any form of denial-of-service attack within the computation process must be prevented with every participant receiving their expected result. Lastly, no participant should have an unfair advantage, and dishonest parties should be able to receive results only if all honest parties receive theirs first. By upholding these principles, secure computation remains reliable, equitable, and resistant to malicious interference.

The most common SMPC-based schemes employed in e-health systems are oblivious transfer, garbled circuits, and secret sharing. Oblivious transfer reinforces security with basic cryptographic mechanisms. The functionality of a 1-out-of-2 oblivious transfer is described. First, a receiver inputs a choice bit which determines which of two same length strings, inputted by the sender, will be transmitted. The sender cannot guess the string received by the receiver and outputs nothing. On the other hand, the receiver outputs the received string while knowing nothing regarding the other string. Oblivious transfer requires a high-demanding public-key cryptography mechanism, leading to the creation of extensions for improving practicality [44]. Garbled circuits enable encrypted function evaluation and ensure that only the final output is revealed without sharing intermediate computation values. It involves two different roles, a garbler and an evaluator, and is represented as a Boolean circuit that consists of XOR and AND gates [45]. In the garbled circuits' process, the garbler assigns two cryptographic keys to each circuit's wire. Each key is represented as a binary value, namely "0" or "1". For every logic gate, a garbled table is constructed via encryption of the output-wire keys by utilizing the corresponding input-wire keys. Afterwards, the garbler transmits the garbler circuit together with their encoded input-wire keys to the evaluator. The evaluator returns the desired function by interacting with the garbler. In many cases, garbler circuits are implemented together with a 1-out-of-2 oblivious transfer protocol in order to ensure that the input-wire key is sent to the evaluator without its disclosure. The final scheme, secret sharing, partitions data into multiple shares and ensures that no individual party can learn the full secret. The distribution among participants ensures that only a predefined subnet of them can reconstruct the original values. Shamir's Secret Sharing is a commonly employed algorithm for sharing the access keys for disclosing a master secret.

Numerous architectures, which implement SMPC-based approaches for securing critical infrastructures, have been proposed throughout the years [46]. designs a SMPC scheme with the purpose of preventing key parameter leakage for precise short-term load forecasting systems. In [43], smart metering data are protected with an SMPC protocol, which employs multiplication triples for computing the production of two secret-shared numbers. Another proposed design is presented in [47], with an SMPC framework with secret sharing protocols being executed. The objective is to provide a solution to the master problem by performing a secure computation on the shares of secret inputs.

Even though SMPC schemes offer many advantages to critical infrastructures, they also present several implementation challenges. The first major drawback is the same as with HE schemes, namely the significant computational and communication overhead added by the employed algorithms [47]. This causes increased latency and resource consumption. Moreover, as multiple interactions between participants are required throughout an SMPC-based protocol, real-time execution becomes difficult with possible scalability issues being added [48].

3.3 Zero-Knowledge Proof (ZKP)

A Zero-Knowledge Proof (ZKP) mechanism requires two parties, specifically a prover and a verifier. The prover aims to prove to the verifier that he has knowledge of a specific piece of information, with the verifier learning nothing more. The verifier is only convinced that the prover's statement is true. It can be employed in biometric systems by ensuring biometric verification through third-party authentication computations without confiding sensitive information. Moreover, it can be applied for proving the knowledge of a secret private key via a corresponding public key and a relation function. Thus, privacy is enhanced by ensuring that no additional information is leaked and only the validity of a statement is displayed. The simple operational flow of a ZKP scheme is presented in Fig. 4.

ZKP schemes can be divided into non-interactive and interactive. In the first schemes, the two parties only interact once for proving the correctness of the statement and do not require any further communication. The most adopted scheme, due to its efficiency in verification is the Zero-Knowledge Succinct Non-Interactive Argument of Knowledge (zkSNARK) [49]. Throughout the years, various works have also been published to enhance this scheme [50, 51]. On the other hand, interactive ZKP schemes involve a sequence of exchanges between prover and verifier. The prover must correspond correctly to specific challenges sent by the verifier to prove knowledge of the secret. A common interactive proof is the GKR protocol (Goldwasser, Kalai, and Rothblum) [52]. This protocol is compatible with tensor structures and suitable for verifying the execution of deep learning models.

In machine learning settings, ZKP schemes are mostly employed for proving the knowledge of model parameters, such as ownership, accuracy, or correctness of prediction [18]. They are not usually employed to train the models, due to the complexity of the procedure. The model owner initially commits to the model before proceeding with the prediction process. In case the prediction programs are expressed as arithmetic circuits, the prover can produce proof based on the

Fig. 4 Operation flow of zero-knowledge proof scheme

machine learning inputs while the verifier checks the correctness of the output with no information disclosure.

Various published papers have implemented these schemes for enhancing communication privacy [53]. proposes an E-participation framework that is based on blockchain technology. It utilizes attribute keys that are generated through a non-interactive ZKP technique and exchanged between participants and the E-participation agency [54]. also provides a ZKP for blockchain-based storage systems. The proposed scheme is responsible for verifying arbitrary SQL queries with fairness for decentralized storage services. In [55], a ZKP for modeling neural networks is presented. Specifically, it provides proof for Rectified Linear Unit (ReLU) activation and its backpropagation, afterwards producing an arithmetic circuit for depicting the entire training process.

Nevertheless, ZKPs display some challenges. First, all attributes must be given to the verifier in order to validate and identify the prover. Thus, efficient sampling is deemed impossible [53]. Finally, for large-scale applications, the computation overhead required can be significant, negatively affecting the real-time processing of the system, especially for resource-constrained IoT devices.

3.4 Federated Learning

Federated Learning (FL) allows the collaboration of multiple institutions, such as hospitals and clinics, with the purpose of training a machine learning model without secret disclosure [56]. Various published works have designed FL-based mechanisms for preserving the privacy of their system [57, 58]. Participants share model updates, including gradients and weights, thus avoiding centralizing sensitive information. As the procedure begins, first, a central server distributes the global model to all institutions. Afterwards, each institution is responsible for performing training on its local dataset and transmitting the generated updates back to the server. Lastly, the central server integrates the updates into the global model. This procedure continues for several rounds.

Even though FL assists in reducing data breaches and is appropriate for large databases, there are several challenges [56]. Model performance is affected by the high computation and bandwidth overhead added due to the frequent model updates and variations in data distribution. Inappropriate model updates can lead to newly created privacy vulnerabilities. Finally, the utility of the model can be significantly decreased due to the vulnerability of FL to poison attacks.

3.5 Anonymization Primitives

Besides the previously mentioned privacy-preserving mechanisms, there are various other less complex security techniques for enhancing e-health systems' privacy.

Differential Privacy (DP) mechanisms insert random noise into the training data or model parameters. Any addition to a single input must result in insignificant changes statistically for the output. Therefore, sensitive information may be leaked but in a control manner with small individual samples being revealed. The smaller the package the better the security achieved. The most popular DF mechanisms rely on specific mathematic equations, such as L-Diversity, Exponential, Laplacian, and Gaussian mechanisms [59–61]. Overall, DF enhances efficiency and mitigates inference attacks but has a limitation in data usability.

Another privacy-preserving scheme is the Trusted Execution Environments (TEEs) mechanism [62]. TEE allows for the program execution to be performed in secure hardware enclaves, therefore efficiently protecting the contents in the memory. The data are transmitted through a secure channel in the enclave, which in turn computes the desired encrypted results. Nevertheless, these mechanisms do not provide a method for protecting the memory address, even though the contents are protected. Thus, they are vulnerable to side-channel attacks.

Data Anonymization is another umbrella term that includes privacy-preserving techniques that eliminate or mask personal identifiable information (PII) from databases [56]. These mitigate privacy risks and specifically minimize the risk of identifying the patients whose data is utilized in AI training. Personal characteristics, such as names, addresses, and medical identifiers, are protected, with re-identification being mitigated. Various mathematical techniques can be employed for data anonymization, including masking, synthetic data generation, and noise injection. However, data anonymization techniques do not guarantee complete security, especially in cases when data are combined with other databases. Lastly, there are also pseudonymization techniques, including Random Number Generators (RNGs), Hash Functions, Message Authentication Codes (MACs), etc., that are employed for replacing identifying information with pseudonyms in an effort to conceal identifiable attributes [63, 64].

4 Conclusions and Future Directions

As AI-based healthcare systems continue to evolve, data privacy and security remain a critical challenge. This chapter serves as a comprehensive guide for healthcare service developers to comprehend and implement privacy-enhancing mechanisms in AI-based e-Health systems. It begins by outlining the framework of healthcare infrastructures that employ AI-based analysis tools, followed by an analysis of privacy vulnerabilities and a threat model affecting data anonymity and security. Principal privacy-enhancing techniques, including Homomorphic Encryption (HE), Secure Multi-Party Computation (SMPC), Zero-Knowledge Proof (ZKP), Federated Learning (FL), Data Anonymization, etc., are extensively discussed. Each method's functionality, integration with AI tools, and effectiveness in handling privacy risks are analyzed, along with real-world implementations that demonstrate their advantages and disadvantages.

The future of privacy-enhancing AI-based e-health systems will be shaped by implementational advancements in cryptographic techniques and regulatory frameworks. One promising direction is the reduction of computational overhead in order to enable real-time processing on resource-constrained healthcare devices. Additionally, hybrid approaches that combine multiple privacy-preserving techniques, such as integrating federated learning with differential privacy, can improve both security and model performance while minimizing data leakage risks. Further research, regarding each mechanism as well as their integration, is required to address adversarial threats, such as model inversion attacks, which can compromise patient data even in privacy-focused AI systems. Moreover, privacy mechanisms based on hardware structures, such as secure enclaves and TEEs, can offer an additional security layer, especially when integrated into the other algorithmic-based privacy mechanisms.

Another significant future direction includes the standardization and regulatory alignment of privacy-enhancing AI-based health frameworks. While GDPR provides basic guidelines, there is a need for globally recognized privacy standards. In recent years, the NIST organization has developed various standardization activities to showcase privacy-enhancing cryptography tools that abide by global regulators and provide high security solutions in traditional infrastructures [23].

Acknowledgments This publication is based upon work from: "Behavioral Next Generation in Wireless Networks for Cyber Security, (BEiNG-WISE)," 2023–2027, COST Action CA22104, supported by COST (European Cooperation in Science and Technology).

References

1. Akram, A., Khan, F., Tahir, S., Iqbal, A., Shah, S. A., & Baz, A. (2024). Privacy preserving inference for deep neural networks: Optimizing homomorphic encryption for efficient and secure classification. *IEEE Access, 12*, 15684–15695.
2. Azure Machine Learning. (2023). Retrieved March 27, 2025, from https://azure.microsoft.com/enau/products/machine-learning
3. Google Prediction API. (2023). Retrtieved March 27, 2025, from https://cloud.google.com/vertexai/docs/predictions/get-predictions
4. Tsantikidou, K., Delimpaltadakis, G., Diasakos, D., & Sklavos, N. (2025). AAL-based smart cane system with security and privacy features for blind and visually impaired individuals. *Microprocessors and Microsystems, 114–115*, 105155.
5. Phillips, M. (2018). International data-sharing norms: From the OECD to the general data protection regulation (GDPR). *Human Genetics, 137*, 575–582.
6. Chenthara, S., Ahmed, K., Wang, H., & Whittaker, F. (2019). Security and privacy-preserving challenges of e-health solutions in cloud computing. *IEEE Access, 7*, 74361–74382.
7. Sklavos, N., Chaves, R., Di Natale, G., & Regazzoni, F. (2017). *Hardware security and trust*. Springer International Publishing.
8. Curzon, J., Kosa, T. A., Akalu, R., & El-Khatib, K. (2021). Privacy and artificial intelligence. *IEEE Transactions on Artificial Intelligence, 2*, 96–108.
9. Tsantikidou, K., & Sklavos, N. (2023). Minimal resource required E-health system with end-to-end authenticated encryption mechanism. In *2023 12th international conference on modern circuits and systems technologies (MOCAST)* (pp. 1–4).

10. Jiang, Z., Van Zoest, V., Deng, W., Ngai, E. C. H., & Liu, J. (2023). Leveraging machine learning for disease diagnoses based on wearable devices: A survey. *IEEE Internet of Things Journal, 10*, 21959–21981.
11. Zhang, L., Wang, X., Wang, J., Pung, R., Wang, H., & Lam, K.-Y. (2024). An efficient FHE-enabled secure cloud–edge computing architecture for IoMT data protection with its application to pandemic modeling. *IEEE Internet of Things Journal, 11*, 15272–15284.
12. Zhang, L., Xu, J., Vijayakumar, P., Sharma, P. K., & Ghosh, U. (2023). Homomorphic encryption-based privacy-preserving federated learning in IoT-enabled healthcare system. *IEEE Transactions on Network Science and Engineering, 10*, 2864–2880.
13. Tsantikidou, K., & Sklavos, N. (2024). Threats, attacks, and cryptography frameworks of cybersecurity in critical infrastructures. *Cryptography, 8*, 7.
14. Guo, H., Li, W., Nejad, M., & Shen, C.-C. (2023). A hybrid Blockchain-edge architecture for electronic health record management with attribute-based cryptographic mechanisms. *IEEE Transactions on Network and Service Management, 20*, 1759–1774.
15. Zografopoulos, I., Ospina, J., Liu, X., & Konstantinou, C. (2021). Cyber-physical energy systems security: Threat modeling, risk assessment, resources, metrics, and case studies. *IEEE Access, 9*, 29775–29818.
16. Tsantikidou, K., & Sklavos, N. (2021). Vulnerabilities of internet of things, for healthcare devices and applications. In *2021 8th NAFOSTED conference on information and computer science (NICS), Hanoi, Vietnam* (pp. 498–503).
17. Xia, K., et al. (2023). Privacy-preserving brain–computer interfaces: A systematic review. *IEEE Transactions on Computational Social Systems, 10*, 2312–2324.
18. Qin, H., He, D., Feng, Q., Khan, M. K., Luo, M., & Choo, K.-K. R. (2024). Cryptographic primitives in privacy-preserving machine learning: A survey. *IEEE Transactions on Knowledge and Data Engineering, 36*, 1919–1934.
19. Li, H., Chen, Q., Zhu, H., Ma, D., Wen, H., & Shen, X. (2020). Privacy leakage via de-anonymization and aggregation in heterogeneous social networks. *IEEE Transactions on Dependable and Secure Computing, 17*, 350–362.
20. Al-Rubaie, M., & Chang, J. M. (2016). Reconstruction attacks against mobile-based continuous authentication systems in the cloud. *IEEE Transactions on Information Forensics and Security, 11*, 2648–2663.
21. Correia-Silva, J. R., Berriel, R. F., Badue, C., de Souza, A. F., & Oliveira-Santos, T. (2018). Copycat CNN: Stealing knowledge by persuading confession with random non-labeled data. In *Proceedings of the international joint conference on neural networks*.
22. Truex, S., Liu, L., Gursoy, M. E., Yu, L., & Wei, W. (2021). Demystifying membership inference attacks in machine learning as a service. *IEEE Transactions on Services Computing, 14*, 2073–2089.
23. NIST. (2025), Privacy-enhancing cryptography (PEC).. Retrieved March 27, 2025, from https:/ /csrc.nist.gov/projects/pec
24. Munjal, K., & Bhatia, R. (2022). A systematic review of homomorphic encryption and its contributions in healthcare industry. *Complex & Intelligent Systems, 9*(4), 1–28.
25. Podschwadt, R., Takabi, D., Hu, P., Rafiei, M. H., & Cai, Z. (2022). A survey of deep learning architectures for privacy-preserving machine learning with fully homomorphic encryption. *IEEE Access, 10*, 117477–117500.
26. Brakerski, Z., Gentry, C., & Vaikuntanathan, V. (2014). (leveled) fully homomorphic encryption without bootstrapping. *ACM Transactions on Computation Theory, 6*, 1–36.
27. Fan, J., & Vercauteren, F. (2012). Somewhat practical fully homomorphic encryption. In *IACR cryptology ePrint archive* (Vol. 144).
28. Cheon, J. H., Kim, A., Kim, M., & Song, Y. (2017). Homomorphic encryption for arithmetic of approximate numbers. In *Proceedings of the international conference on the theory and application of cryptology and information security*.
29. PALISADE homomorphic encryption software library—An open-source lattice crypto software library. (2015). Retrieved March 27, 2025, from https://palisade-crypto.org/

30. Halevi, S., & Shoup, V. (2014). Algorithms in Helib. In *Proceedings of the Annual Cryptology Conference*.
31. Microsoft SEAL (Release 3.6). (2020). Retrieved March 27, 2025, from https://github.com/Microsoft/SEAL
32. Al Badawi, A., et al. (2022). OpenFHE: Open-source fully homomorphic encryption library. In *Cryptology archive, paper 2022/915*. Retrieved March 27, 2025, from https://eprint.iacr.org/2022/915
33. Tanim, S. A., Aurnob, A. R., Mridha, M. F., Safran, M., Alfarhood, S., & Che, D. (2024). Secure federated learning for Parkinson's disease: Non-IID data partitioning and homomorphic encryption strategies. *IEEE Access, 12*, 127309–127327.
34. Pung, R., et al. (2022). Using high-resolution contact networks to evaluate SARS-CoV-2 transmission and control in large-scale multi-day events. *Nature Communications, 13*, 1956.
35. Al Badawi, A., Jin, C., Lin, J., Mun, C. F., Jie, S. J., Tan, B. H. M., et al. (2021). Towards the AlexNet moment for homomorphic encryption: HCNN the first homomorphic CNN on encrypted data with GPUs. *IEEE Transactions on Emerging Topics in Computing, 9*, 1330–1343.
36. Khan, T., Bakas, A., & Michalas, A. (2021). Blind faith: Privacy-preserving machine learning using function approximation. In *Proceedings of the IEEE symposium on computers and communications (ISCC)*.
37. Kim, D., & Guyot, C. (2023). Optimized privacy-preserving CNN inference with fully homomorphic encryption. *IEEE Transactions on Information Forensics and Security, 18*, 2175–2187.
38. Tsantikidou, K., & Sklavos, N. (2024). Lightweight privacy architecture for homomorphic operations in e-health systems. In *2024 IEEE international conference on internet of things and intelligence systems (IoTaIS), Bali, Indonesia* (pp. 290–295).
39. Su, Y., Yang, B., Yang, C., & Tian, L. (2020). FPGA-based hardware accelerator for leveled ring-LWE fully homomorphic encryption. *IEEE Access, 8*, 168008–168025.
40. Behera, S., & Prathuri, J. R. (2022). Design of Novel Hardware Architecture for fully homomorphic encryption algorithms in FPGA for real-time data in cloud computing. *IEEE Access, 10*, 131406–131418.
41. Nguyen, T. T., Kim, J., & Lee, H. (2023). CKKS-based homomorphic encryption architecture using parallel NTT multiplier. In *2023 IEEE international symposium on circuits and systems (ISCAS), Monterey, CA, USA*.
42. Syafalni, I., Jonatan, G., Sutisna, N., Mulyawan, R., & Adiono, T. (2022). Efficient homomorphic encryption accelerator with integrated PRNG using low-Cost FPGA. *IEEE Access, 10*, 7753–7771.
43. von der Heyden, J., et al. (2025). Privacy-preserving power flow analysis via secure multi-party computation. *IEEE Transactions on Smart Grid, 16*, 344–355.
44. Nielsen, J. B. (2007). Extending oblivious transfers efficiently-how to get robustness almost for free. In *Proc. IACR cryptology ePrint archive*.
45. Yao, A. C.-C. (1986). How to generate and exchange secrets. In *Proceedings of the 27th annual IEEE symposium on foundations of computer science*.
46. Si, C., Wang, H., Chen, L., Zhao, J., Min, Y., & Xu, F. (2024). Robust co-modeling for privacy-preserving short-term load forecasting with incongruent load data distributions. *IEEE Transactions on Smart Grid, 15*, 2985–2999.
47. Tian, N., Guo, Q., Sun, H., & Zhou, X. (2022). Fully privacy-preserving distributed optimization in power systems based on secret sharing. *iEnergy, 1*, 351–362.
48. Krishna Prakasha, K., & Sumalatha, U. (2025). Privacy-preserving techniques in biometric systems: Approaches and challenges. *IEEE Access, 13*, 32584–32616.
49. Gennaro, R., Gentry, C., Parno, B., & Raykova, M. (2013). Quadratic span programs and succinct NIZKs without PCPs. In *Annual international conference on the theory and applications of cryptographic techniques*.
50. Bünz, B., Bootle, J., Boneh, D., Poelstra, A., Wuille, P., & Maxwell, G. (2018). Bulletproofs: Short proofs for confidential transactions and more. In *Proceedings of the IEEE symposium on security and privacy (SP)*.

51. Groth, J., Kohlweiss, M., Maller, M., Meiklejohn, S., & Miers, I. (2018). Updatable and universal common reference strings with applications to Zk-SNARKs. In *Proceedings of the annual international cryptology conference*.
52. Goldwasser, S., Kalai, Y. T., & Rothblum, G. (2008). N.: Delegating computation: Interactive proofs for muggles. In *Proceedings of the 40th annual ACM symposium on theory of computing*.
53. Seo, J., Lee, J., Joo, Y., Lee, K., Sugumaran, V., & Park, S. (2025). A Blockchain-based E-participation framework utilizing zero-knowledge proofs with guaranteed sampling and differential reward mechanisms. *IEEE Access, 13*, 25752–25764.
54. Wang, H., Guo, Y., Bie, R., & Jia, X. (2024). Verifiable arbitrary queries with zero knowledge confidentiality in decentralized storage. *IEEE Transactions on Information Forensics and Security, 19*, 1071–1085.
55. Sun, H., Bai, T., Li, J., & Zhang, H. (2025). zkDL: Efficient zero-knowledge proofs of deep learning training. *IEEE Transactions on Information Forensics and Security, 20*, 914–927.
56. Nankya, M., Mugisa, A., Usman, Y., Upadhyay, A., & Chataut, R. (2024). Security and privacy in E-health systems: A review of AI and machine learning techniques. *IEEE Access, 12*, 148796–148816.
57. Dong, C., et al. (2024). Privacy-preserving and byzantine-robust federated learning. *IEEE Transactions on Dependable and Secure Computing, 21*, 889–904.
58. Chronis, C., et al. (2024). A survey on the use of federated learning in privacy-preserving recommender systems. *IEEE Open Journal of the Computer Society, 5*, 227–247.
59. McSherry, F., & Talwar, K. (2007). Mechanism design via differential privacy. In *Proceedings of the 48th annual IEEE symposium on foundations of computer science*.
60. Dwork, C., McSherry, F., Nissim, K., & Smith, A. (2006). Calibrating noise to sensitivity in private data analysis. In *Proceedings of the third conference on theory of cryptography*.
61. Dwork, C., et al. (2014). The algorithmic foundations of differential privacy. *Foundations and Trends® in Theoretical Computer Science, 9*, 211–407.
62. McKeen, F., et al. (2013). Innovative instructions and software model for isolated execution. In *Proceedings of the 2nd international workshop on hardware and architectural support for security and privacy*.
63. Tsantikidou, K., & Sklavos, N. (2024). An area efficient architecture of a novel chaotic system for high randomness security in e-health. *IEEE Computer Architecture Letters, 23*, 104–107.
64. Tsantikidou, K., & Sklavos, N. (2024). A compact pipelined architecture of chaotic logistic map for e-health applications. In *2024 13th international conference on modern circuits and systems technologies (MOCAST), Sofia, Bulgaria* (pp. 1–4).

Machine Learning Approaches to Detect Psychological Patterns in Social Media Content

Kapil Sharma, Neha Gupta, Shruti Sharma, and Vandit Hedau

1 Introduction

1.1 Social Media's Effect on Mental Health and Machine Learning-Based Predictive Modelling

Social media's widespread use has raised worries near its possible impact on users' mental health. With an emphasis on mediating factors like social isolation, smartphone addiction, and social capital—both linking and bonding types—this study investigates the assembly between social media use and mental health. There is a chance to gain valuable insights into users' psychological behaviours because of the huge volumes of data generated on social media sites like Facebook, Instagram, and Twitter.

This study's main goal is to create analytical models by analysing social media content and evaluating its relationship to psychological health issues using machine learning techniques. To find trends in post content, sentiment, and engagement metrics, algorithms similar Random Forest (RF), Logistic Regression (LR), and Support Vector Machines (SVM) are used as a model. These trends could offer useful indicators of mental health conditions like stress, anxiety, and depression.

Posts on social media provide ongoing data for psychological evaluation since they are real-time reflections of users' feelings, ideas, and responses what actually they have in their mind. Sentiment analysis recognizes changes in emotional

K. Sharma · N. Gupta (✉)
Symbiosis University of Applied Sciences, Indore, India
e-mail: Kapil.sharma@suas.ac.in; Neha.gupta@suas.ac.in

S. Sharma · V. Hedau
DAVV, Indore, India

C. Rompante Cunha et al. (eds.), *Transformative Cloud Computing, IoT and Extended Reality*, Signals and Communication Technology,
https://doi.org/10.1007/978-3-032-03554-7_13

tone, while natural language processing (NLP) methods allow for the methodical assessment of textual content. Combining text analysis and machine learning offers a fresh way to understand the psychological effects of social media use.

In addition to outlining the research goals, this chapter highlights the importance of machine learning and its models in forecasting and predicting outcomes related to mental health. Moreover, covered are ethical issues like user privacy and responsible data use. This study intends to support early psychological distress detection and enhanced mental health interventions in the digital age by fusing data-driven insights with ethical considerations.

Social media use is increasing enormously, which has raised questions about possible effects on users' psychological health. The main goal of this research is to explain how social media use touches psychological health. It offers a more thorough analysis of the phenomena by considering a number of mediators, such as social isolation, smartphone addiction, and social capital types (such as bridging and bonding social capital). With the cumulative use of social media platforms, there is a wealth of data available that can provide insights into individuals' psychological behaviour. This research explores the application of machine learning techniques, including Random Forest (RF) classifier, Logistics Regression (LR), and Support Vector Machine (SVM), to predict psychological behaviour using social media posts. By analysing the content and sentiment of users' posts, along with other metadata such as frequency of posts and engagement metrics, these algorithms can uncover patterns and correlations that offer valuable predictive insights (Fig. 1).

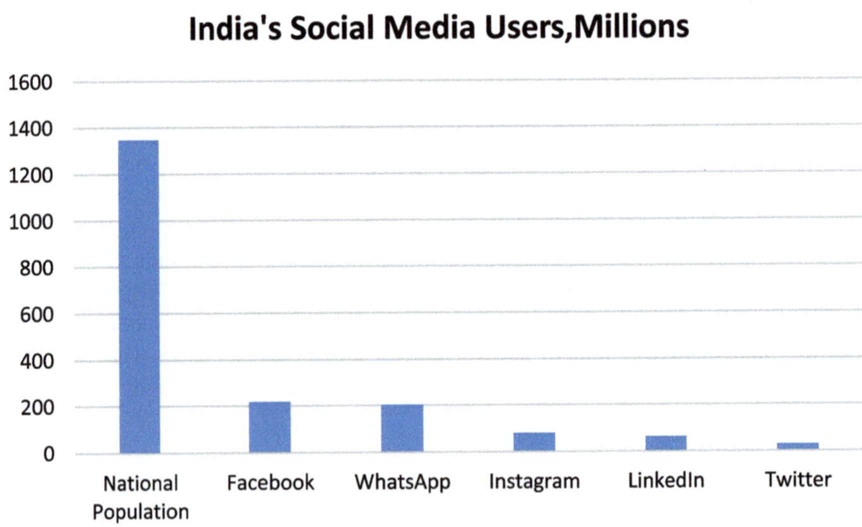

Fig. 1 Popular social media platforms in India by user base (in millions) [2]

1.2 Prediction of Psychological Behaviour Using Social Media Posts

Social media has become a part of life in the last few years. It now offers unique opportunities to analyse and forecast behaviour within the field of psychology. The vast amount of data offered by users of Twitter, Facebook, and Instagram has become a valuable resource when looking into the emotional and psychological status of a person. If these posts are texted, they can be analysed with the help of NLP and machine learning tools to unveil invaluable pointers into the state of one's mental well-being, uncover psychosomatic ailments, and even forecast psychosocial issues. Social media data is significantly helpful in measuring psychological behaviour due to the content being delivered in real time.

A social media post reflects the before and current status of a person's feelings, emotions, and thoughts. The absence of emotion in sociological texting is dictated by a specific set of words in the text, which, when taken in large quantities, leads to stress, depression, or anxiety. From another point of view, one can associate social media with the very unwanted actions of the harmful kind of sociological robot, which leaves traces of traces of people in everything done or accomplished done with feeling (Fig. 2).

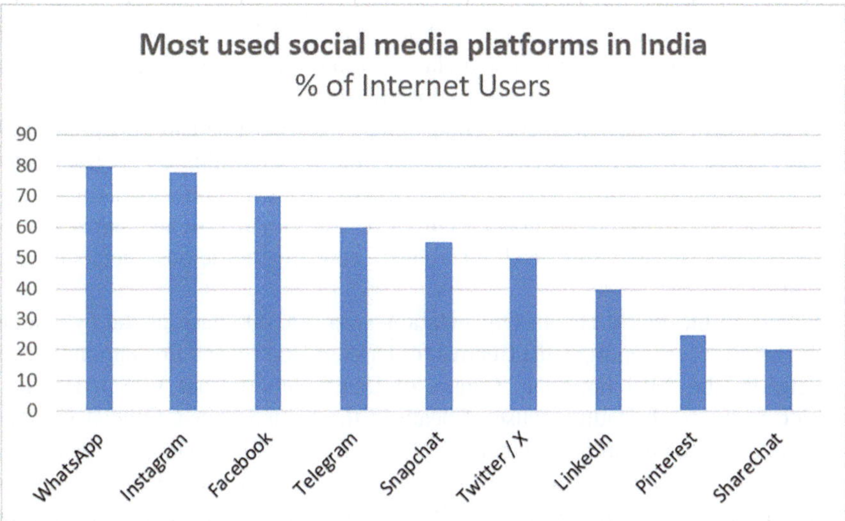

Fig. 2 Most used social media platforms in India (% of internet users) [4]

Social media has become an indispensable aspect of modern life, offering researchers an unprecedented opportunity to analyse digital footprints and predict psychological behaviour. The extensive data generated on stages such as Twitter, Facebook, and Instagram provide a rich resource for assessing emotional and mental states. Machine learning and NLP techniques can extract meaningful patterns from user-generated content, enabling early detection of psychological distress and behavioural trends.

One of the major compensations of social media data is its real-time nature, capturing immediate emotional responses and thoughts. Linguistic patterns, sentiment shifts, and keyword frequency in posts can serve as indicators of anxiety, depression, or stress. Sentiment analysis, in particular, plays a crucial role in classifying negative emotional states that may signal underlying mental health concerns.

Interaction metrics—such as social media post frequency, engagement levels (likes, comments, shares), and changes in activity—further contribute to psychological assessment. Sudden declines or spikes in social media participation may reflect mood fluctuations or emotional distress of the individual. Combining these behavioural indicators with textual analysis enhances the accuracy of psychological state predictions.

Machine learning models, counting Support Vector Machine (SVM), Logistic Regression (LR), and deep learning (DL) architectures, have demonstrated efficacy in classifying mental health circumstances from social media data. Keyword extraction and topic modelling can reveal recurring themes related to loneliness, social anxiety, or depressive symptoms. Supervised learning approaches, trained on annotated datasets, enable automated classification of social media posts based on psychological criteria, facilitating early intervention.

In spite of these advancements, ethical challenges must be carefully addressed. Privacy concerns, data security, and the potential for algorithmic bias necessitate stringent safeguards. Ensuring responsible data usage and minimizing harm to users are paramount. Additionally, model accuracy depends on diverse and representative datasets, underscoring the need for robust, ethically conscious methodologies (Fig. 3).

Giving to the World Health Organization, over 800,000 persons die by suicide each year. A recent report by the Centres for Disease Control and Prevention (CDC), specifically the 2019 Important Causes of Death Report from WISQARS, lists suicide as the tenth leading cause of death in the United States. Official US data indicate that one person dies by suicide every 11.1 min. In Canada, 2017 statistics show that 4157 individuals died by suicide, making it the ninth leading cause of death in the country. During the COVID-19 pandemic, clinical psychologists and researchers observed a rise in mental health issues, often amplified by social media exposure during lockdowns. The pandemic has left lasting effects on public mental health due to economic uncertainty and social isolation. Suicide not only has severe emotional consequences but also impacts the physical and economic well-being of society. Early prediction of suicide risk can help lower suicide rates by encouraging timely preventive action. A report from August 2021 revealed that 1.6 million people in England were waiting for mental health services, and an estimated eight

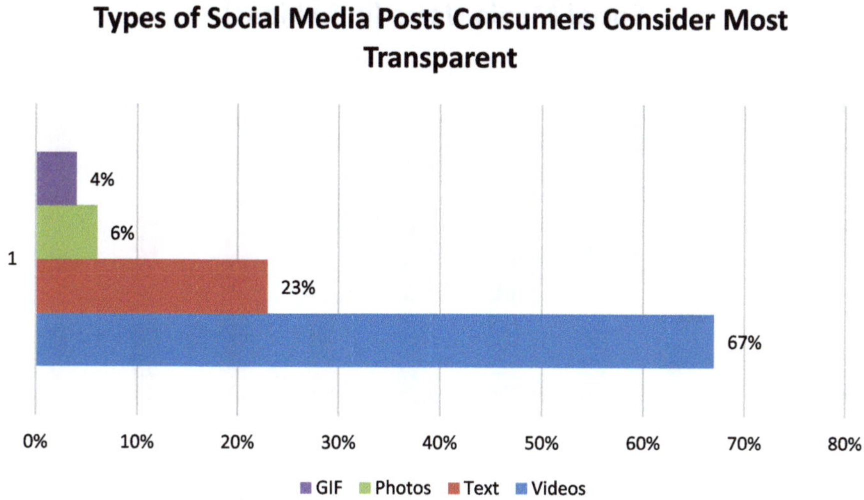

Fig. 3 Types of social media posts consumers consider most transparent (Adapted from Statista, 2024)

million were unable to receive specialized care due to not meeting the criteria for treatment. This highlights the importance of automating mental health detection using social media data, where users freely express their thoughts, emotions, and beliefs. These posts, often containing diverse, unstructured, and informal content, are readable by humans but challenging for automated systems to interpret. WHO ranks unhappiness as one of the most devastating diseases in the World [1]. In addition, about two thirds of depressed people do not seek suitable treatments, which lead to major consequences. Medical science relies on asking the patients questions about their situations, which does not identify depression in a precise way [2]. The patient must attend more than one session during a period of 2 weeks. The classification of a not depressed condition as a depressed is a False Positive problem [3]. However, researchers found that the Electronic Health Record (EHR) structures are not optimally designed to handle integrating behavioural health and primary care. EHRs lack to support documenting and tracking data for behavioural health circumstances such as depression.

Figure 4 shows number of social media users in India from 2015 to 2019 in blue and it also shows forecasted data of social media users from 2020 to 2023 in orange. A statistic made by researcher eMarketer [5] has shown the number of social media users in 2015 was 142 million and it is increasing day by day. According to eMarketer, this figure will be 448 million in 2023 in India. Most of the people use social media to rapid their feelings, emotions, and what they are doing on their daily routines. Rich bodies of work have been approved that social media is an open area for many people to express their negative emotions by division information which reflects those emotions [6]. Many researchers have been positively proving that social media has been successfully used to maintain people's mental health. By

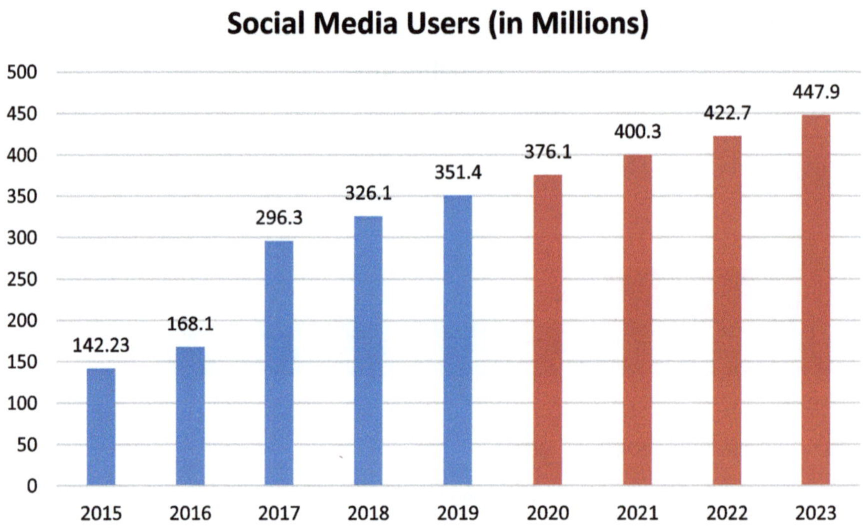

Fig. 4 Number of social network users in India from 2015 to 2023 (in millions) [4, 10]

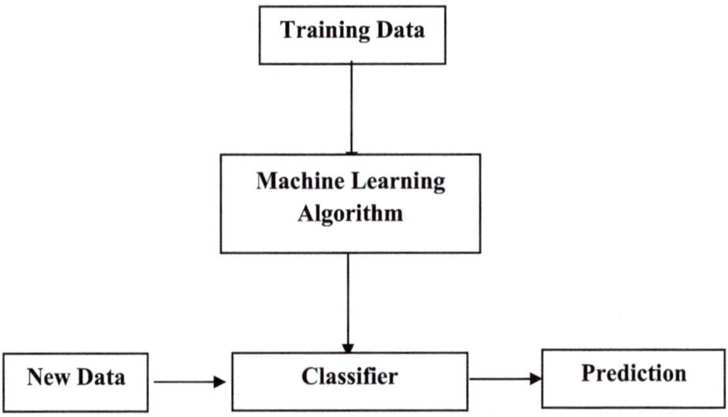

Fig. 5 Flow diagram

mining the social media posts of users, we may get a broad image of psychological behaviour. From user's profile in Social media, we can gather all the information that relates to person's mood, activities, sleep hours, thinking style, interactions, guilt feeling, worthlessness, loneliness, and helplessness. Recovering such behavioural attributes shows indications of depression on the social media users, which could be used to predict if the user is unhappy or not. Psychiatrists, parents, and friends could track the user depression by the proposed tool, which will save the time before the depressed user could get into major despair phase [7] (Fig. 5).

1.3 Sources of Data

To understand and predict psychological behaviour built on social media activity, data was gathered from Kaggle (referenced) 1 lakhs+ posts, ensuring a comprehensive analysis:

1.4 Primary Data

Primary data was collected from social media platforms, Twitter (Kaggle), which serve as the foundation for behavioural analysis. User posts, comments, and engagement activities (likes, shares) were retrieved using Kaggle while adhering to the privacy policies and regulations of the respective platforms [8].

1.5 Secondary Data

Existing studies, datasets, and research papers related to social media use and psychological behaviour were reviewed. Sources include academic journals, published articles, and publicly available social media datasets [9]. These sources provided insights into patterns of user engagement and psychological markers associated with certain behaviours.

1.6 Models

This mathematical model participates Logistic Regression, Support Vector Machine (SVM), and Random Forest Classifier to predict psychological behaviour based on social media posts. Below are the formulations for each model:

1.6.1 Logistic Regression

Logistic Regression models the likelihood that a given input (social media post) belongs to a particular class.

Model Formulation

Given a feature vector x derived from a social media post, Logistic Regression predicts the probability $p(y = 1|x)$ that the post exhibits a particular psychological state.

$$p\,(y = 1\,|x\,) = \sigma\left(\omega^{\top}x + b\right) = \frac{1}{1 + \exp\left(-\left(w^{\top}x + b\right)\right)} \tag{1}$$

where

- $\boldsymbol{w}$ is the weight vector.
- $\boldsymbol{x}$ is the feature vector.
- $\boldsymbol{b}$ is the bias term.
- $\boldsymbol{\sigma(z)}$ is the sigmoid function,
- $\boldsymbol{\sigma(z)} = 1/1 + \mathrm{e}^{-z1}$.

1.6.2 Support Vector Machine (SVM)

SVM seeks to find the optimal hyperplane that separates data points of dissimilar classes with the maximum margin.

Model Formulation

For a linear SVM, the decision boundary is given by:

$$f(x) = w^{\top}x + b \tag{2}$$

The goal is to find w and b that maximize the margin $2/\|w\|$ between the classes, subject to the constraints:

$$y_i\left(w^{\top}x_i + b\right) \geq 1 - \xi_i, \quad \text{for all } i \tag{3}$$

where

- $y_i \in \{-1, 1\}$ represents the class labels.
- $\boldsymbol{\xi_i} \geq 0$ are slack variables to allow for misclassification in the case of non-linearly separable data.
- The margin maximization problem can be formulated as a quadratic optimization problem:

$$\min_{w,b,\{\xi_i\}} \quad \tfrac{1}{2}\|w\|^2 + C\sum_{i=1}^{N}\xi_i$$
$$\textbf{subject to}: y_i\left(w^{\top}x_i + b\right) \geq 1 - \xi_i$$

where C is a regularization parameter.

1.6.3 Random Forest Classifier

Random Forest is an ensemble of decision trees, where each tree is trained on a random subset of the data and features [10]. The final prediction is made by aggregating the predictions of all trees.

Decision Tree Model

Separately decision tree T_m in the forest classifies a sample by traversing the tree from the root to a leaf node, where a decision is made based on a series of binary splits on the features [11].

Random Forest Formulation

Given MMM decision trees, the final prediction y^{\hat}{y}y^ for a sample is the mode of the predictions from all trees:

$$\hat{y} = \text{mode}\ (\{T_1(x), T_2(x), \ldots, T_M(x)\}) \tag{4}$$

Gini Impurity

Respectively decision tree in the Random Forest is built by minimizing the Gini impurity at each split:

$$G(p) = 1 - \sum_{k=1}^{K} p_k^2 \tag{5}$$

where

- p_k is the proportion of samples belonging to class k.
- K is the number of classes (Table 1).

2 Conclusion

This study validates the effectiveness of machine learning in predicting psychological behaviour using communal media data, achieving *77.36% accuracy*. Algorithms like *Logistic Regression and SVM* proved particularly robust, enabling scalable analysis of emotional and behavioural patterns for early mental health insights.

Table 1 Some posts that indicate depression

Depression indicative posts
"Hey bro, what happen, is everything fine or not?" no .... Nothing is fine, I don't appreciate why only me...Why people always cheat only me ...
Everybody taunts only on me .... Why? I'm fed up of this.
I wish I have someone that cares for me, understands me, love me ... I'm sad.
I lost my interest in all activities. I don't like anything. I want to be alone...
Why I'm not a first importance for a people .... This makes me cry.
How is that possible. How I can fail. I don't know what to do. There is no reason for me to be alive ....
I identified with depression last week.
I'm fed up of all this, why can't people leave me alone, why they delay all the time.

By leveraging digital footprints, this research underscores AI's potential to support proactive mental health monitoring and personalized interventions.

However, ethical challenges—such as *data privacy, algorithmic bias, and interpretability*—must be addressed to confirm responsible deployment. Future work should focus on:

1. *Enhancing model performance* with deep learning (e.g. transformers) and multimodal data (images, temporal activity).
2. *Expanding feature sets* to capture nuanced linguistic and behavioural cues.
3. *Developing ethical frameworks* for transparent, secure, and bias-mitigated AI applications in mental health.

These advancements could transform social media analytics into a reliable tool for psychological assessment, bridging gaps in traditional healthcare while prioritizing user trust and safety.

References

1. World Health Organization. (2023). Depression and other common mental disorders: Global health estimates (2023). https://www.who.int/publications/i/item/depression-global-health-estimates
2. Statista. (2024). Number of social network users worldwide from 2018 to 2025. https://www.statista.com/statistics/278414/number-of-worldwide-social-network-users/
3. Liu, X., Rivera, S. C., & Lipson, S. M. (2024). Digital phenotyping for depression: A meta-analysis of machine learning approaches. *Nature Mental Health, 2*(1), 45–60. https://doi.org/10.1038/s44220-023-00001-5
4. Statista. (2024). India social media users 2024. Retrieved May 14, 2025, from https://www.statista.com/statistics/278414/number-of-worldwide-social-network-users/
5. Pew Research Center. (2024). Teens, social media, and mental health: A 2024 landscape. https://www.pewresearch.org/internet/2024/02/20/teens-social-media-and-mental-health-2024/
6. Marcus, M., Yasamy, M., Ommeren, M., Chisholm, D., & Saxena, S. (2012). *Depression: A global public health concern*. WHO Dataset.

7. Hussain, J., Ali, M., Bilal, H., Afzal, M., Ahmad, H., Banos, O., & Lee, S. (2015). SNSW based predictive model for depression. In *The 13th international conference on smart homes and health telematics, ICOST 2015* (pp. 349–354).
8. Wakefield, J., Schmitz, M., & Baer, J. (2010). Does the DSM-IV clinical significance criterion for major depression reduce false positives? The National Comorbidity Survey replication. *American Journal of Psychiatry, 167*(3), 298–304.
9. Rowe, J. (2015). For behavioral health, researchers cite need for new IT. Retrieved October 18, 2015, from http://www.himssfuturecare.com/blog/behavioral-health-researchers-cite-need-new-it
10. Statista. (2015). Social network users in 2014 and 2018. Retrieved October 20, 2015, from http://www.statista.com/statistics/278341/number-of-social-network-users-in-selected-countries/
11. Frances, A., Pincus, H., & First, M. (1994). Major depressive episode. In *Diagnostic and statistical manual of mental disorders* (4th ed., pp. 326–327). American Psychiatric Association.

Index

C. Rompante Cunha et al. (eds.), *Transformative Cloud Computing, IoT and Extended Reality*, Signals and Communication Technology,
https://doi.org/10.1007/978-3-032-03554-7

MIX
Papier aus verantwortungsvollen Quellen
Paper from responsible sources
FSC® C105338

If you have any concerns about our products,
you can contact us on
ProductSafety@springernature.com

In case Publisher is established outside the EU,
the EU authorized representative is:
Springer Nature Customer Service Center GmbH
Europaplatz 3, 69115 Heidelberg, Germany

Printed by Libri Plureos GmbH
in Hamburg, Germany